Nanotechnology and Nanomaterials for Energy

Series Editor
Pierre-Noël Favennec

Nanotechnology and Nanomaterials for Energy

Pierre Camille Lacaze
Jean-Christophe Lacroix

ISTE

WILEY

First published 2021 in Great Britain and the United States by ISTE Ltd and John Wiley & Sons, Inc.

ISTE Ltd
27-37 St George's Road
London SW19 4EU
UK

www.iste.co.uk

John Wiley & Sons, Inc.
111 River Street
Hoboken, NJ 07030
USA

www.wiley.com

Library of Congress Control Number: 2021941647

British Library Cataloguing-in-Publication Data
A CIP record for this book is available from the British Library
ISBN 978-1-78630-497-1

MIX
Paper from responsible sources
FSC® C013604

Contents

Introduction ix

Part 1. Nanomaterials and Nanotechnologies 1

Chapter 1. Carbon-based Nanomaterials 3

1.1. Fullerenes 4
1.1.1. Properties of fullerenes 5
1.2. Carbon nanodiamonds 11
1.2.1. Principal techniques used in creating nanodiamonds 11
1.2.2. Key properties of nanodiamonds 13
1.3. Carbon dots or carbon quantum dots 16
1.3.1. CQD production methods 16
1.3.2. Fluorescence properties of CQDs 18
1.3.3. CQD applications 21
1.4. Carbon nanotubes 21
1.4.1. Chirality of carbon nanotubes 24
1.4.2. Mechanistic models of CNT growth 26
1.4.3. CNT arrays aligned horizontally or perpendicularly to a planar substrate 31
1.4.4. Key properties and applications of CNTs 34
1.4.5. Conclusion 37
1.5. Graphene 37
1.5.1. Electrical properties of exfoliated graphene 38
1.5.2. Graphene production techniques 41
1.5.3. Applications of graphene and graphene derivatives 51
1.5.4. Conclusion 62

1.6. Graphene quantum dots. 63
1.6.1. GQD production methods 63
1.6.2. Properties and applications of GQDs 66
1.6.3. Graphdiyne: a new alternative to graphene. 72
1.7. Conclusions and perspectives of carbon-based nanomaterials 77

Chapter 2. Inorganic Nanomaterials 79

2.1. Metallic nanoparticles. 80
2.1.1. Gold nanoparticles (Au NPs) 81
2.1.2. Core-shell type bimetallic nanoparticles 83
2.2. Metal nanoclusters. 87
2.2.1. Production methods for gold nanoclusters 88
2.2.2. Structure and stability criteria of Au NC 90
2.2.3. Luminescence properties of Au NCs 91
2.2.4. Applications using the luminescent properties of Au NCs. 95
2.2.5. Conclusion. 97
2.3. Semiconductor quantum dots 97
2.3.1. Development of colloidal QDs 98
2.4. Two-dimensional inorganic lamellar nanosheets. 103
2.4.1. Transition metal dichalcogenides 104
2.4.2. Conclusion. 113
2.5. Hybrid metal-organic frameworks 113
2.5.1. MOF production 113
2.5.2. Potential applications of MOFs. 119
2.5.3. Conclusions 128
2.6. Conclusions on inorganic nanomaterials. 129

Part 2. Nanotechnology and Nanomaterials for Energy 131

Chapter 3. Energy Storage 133

3.1. Worldwide energy use 133
3.2. Energy storage systems. 135
3.2.1. Non-chemical/electrochemical storage 135
3.2.2. Chemical and electrochemical storage systems 136
3.2.3. Rechargeable batteries. 139
3.2.4. Supercapacitors 184
3.2.5. Pseudocapacitors 189
3.3. Conclusions on energy storage 193

Chapter 4. Energy Conversion 195

4.1. Photovoltaics 196
4.1.1. General principles of the photovoltaic process. 197
4.1.2. Photovoltaic technologies 200
4.2. Electroluminescence, lighting and display. 225
4.2.1. Inorganic light-emitting diodes. 226
4.2.2. Organic light-emitting diodes. 233
4.2.3. QDot light-emitting diodes 244
4.3. Conclusions on energy conversion 249

Chapter 5. Electro- and Photocatalysis 251

5.1. Water splitting 252
5.2. Electrolysis techniques 253
5.3. HER and OER processes in water splitting 257
5.3.1. HER in an acidic medium 257
5.3.2. HER in alkaline media 274
5.3.3. Conclusions on HER reactions 279
5.3.4. Catalysts for oxygen evolution reaction. 279
5.4. Photoelectrochemical water splitting 294
5.4.1. Heterogeneous photocatalysts 297
5.4.2. Photocatalytic systems with two SC heterojunctions 298
5.4.3. Conclusions 302
5.5. Fuel cells 302
5.5.1. Operating principle of a fuel cell. 303
5.5.2. Choice of O_2 reduction catalysts 306
5.5.3. Conclusions on electrocatalysis and photocatalysis 310

Conclusion 313

References 317

Index 359

Introduction

The term "nanomaterials", by general consensus, is used to designate both nano-objects and nanostructured materials. To qualify as a nanomaterial, the dimensions of an object or of a structural element of a macroscopic material must be less than about one hundred nanometers. In addition to this first, generic definition, a distinction is made between zero-dimensional nanoparticles (0D-NP); nanowires (NW) and nanotubes (NT), which are one-dimensional (1D-NP); nanosheets and multi-sheets, which are two-dimensional (2D-NP); and, finally, three-dimensional (3D) nanostructured materials. Further distinctions are made between these forms based on their material composition: carbonaceous, inorganic or hybrid.

I.1. History of nanotechnology

Nanotechnologies first came into public view in the year 2000. This was the year the National Nanotechnology Initiative[1], or NNI, was launched in the United States, propelled by the NSF (National Science Foundation), a large number of academic societies and multiple federal departments. The new and innovative properties of nanomaterials discovered over the course of the 1980s and 1990s opened a vast range of perspectives in terms of applications, attracting the attention of researchers and industrialists, who began to devote considerable time and attention to exploiting these new opportunities. These initiatives attracted almost unprecedented support from public authorities, and President Clinton, visiting Caltech in January 2000, specifically designated this initiative as a national developmental priority. Different fields of application were defined in collaboration with several government agencies, who also provided funding; this new area of research was seen as a likely

1 The term nanotechnology was first used in the 1970s by the Japanese scientist Noria Tanegouchi to denote the body of research concerning materials with dimensions in the tens of nanometers at most.

source of significant scientific progress with the potential to benefit the whole of society.

Why, then, did nanotechnology attract such levels of attention from the US government in the early 2000s? Without going into too much detail, it is important to note that the 1980s and 1990s were marked by a number of important scientific discoveries relating to extremely small elements; this went hand-in-hand with technological developments which greatly increased the capacity to observe the state and performances of materials.

Transmission electron microscopy (TEM) was first commercialized by Siemens and Ernst Ruska[2] and continued to develop over the following decades. By the 1990s, TEM performance had reached a level where it was possible to observe objects on a nanometric scale [RUS 86]. Scanning electron microscopy (SEM) developed in parallel, although commercial applications appeared at a later stage; this technique is based on the analysis of reflected secondary electrons. The first apparatus was launched onto the market by Cambridge Scientific Instruments in the 1960s. SEM makes it possible to observe the texture of material surfaces at nanometric level, constituting a breakthrough in terms of observing nano-objects in 3D.

The decisive breakthrough into the "nanoworld", however, was not made until the mid-1980s. Rohrer and Binnig's invention of near-field microscopy [ROH 86], including scanning tunneling microscopy (STM) and atomic force microscopy (AFM), marked a further step in the observation of sub-nanometric objects: the long-pursued goal of atomic-level observation had finally been attained. These new microscopy techniques, which now play a central role in the study of nanomaterials, were increasingly used to produce high-resolution images. Compared to classic TEM and SEM techniques, they also present a practical advantage in that they do not need to be used in a vacuum, and can indeed be used in "everyday" atmospheric conditions.

These technological innovations in the field of observation went hand-in-hand with the discovery of new materials, on a nanometric scale, which displayed entirely new physical and chemical properties due to their very smallness. Thus, in 1985, Kroto *et al.* (Nobel Prize winners in 1996) discovered fullerenes: new carbon nanomaterials with exceptional properties [KRO 85]. The first fullerenes were obtained in the form of traces; they are now produced commercially on a large scale.

2 Ernst Ruska was the first to suggest using electron beams to observe microscopic objects via a transmission mechanism. Ruska produced the first commercial electron microscope, in collaboration with Siemens, in 1933. He went on to devote his whole scientific career to improve this technique, which earned him a Nobel Prize in 1986.

Fullerenes consist of a spherical assembly of carbon atoms, assembled in hexagon and pentagon shapes. Often considered as "electron wells", they have particularly interesting chemical properties and are widely used in various domains, notably in plastic photovoltaic cells; their application in this field in the 1990s resulted in significant gains in energy yield [YU 95].

A few years later, in Japan, Lijima [LIJ 91] discovered the tubular structure of carbon nanotubes (CNTs), then of single wall carbon nanotubes (SWCNTs) [BET 93], which also consist of a hexagonal assembly of carbon atoms. These nanotubes vary from one to several nanometers in diameter, and have a length of several hundred nanometers. As we shall see later, the mechanical properties of these nanotubes largely surpass those of steel; furthermore, they also possess exceptional electronic properties.

These discoveries rapidly attracted the attention of scientific agencies in the United States. An ambitious program of fundamental and applied research into nanotechnologies was launched in the year 2000, with both scientific and industrial support. A series of directives and recommendations issued by the NNI attracted attention on the world stage, and a number of other nations launched their own initiatives, highlighting the strategic importance of the domain.

The scale of technological research increased considerably from the early 2000s on, and the number of publications in connection with nanomaterials or nanotechnologies continues to grow. While the creation and use of nanometric elements in the field of electronics are well established, other fields, including chemistry, the energy sector, biology and medicine, stand to make considerable gains from the progression of nanotechnologies. New fields of investigation are opened up on a regular basis, and their results look highly likely to revolutionize both theoretical knowledge and practical applications in these domains.

I.2. Outline of this book

Our main aim in this book is to highlight new breakthroughs and areas of research in nanotechnologies which have appeared in recent years, notably since 2010; however, we shall begin by presenting a brief overview of earlier discoveries, essential to understanding later work. Many of the new findings presented here have yet to be used commercially, but their interest in terms of research and potential future applications is immense.

We have chosen to focus on problems and solutions relating to the energy sector. Our discussion is split into two parts: first, a description of nanomaterials and their properties, and second, a discussion of the ways these materials are, or may be, used in the energy sector for storage and conversion, electrocatalysis and photocatalysis.

Chapter 1 is devoted to the subject of carbon nanomaterials, and includes a description of recent preparation methods, properties and major applications. Some of these materials (fullerenes, carbon nanotubes, nanodiamonds, etc.) largely predate the year 2000. However, they are yet to reveal all of their secrets, notably in terms of the remarkable properties they possess and in new contexts of application. Two-dimensional families of nanomaterials, derived from graphene, are a far more recent discovery; graphene itself looks set to become the star material of the 21st century. More recently still, the family of 2D carbon composites has expanded to include materials such as graphdiyne, with a huge range of potential applications.

Chapter 2 concerns the family of inorganic nanomaterials. This field, too, has undergone considerable development in recent years, and new properties have emerged with the discovery of materials at an increasingly small level. Atom clusters, which contain from ten to a few hundred atoms and are smaller than 1.5 nm in size, possess physico-chemical properties which set them apart from larger NPs; this is a result of their spatial confinement. Quantum dots (QD), which are generally made up of binary alloys of semiconductors (SC), constitute another important family of nanomaterials and are particularly notable for their luminescent properties. The latest set of 2D materials include dichalcogenides of transition metals, along with a significant number of inorganic lamellar materials; the interest of these materials is already apparent, particularly in terms of their applications in the energy sector.

The second part of this book constitutes a critical study of new problems in the domain of energy, notably in terms of energy transition and preserving the environment. The fight to reduce greenhouse gas emissions raises two major problems: those of storing and converting energy. Nanotechnologies and nanomaterials stand to make a decisive contribution to solving these issues.

Chapter 3 is devoted to the key question of energy storage, a major challenge linked to the development of renewable energies. This field encompasses large-capacity static storage, the production of batteries and supercapacitors for electric vehicles, and also microbatteries, designed to provide power for small portable electronic devices. All of these different systems are subject to different constraints, leading us to focus on different design choices.

Chapter 4 concerns energy conversion, notably in terms of photovoltaic sources and lighting. Solar power is nothing new, but huge progress has been made in recent years, notably with the development of organic photovoltaics and that of perovskites; these solutions provide a genuine alternative to inorganic silicon-based systems, and, importantly, are more cost-effective. Lighting and display devices constitute another important subject in the energy field, notably in connection with the general phenomenon of electroluminescence. The overarching goal is to reduce energy consumption while providing a better quality of lighting than that obtained using incandescent bulbs or neon tubes. Considerable advances have been made over the last few years; as before, parallel developments in inorganic and organic nanomaterials have found practical applications in this area. The development of new forms of memory electronics (not covered in this book) is another important area of research, where once again the reduction of energy consumption is a major goal.

Chapter 5 concerns electrocatalysis and photocatalysis. Electrocatalysis relates primarily to the development of electrolysis cells (water splitting) and fuel cells, a main focus for future advances in converting chemical energy into electrical energy. Hydrogen is a particularly promising source of fuel: it is non-polluting, and constitutes a highly efficient means of transporting and storing energy. One especially promising area of research relates to the production of hydrogen by electrolyzing water, powered by a photovoltaic cell. This development is connected to the broader aim of developing systems which use non-precious metals to catalyze oxygen evolution reactions (OER) and oxygen reduction reactions (ORR). This area of research is essential for the optimization of water splitting and H_2/O_2 fuel cells. Extensive research is also underway on the subject of photocatalysis: the aim in this case is to optimize the use of solar energy for carrying out chemical transformations. Photocatalysis can also be combined with electrolysis reactions (photo-electrocatalysis) in order to transform ordinary compounds into high-added-value compounds, or into energy products which may be used either for storage or as a fuel source.

I.3. Dedication

We dedicate this book in memory of Jacques-Emile Dubois, eminent member of the Science Faculty of the University of Paris and founder of the ITODYS laboratory, who would have celebrated his 100th birthday this year, were he still among us. Professor Dubois was a pioneer and outstanding practitioner of the multidisciplinary approach to science in the realm of Physical Organic Chemistry. He blazed a trail in his revolutionary application of Computer Science to Chemistry, as well as in launching the study of phenomena at the solid-liquid interface. His drive and enthusiasm for uncovering and delving into virgin areas of research were and remain a source of inspiration to many.

PART 1

Nanomaterials and Nanotechnologies

1

Carbon-based Nanomaterials

Fullerenes, carbon nanotubes (CNTs), nanodiamonds (NDs) and carbon quantum dots (CQDs) are nothing new, and were discovered long before the year 2000; these discoveries were followed, in 2005, by that of graphene. Due to their remarkable physico-chemical properties, fullerenes, CNTs and graphene have given rise to innumerable applications, and their use has progressed continually; nevertheless, the use of graphene has increased much faster than that of fullerenes and CNTs (Figure 1.1).

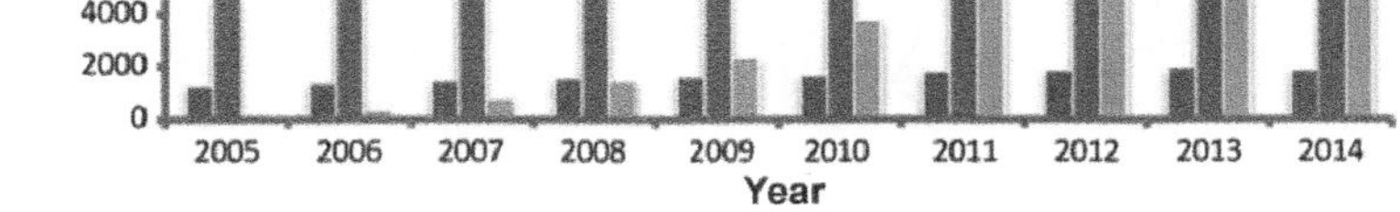

Figure 1.1. *Number of publications on the subject of fullerenes, CNTs and graphene from 2005 to 2014. Based on [LI 15]*

The number of publications concerning these three families of compounds between 2005 and 2014 clearly shows an "explosion" in the number of articles on the subject of graphene; over the course of 7 years, from 2007 to 2014, this went

For a color version of all the figures in this chapter, see: www.iste.co.uk/lacaze/nanotechnology.zip.

from around 50 to over 16000, exceeding scientific production on the subject of CNT, discovered a whole decade earlier than graphene.

Moreover, the implication of Asian countries (notably China, South Korea and Japan) in this research is notable, representing more than 50% of total publications between 2005 and 2014 (China alone accounts for 36% of production); the United States accounts for just over 22%, and the rest of the world for 27%.

1.1. Fullerenes

Since Kroto *et al.*'s [KRO 85] discovery of buckminsterfullerene C_{60} (henceforth C_{60} fullerene) in 1985, where they demonstrated that it had a balloon-like structure made up of an assembly of sp^2 carbon atoms arranged in the form of pentagons (12) and hexagons (20), a whole area of chemistry has developed in connection with fullerenes; a whole host of applications have also emerged in biology, medicine, electronics and energy.

Following on from C_{60} and C_{70}, the two most stable and widespread forms (Figure 1.2), new fullerenes with higher carbon numbers have subsequently been synthesized.

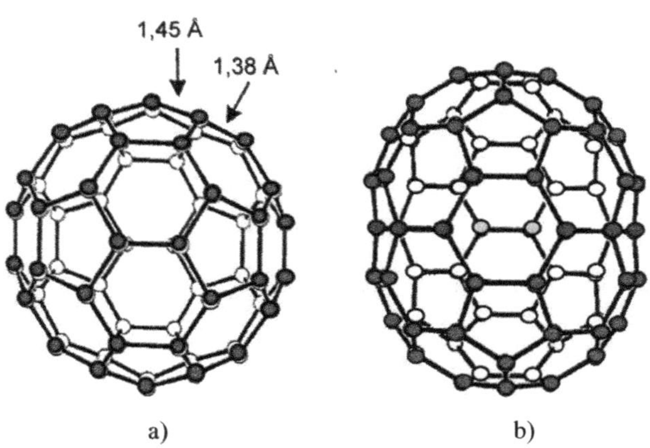

Figure 1.2. *Fullerenes: C_{60} (a) and C_{70} (b)*

COMMENT ON FIGURE 1.2.– *a) The C_{60} fullerene consists of an assembly of 10 pentagonal and 20 hexagonal carbon rings. The red and white symbols represent the C atoms in the foreground and background of the projection plane. b) The structure of the C_{70} fullerene is equivalent to that of two C_{60} hemispheres joined by*

10 C atoms (six shown in dark blue, four in light blue, two of which cannot be seen) at the midplane of the molecule. Adapted from [BAL 98a].

The stability of these fullerenes results from the arrangement of hexagons and pentagons. The IPR (Isolated Pentagon Rule) notably states that a fullerene, with a given number n of carbon atoms, will be stable if each pentagon is surrounded by hexagons and has no contact with another pentagon. In the case of the C_{60} fullerene, there are 1812 geometric isomers of which only one (with symmetry I_h) verifies the IPR. The energy difference between the most stable (noted 1) and the least stable (noted 1812) isomers in the series is of the order of 540 kcal/mol [SUR 17]. This energy largely results from the tension exerted by the curvature of the bonds.

1.1.1. *Properties of fullerenes*

Given their cage-like carbon structure, featuring only covalent bonds between sp^2 carbons, fullerenes are naturally hydrophobic and therefore insoluble in aqueous media. In order to use fullerenes for biological purposes, transformations must be carried out to make them more hydrophilic; this is achieved by grafting hydroxylated groups (COOH or OH), corresponding to exo-functionalization reactions, notably onto C_{60} and a number of other fullerenes (C_{70}, C_{80}, etc.) which are produced on a commercial scale. However, these latter fullerenes are expensive due to the difficulties inherent in large-scale production.

The presence of a cavity suggests that it might be possible to encapsulate chemical species. This is of interest with regard to the design of various markers, such as those used in medical imaging. The encapsulation of metals would result in a whole new category of nanomaterials, corresponding to the formation of endometallofullerenes (EMF), whose stability is a function of the fullerene and the small number of metal atoms encapsulated within the cage. The electrophilic character of fullerenes and their strong electronic affinity also mean that they are strong electron receivers, and are sometimes considered as “electron sinks”.

1.1.1.1. *Electrophilic and antioxidant properties of fullerenes*

The electrophilic character of fullerenes was established as early as the 1990s. This discovery suggested the possibility of forming anions with six negative charges, corresponding to the acceptance of six electrons by the LUMOs (lowest unoccupied molecular orbitals) of the fullerene. This hypothesis was confirmed by the existence of potassium salts (K_6C_{60}) in crystalline form.

More evidently still, the progressive six-step electrochemical reduction of C_{60} and C_{70} demonstrates the possibility of obtaining six stable reduction states of fullerene [XIE 92] (Figure 1.3).

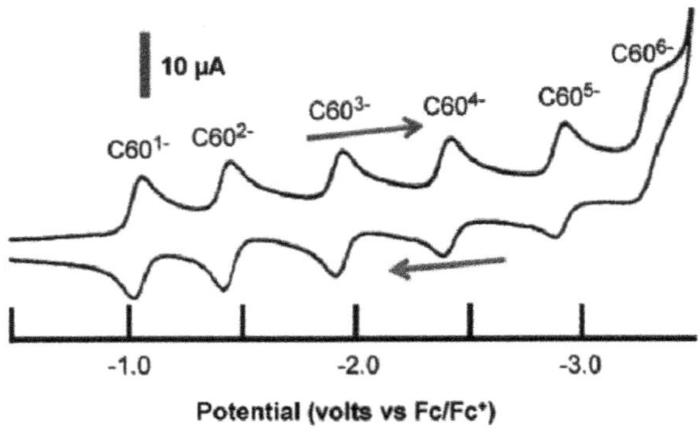

Figure 1.3. *Voltametric curves showing redox of C_{60} in a low temperature acetonitrile/toluene solution (–10°C)*

COMMENT ON FIGURE 1.3.– *The reduction curve obtained by a potential sweep from 0 to –3.5 V (blue arrow) shows six distinct reduction peaks, each corresponding to the gain of an electron. The reverse oxidation curve (red arrow) is clearly defined and also shows six oxidation peaks, corresponding to the six successive anionic states; this proves the stability and reversibility of these six redox states. Sweep speed: 100 mV/s. Potentials given against the ferrocene/ferricenium$^+$ (Fc/Fc$^+$) redox couple. Adapted from [XIE 92].*

The electronic affinity of fullerenes means that they can be used as electron carriers in photovoltaic cells, achieving energy yields of the order of 10% (see section 4.1). Fullerenes also have numerous applications in the biomedical field. C_{60} and its derivatives, in combination with various hydrophilic groups with relevant biological properties, have been used in cancer and AIDS therapies (anti-retroviral therapy against HIV-1). Functionalized by carbohydrate chains, fullerenes have also proven to be good antibacterial agents. Their high affinity for radicals, which results from radical addition on the numerous double bonds of the fullerene, effectively makes them act as "sponges" for these radicals. As powerful antioxidants, fullerenes are used in biology to neutralize radical oxygen species (ROS) such as the $O_2^{\bullet-}$ superoxide ion, $HO^{\bullet}$ hydroxyls, and hydrogen peroxide, H_2O_2, which are particularly damaging to DNA and certain proteins. They are also used in "niche" cosmetic products, including creams based on liposoluble C_{60} fullerenes, marketed as antioxidants, with a power comparable to, or better than, that of vitamin E-based products [ACQ 17]. All these properties, used in a medical context, have been described in well-documented papers [BAK 07, LAL 13, CAS 17].

1.1.1.2. *Chemical reactivity and exo-functionalization*

As fullerenes are considered to be electrophilic reagents, a whole chemistry of functionalization, corresponding to nucleophilic or radical additions or even cycloaddition reactions of the Diels-Alder and Prato type[1], can be envisaged. This has been described in detail by Taylor and Walton [TAY 93], and more recently by Georgakylas *et al.* [GEO 15] in the case of C_{60} and C_{70}.

Among fullerene derivatives, the fullerenols and carboxyfullerenes obtained by grafting hydroxyl or carboxyl groups onto the surface of fullerenes present an advantage in comparison with simple fullerenes in that they are soluble in aqueous and biological media. This is the case of hexacarboxylated fullerenes (carboxyfullerenes), which are fullerenes carrying three pairs of carboxylic acids, comprising a mixture of two stereoisomers, C3-C_{60} and D3-C_{60} (Figure 1.4)[2].

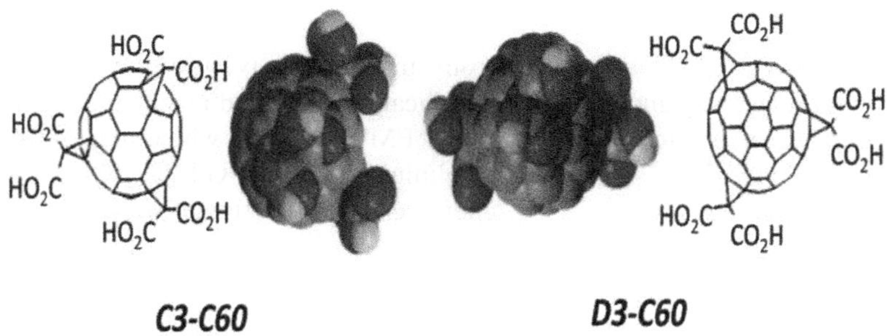

Figure 1.4. *Structure of two carboxyfullerene stereoisomers, C3-C_{60} and D3-C_{60}. Based on [DUG 97]*

Like simple fullerenes – but with the advantage of aqueous solubility – these derivatives have strong antioxidant properties; they notably have the capacity to destroy the peroxide ion $O_2^{\bullet-}$, a toxic subproduct of cell metabolism. This property makes them excellent neuroprotectors, and research in this area is still ongoing [DUG 97, ALI 04, GHA 05, YE 15b].

---

1 The Prato reaction is a cycloaddition of a nitrogen ylide to an ethylenic double bond, resulting in an (R)N-pyrrolidine or, in the case of fullerenes, an (R)-pyrrolidino fullerene. The nitrogen ylide is obtained *in situ* by the action of formaldehyde on an amino acid (RNH-CH_2-COOH + $CH_2O \rightarrow H_2C = N^+(R)–CH_2^- + CO_2 + H_2O$).

2 These C3-C_{60} and D3-C_{60} isomers, obtained by reacting ethyl 2-bromo malonate with C_{60} in the presence of NaH, differ in that they belong to symmetry groups C3 and D3.

1.1.1.3. *Endometallofullerenes*

The idea that the interior of the fullerene "cage" to trap atoms came hot on the heels of the discovery of fullerenes themselves. As early as 1985, Heath *et al.* [HEA 85] identified the first fullerene made up of 60 carbon atoms and containing a lanthanum atom[3] using mass spectrometry. A few years later, the same authors isolated several metallofullerenes with 60, 70, 74 and 82 carbon atoms, the last of which, La@C_{82}[4], is the only one which is stable in contact with air [CHA 91].

One remarkable feature of these new compounds is that the metal is not released when the compound is placed in a biological medium; this constitutes a considerable advance in comparison with classic metal chelates. This stability probably results from an electron transfer between the La atom and the fullerene, creating a La^{3+}@C_{82}^{3-} ion pair, and presumably also to the fact that the mesh of the fullerene is small enough to prevent the La^{3+} ion from diffusing outward.

This discovery, which opened promising perspectives in the medical field (notably for diagnostic and therapeutic applications) resulted in an almost immediate upsurge in research into synthesizing new EMF, some of which are useful in the field of oncology. For instance, the gadolinium fullerenol Gd@$C_{82}(OH)_{22}$, initially used as a contrast agent in nuclear magnetic resonance imaging (MRI), has also been shown to have strong anti-cancer properties, different from those of simple fullerenes; furthermore, it retains low cytotoxicity [KAN 14][5].

1.1.1.4. *Endocluster fullerenes*

This new family of metallofullerenes was isolated for the first time in 1999 by Dorn and coworkers [STE 99]; as before, these molecules are obtained by producing an electrical discharge between graphite electrodes, containing a metal oxide[6], in a

3 The lanthanum fullerene was synthesized by laser ablation of a disk of graphite impregnated with lanthanum salts (La Cl_3).

4 The notation La@C_{60} indicates that the lanthanum ion is contained inside the cage formed by the fullerene.

5 The gadolinium ion (Gd^{3+}), used as a contrast agent in MRI, is unfortunately extremely toxic. This toxicity is greatly reduced when the Gd is encased in a hydrophilic modified fullerene, i.e. one which is soluble in biological media. A commercial product made by Bayer, Magnevist®, corresponds to the EMF C_{60} gadolinium EMF (Gd@$C_{60}[C(COOH_yNa_{1-y})_2]_{10}$ into which a $[C(COOH_yNa_{1-y})_2]_{10}$ chain is grafted [LAU 07].

6 The most widespread method consists of using the electrical arc discharge to vaporize graphite. Other methods based on laser ablation, radiofrequency heating or heating using a resistor in an oven, ion implantation through the cage or opening the cage to insert ions have also been used to synthesize ECFs.

reactor. The difference here lies in the inclusion of a small quantity of dinitrogen in the reactor (Figure 1.5).

The remarkable result of this reaction is that it produces a new type of fullerene in which a cluster corresponding to a trimetallic nitride is embedded. The most abundant endocluster fullerenes (ECF) of this family is obtained using scandium. It has the formula $Sc_3N@C_{80}$ and consists of an 80-atom cage, into which scandium nitride with a symmetrical plane structure is inserted. A detailed description of the various functionalization reactions carried out on the surface of the cage of metallocluster fullerenes, which sometimes present higher levels of reactivity compared to simple fullerenes, can be found in a number of recent articles [GOO 17, YAN 17, BAO 18].

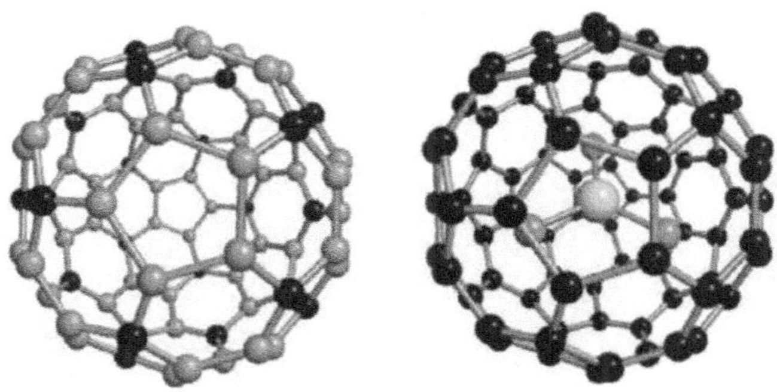

a) b)

Figure 1.5. *Structure of the C_{80} fullerene and of the ECF $Sc_3N@C_{80}$. a) C_{80} fullerene (symmetry Ih, symmetry axis of order 5) showing a 5 C ring (light blue) surrounded by five rings of 6 C atoms (dark blue). b) Structure of the $S_{c3}N@C_{80}$ endocluster fullerene. Based on [STE 99]*

An additional family of EMF has been developed by replacing nitride clusters with clusters of metal dicarbides of the form M_2C_{2n}, which can exist in the form $M_2C_2@C_{2n-2}$. Depending on the atomic volume of the metal, the size of the fullerene cages can vary from 72 to 88 carbon atoms (in the case of scandium) and from 90 to 104 atoms (lanthanum). More recently still, a series of "giant" fullerenes has been discovered involving yttrium carbide clusters: the size of the cages in this case can be up to 130 carbon atoms. The formula for these clusters is Y_2C_{2n} where 2n is from 92 to 130 inclusive [PAN 18].

All these compounds have immense potential applications, but the fact that they are currently obtained in only very small quantities (a few tens of mg) and in the form of mixtures, requiring painstaking separation using preparative chromatography, obviously limits their practical use. In contrast with simple fullerenes (C_{60} and C_{70}) and their exo-functionalized derivatives, which have already been subject to extensive research in connection with biomedical applications [CAS 17], ECF and EMF are generally studied in the context of fundamental research, aimed at better understanding the behavior of clusters of a small number of atoms, confined in a constrained space and interacting with carbon atoms.

1.1.1.5. *Onion-like fullerenes*

Onion-like fullerenes (OLFs) are a family of multi-shell fullerenes consisting of a concentric, nested assembly of multiple fullerenes of no more than around 10 nm. Various methods are used to create OLFs; the most common approach involves heat annealing, a powder made up of carbon nanodiamonds (NDs; see section 1.2) to produce large quantities of these OLFs [ZEI 16] (Figure 1.6).

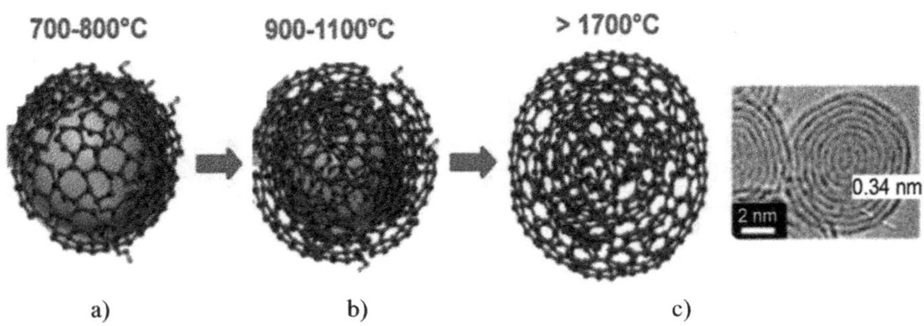

Figure 1.6. *Progressive transformation of nanodiamonds (ND) into OLF. a) Beginnings of the formation of a fullerene structure at 700–800°C on the surface of the ND. b) Formation of multiple concentric layers of fullerenes at 900–1100°C. c) Terminal phase in obtaining an OLF at T > 1700°C and the corresponding TEM image, showing multiple concentric carbon layers. Adapted from [ZEI 16]*

The transformation of NDs (which are commercially available) into OLFs essentially consists of a gradual change from sp^3 to sp^2 carbons. This transformation, which takes place in vacuum or in the presence of inert gases, starts to take place at temperatures around 1300°C and is completed at temperatures of up to 1800–2000°C. Under these conditions, we obtain OLFs which are homogeneous in size, about 5 nm in diameter, but in agglomerated form. Stable solutions can be obtained after disaggregation and an acid treatment, and these solutions are used in manufacturing conductive inks [VAN 17].

These compounds have electrical conduction properties far superior to those of activated carbon[7] and are thus highly valuable for creating supercapacitors with extremely fast response times [PEC 10]. They are eminently suited for use as power devices for energy storage, notably in the realization of pseudo-supercapacitors (see Chapter 3), combining a high charge/discharge speed with larger amounts of stored energy than those found in simple supercapacitors [VAN 17]. A very large number of alternative preparation techniques, which do not use NDs, may be used; these involve degrading various low-cost carbon products, and this makes them more economically competitive in relation to activated carbon [ZEI 16].

1.2. Carbon nanodiamonds

Although NDs were first discovered in the 1960s[8], they were not widely used until the late 1990s, once a detonation-based production technique had been firmly established [GRE 88]. These nanomaterials are now produced on a commercial scale, and are used both for improving the mechanical properties of plastics and for diagnosis and treatment in the field of medicine.

1.2.1. *Principal techniques used in creating nanodiamonds*

Owing to growing awareness of their importance, many different techniques have been developed for producing NDs. The most widespread approaches currently include detonation in a closed chamber, laser ablation and grinding using diamond microcrystal microbeads at high temperature and high pressure [MOC 12]. Their size is between 4 and 5 nm. Nanodiamonds can also be produced using CVD (chemical vapor deposition) techniques, in the form of films of varying thicknesses, between 20 nm and 5 µm [BUT 08]. The detonation method consists of introducing a mixture of two explosive compounds, such as the trinitrate derivatives trinitrotoluene (TNT-$C_7N_3O_6H_5$), trinitrobenzene (TNB-$C_6N_3O_6H_3$) or hexogen

---

7 Activated carbon (or activated charcoal) is a variety of porous carbon, available in the form of micronic powder or granules. It is obtained by pyrolyzing waste natural products, such as olive pulp, making it a particularly cheap product. Its high porosity and very high molecular adsorption power make it ideal for liquid filtering and gas purification operations; it is also an ideal substance for electrodes used in constructing supercapacitors.

8 First discovered by Danilenko *et al.* [DAN 04b] in 1963 but ignored by the scientific community, NDs were synthesized for the second time in 1982 by another Russian team, in the context of research aimed at synthesizing diamonds using detonations in a constrained space. Diamond-containing soots were initially considered as an insignificant subproduct; they finally began to attract attention once the possibilities for functionalization became apparent, and as potential uses in the biomedical sphere began to come to light.

($C_3N_6O_6H_6$), which do not release oxygen on explosion, into a closed metal chamber. The reaction takes place in a humid atmosphere of nitrogen and carbon dioxide (Figure 1.7).

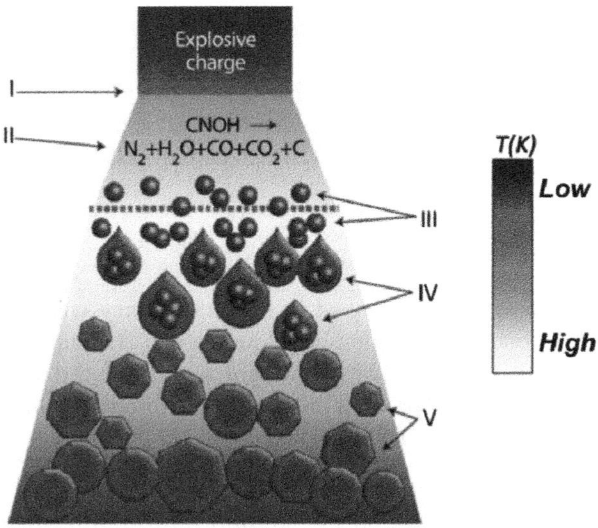

Figure 1.7. *Simplified diagram showing the detonation technique for nanodiamond production*

COMMENT ON FIGURE 1.7.– *I) Front of the shockwave resulting from the explosion. II) Products resulting from the explosion of a combination of TNT/TNB. III) Region in which the temperature (from 2200 to 4500 K) and pressure (from 13.5 to 16.5 GPa, i.e., approximately 150000 atmospheres) correspond to the formation of carbon clusters. IV) Coagulation of clusters in the form of nanodroplets. V) Crystallization, growth and agglomeration of NDs. Adapted from [MOC 12].*

As soon as the explosion occurs, the temperature and pressure immediately increase, establishing within a confined space the conditions needed for carbon nanoclusters to form. These nanoclusters become ND (in soot form) on reaching the coldest parts of the reactor [DAN 05]. The soot formed after the explosion of the charge may contain up to 75% ND, ranging from 4 to 5 nm in size; the remainder is made up of graphitic carbon (between 25% and 75% by weight, depending on the conditions), metals and oxides (between 1% and 8 % by weight), and nitrogenous derivatives. The yield in terms of carbon-based products is around 4%–10% of the mass of the explosive charge[9].

Nanoparticles (NPs) are essentially composed of sp^3 carbons, but they may also be partially covered by layers of graphitic carbon (sp^2) on the surface, or functionalized by various chemical groups resulting from reactions with dangling bonds on the surface.

1.2.2. *Key properties of nanodiamonds*

From a mechanical perspective, the properties of NDs are very similar to those of pure diamonds. They are characterized by extreme hardness and a very high Young elasticity modulus, making them ideal for use in polishing hard surfaces such as ceramics. Their high level of chemical stability means that they can also be used in very hostile environments.

Fluorescence and biocompatibility are two further interesting properties, opening the way for biomedical applications in both diagnostics and treatment, due to the ease of surface functionalization.

1.2.2.1. *Fluorescent nanodiamonds*

Fluorescent nanodiamonds (FNDs) are a new family of nanomaterials, ranging in size from 35 to 100 nm [HSI 16] and characterized by the presence of a structural defect inside the crystal (Figure 1.8a) [CHA 08].

This defect, named NV, corresponds to a coupling between a vacancy V (absence of a carbon atom) and a nitrogen atom N, adjacent to the vacancy. It can be created easily by irradiating ND crystallites with a beam of helium (He^+) ions [CHA 08] or protons (H^+)[10]. Nanodiamonds treated in this way (equivalent to n-doped NDs) emit

---

9 Raman spectrometry may be used to identify the main functional groups found on the surface of NDs. The core of the ND is characterized by a high-intensity peak centered at 1328 cm^{-1}; this is characteristic of sp^3 carbons. The presence of graphitic carbons, sp^2, along with that of oxygenated functional groups resulting from oxidation on contact with air is shown by the presence of a wide spectral band located between 1500 and 1800 cm^{-1}; this results from the addition of the graphitic carbon band, centered at 1590 cm^{-1}, a band due to the vibration of the angular deformation of OH at 1640 cm^{-1}, and a plateau resulting from the presence of carboxyl groups between 1740 and 1760 cm^{-1}.

10 The sub-micrometer diamond powders used to make FNDs are obtained at high temperature and high pressure, and contain about 100 ppm of impurities in the form of nitrogen atoms dispersed throughout the diamond structure. Under the effect of ion

stable red fluorescence when stimulated by a laser. The intensity of the fluorescence is a function of the concentration of NV, which increases as the energy of the He^+ ion beam used to create the defect is strong (NV concentration may vary between 10 and 30 ppm in relation to the number of carbon atoms; in the absence of treatment, their concentration is lower than 1 ppm)[11]. The biocompatibility and high stability of FNDs, along with their non-toxic nature and high-intensity red fluorescence, make them ideal for use in the biomedical sphere; they are a helpful replacement for inorganic quantum dots, which are usually made up of toxic elements (Cd, Se, Pb, etc.).

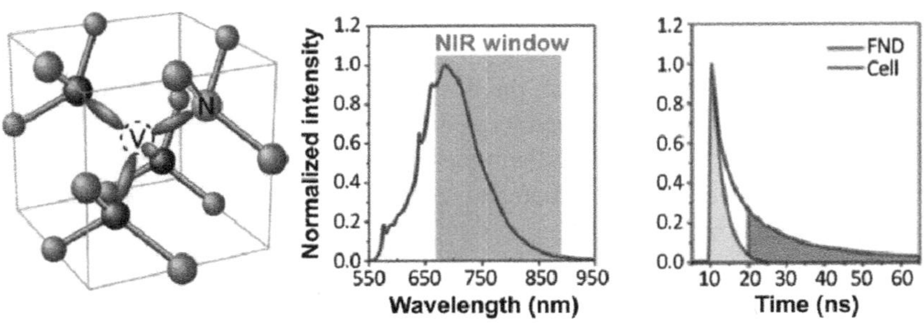

a) b) c)

Figure 1.8. *Fluorescence of ND nanocrystallites containing NV defects*

COMMENT ON FIGURE 1.8.– *a) Structure of an NV defect resulting from the proximity of a nitrogen atom (impurity due to ND preparation) to a vacancy corresponding to the absence of a carbon atom. The lobes shown in purple correspond to the dangling bonds of the carbon atoms neighboring the vacancy V. b) Fluorescence spectrum of FND (size: 35 nm) obtained by laser irradiation at 532 nm. The pink rectangle represents the window for which the autofluorescence of biological tissue is negligible. c) Time difference between the fluorescence lifetimes of an FND (pink) and that of a biological cell due to the presence of endogenous fluorophores (green). The use of a 10 ns shift between excitation and detection neutralizes the effect of the underlying fluorescence of the biological medium. Adapted from [HSI 16].*

---

bombardment, vacancies (V) are produced; their migration and proximity to nitrogen atoms (N) leads to the formation of NV centers.

11 NV centers can exist in two forms: NV° and NV-. The preponderant NV- form is responsible for all magneto-optical properties due to its different spin numbers $m_s = 0$ and ± 1. NVs are characterized by a strong absorption band centered at 550 nm, and an intense fluorescence band around 685 nm, with high quantum yield (70%) [SCH 14].

1.2.2.2. *Boron-doped diamonds*

Nanodiamonds made up of a core of sp^3 carbons are insulators, with resistivity reaching values of the order of 10^{12} Ω cm^{-1}; the band gap, typically that of an insulating material, is equal to 5.47 eV [HE 19].

These characteristics are completely reversed when NDs are doped with boron. Unlike nitrogen doping, which does not exceed concentrations of 30 ppm, boron doping can be very high: it is possible to obtain concentrations of up to 30000 ppm, transforming the boron-doped nanodiamonds (BDD) into conductive materials suitable for use in electrochemistry. In cases of high boron doping (10^{20}–10^{21} atoms/cm^3), the resistivity of BDD is very low, corresponding to that of a semi-metal (10^{-3} Ω cm), meaning that this new material is suitable for use as electrodes. Lower levels of doping, around 10^8 atoms/cm^{-3}, give much higher resistivities of around 10^4 Ω cm [PAN 05].

BDDs containing very low levels of graphitic carbon constitute an exceptional material for electrodes [GAR 15a, HE 19], giving an electroactivity window, for both oxidation and reduction, which is much higher than that of platinum or glassy carbon. The reaction corresponding to the reduction of H^+ ions in an acidic medium with the release of dihydrogen occurs at −1.2 V compared to an SHE (standard H_2/H^+ reference electrode), and the reaction corresponding to the oxidation of water with the release of dioxygen occurs at 2.4 V/SHE, that is, in both cases, there is considerable overvoltage compared to thermodynamic equilibrium voltages[12]. This property, which is *a priori* counterproductive from a catalytic point of view, has been used for the oxidation of organic pollutants in water [PAN 05].

Several research groups currently consider that electrodes made of BDD films are the best available option for the destruction of organic materials [HE 19]. Their high resistance to corrosion, non-fouling of their surface due to their hydrophobic character and the very high overvoltage required for oxygen production mean that large quantities of $^{\bullet}OH$ radicals are produced and adsorbed on the electrode surface in the form of BDD-$OH^{\bullet}$. These highly reactive radical species have the effect of significantly increasing the oxidizing power of the electrodes with respect to organic compounds. Given their very high oxidizing power, their efficiency in wastewater treatment is much better than that obtained with platinum, graphite or even DSA

---

12 The reduction of H^+ ions on BDD in acid medium with the release of dihydrogen begins at −1.2 V/SHE (Standard Hydrogen Electrode) compared to −0.2 V on a glassy carbon electrode (GC); the reaction corresponding to the oxidation of water with the release of O_2 occurs at 2.4 V/SHE, instead of 1.5 V on GC. It should be noted that these properties, while they are helpful in terms of the oxidation of organic pollutants, represent a disadvantage in terms of energy for water splitting, which corresponds to the production of dioxygen and dihydrogen.

(a "dimensionally stable anode" by DeNora) electrodes, which are usually used for this kind of application [GAR 15b].

BDD electrodes can also be produced in the form of porous films, increasing the active surface of the electrodes and thus improving electrochemical performance. However, this raises a number of difficulties, principally in terms of the adhesion of the BDD film to the substrate (usually Ti or Si); this must be improved before the film can be used for long-term electrolysis operations [GAR 15b].

1.3. Carbon dots or carbon quantum dots

Carbon dots (CDs) or carbon quantum dots (CQD) are smaller than 10 nm in size, and consist of a core of graphitic-type sp^2 carbons, corresponding to fragments of graphene or graphene oxide (GO), linked together by sp^3 carbons. Their surface contains significant quantities of oxygen, essentially in the form of COOH, carbonyls and hydroxyl groups [LIM 15] (Figure 1.9). Discovered by chance during the preparation and purification of CNTs [XU 04], CDs rapidly attracted attention for their fluorescent properties, which mean they have the potential to be used as reagents for biomedical imaging. As in the case of NDs, their non-toxic nature presents an additional advantage, notably making them preferable to inorganic, heavy metal semiconductor-based quantum dots (QD) in many cases.

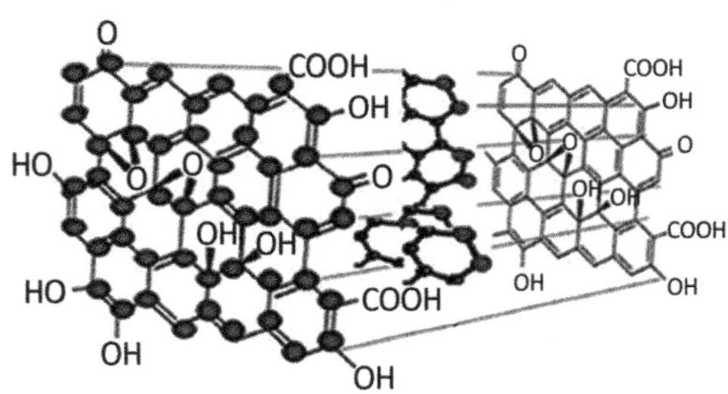

Figure 1.9. *Structure of carbon quantum dots (CQD) showing the association of graphene oxide and graphene fragments. Adapted from [DEM 13] and [LIM 15]*

1.3.1. *CQD production methods*

CQDs were first discovered in carbon soot. Produced by an electric arc discharge between graphite electrodes, CQDs were identified and isolated during electrophoresis separation of the various carbon NPs contained in the soot, thanks to their fluorescence [XU 04]. Interest in these new fluorescent nanomaterials soon resulted in the development of other methods, such as laser ablation of carbon-based precursors [SUN 06, HU 09].

Many different production methods have been described in the literature [WAN 14b, FER 15, LIM 15]. As fluorescence is a key property of CQDs, the main objective is generally to obtain the best possible fluorescence properties at the lowest cost; different approaches result in different specific properties. The most common production methods involve pyrolysis, applied to hydroxylated or easily oxidizable chemical compounds (amino products), carried out by microwave irradiation of the precursor product to create carbonization. This is the simplest and least expensive method.

For instance, the carbonization of citric acid in the presence of amines, used as "passivation" agents[13], produces (after microwave irradiation) CQDs with very high monodispersity and a size distribution between 2.3 and 3 nm. Laser irradiation of a solution of these QDs at 360 nm produces an intense blue fluorescence, centered around 450 nm, with a high quantum yield (QY) of 30% [ZHA 12].

Another way of treating a hydroxylated organic precursor consists of successively using a highly concentrated acid (e.g., H_2SO_4) for its dehydrating power, followed by an acid with a strong oxidizing power (HNO_3), which breaks down bonds, leading to the formation of CQDs. Passivation of the CQDs using an amino compound is then used to increase the intensity of fluorescence of the CQDs [WAN 14b]. The same result can be achieved in a single step by heating polyethyleneimine (PEI), in contact with concentrated nitric acid, to 120°C with reflux [SHE 13a]. This technique has the advantage of being readily available and applicable to many organic precursors, but does not allow for very precise control of CQD sizes [WAN 14b].

A variant of alcohol degradation by electrochemical oxidation in a strongly basic medium has also been proposed, and this also results in fluorescent CQDs. This technique has the advantage of simplicity, insofar as CQDs can be obtained in a single step; it is possible to control CQD size using this approach, but unfortunately only a few compounds can be transformed using this method [LI 10a].

---

13 Citric acid is often chosen as an organic precursor because of the ease with which it can be dehydrated and degraded, a result of the presence of carboxyl and hydroxyl groups. The "passivation" amine probably acts as a nitrogen doping agent for the CQD core, and is thus responsible for an increase in the intensity of fluorescence; this effect is similar to that obtained by N-doping nanodiamonds.

The hydrothermal route has also been widely developed, and allows high-quality CQDs to be obtained at relatively low cost. This method consists of heat treating an organic precursor, placed in a sealed reactor, in the presence of high-temperature steam. The carbonization reaction can be carried out by choosing organic precursors such as glucose, citric acid or orange juice, along with a whole series of easily degradable natural compounds [WAN 14b].

When the organic precursor is pyrolyzed in a confined space, it becomes effectively possible to control the size of the CQDs. The use of nanoporous materials such as silica (in the form of microspheres) means that organic precursors can be inserted into the pores, which then act as nanoreactors. Hydrophilic CQDs are thus obtained by pyrolysis in air of silica microspheres (average diameter 1.3 μm, pore size 3.6 nm), soaked in salts and citric acid, at 300°C for 2 hours. After dissolving the silica using hydrofluoric acid, dialysis is used to recover CQDs, which are almost mono-disperse (sized between 1.5 and 2.5 nm). These are characterized by excellent photostability and high QYs in terms of fluorescence (23%) [ZON 11].

CQDs can be given a variety of different properties by doping or functionalizing the surface; one reason to do this is to increase the intensity of fluorescence. The ultimate goal is to match the QYs obtained using certain organic dyes or inorganic QDs such as CdSe, for which QY can reach 80% [LIM 15].

1.3.2. *Fluorescence properties of CQDs*

Unlike inorganic QDs, for which absorption is determined by the band gap, CQDs are characterized by a very broad absorption band resulting from plasmonic absorption of the π orbitals[14]. This absorption band covers a large part of the UV-visible spectrum and extends into the near infrared, thus intercepting a large part of the solar spectrum [FER 15].

Two main emission mechanisms, which are not always easy to identify, are responsible for CQD fluorescence. The first results from transitions between energy levels in domains composed of conjugated sp^2 carbons; the second, less evident, relates to surface defects present in CQDs. A third type of mechanism has also been identified, corresponding to an up-conversion of photons, where an emission is produced in the visible spectrum as a result of excitation in the red spectrum.

14 Plasmonic absorption corresponds to a collective excitation of delocalized π-electrons on multiple benzene rings.

It is widely accepted that photoexcitation of CQDs results in a separation of charges in the core carbons, and that their recombination with surface defects results in fluorescence. This also explains the high degree of variation in emissions, which is not strictly related to a confinement effect, specific to mineral semiconductor QDs (Figure 1.10).

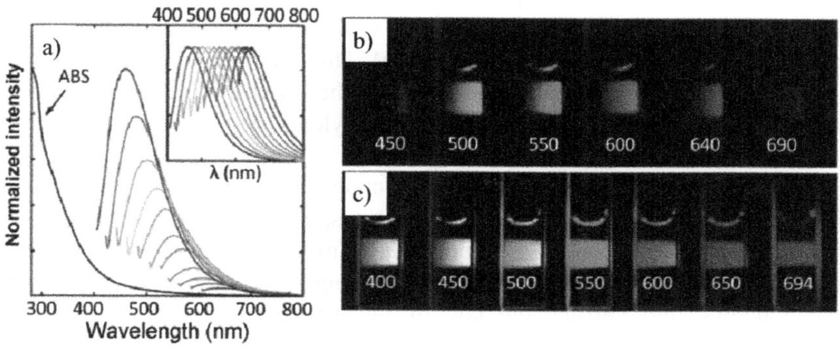

Figure 1.10. *Luminescence of an aqueous solution of CQDs produced by laser ablation of graphite powder*

COMMENT ON FIGURE 1.10.– *a) Absorption (ABS) and luminescence spectra obtained by excitation at different wavelengths (from 400 to 600 nm in 20 nm increments – from left to right) of poly(N-propionylethyleneimine-co-ethyleneimine) (PPEI-EI)-functionalized CQDs. Intensities are normalized with respect to quantum yields; the inset figure shows normalization with respect to peak intensity. b) Photographs of the luminescence of polyethylene glycol (PEG)-functionalized CQD solutions with a molar mass of 1500 and two amino groups – CH_2NH_2 (PEG_{1500N}) – at chain ends. Irradiation was applied at 400 nm, and the photographs are obtained after passing the emitted light through color filters of the indicated wavelengths (in nm). c) The same CQD solutions, photographed directly after irradiation at the indicated wavelengths (in nm). Adapted from [SUN 06].*

One of the most striking features of the fluorescence emission of CQDs is the way the emission spectrum depends on that of the source of excitation. Different explanations have been put forward for this dependence. The quantum size effect, defects and the surface states present in the CQDs, the presence of fluorophores with different degrees of conjugation and the radiative recombination of electron-holes due to clusters of sp^2 carbons enclosed in a matrix of sp^3 carbons all play a determining role, and may be responsible for fluorescence [WAN 10]. In general, the fluorescence QYs of most CQDs are in the range of 10%; however, in some cases, high yields of around 80% have been observed. These appear to be connected to the presence of nitrogen, resulting from the precursors used during pyrolysis (citric acid and ethylene diamine) [ZHU 13].

Finally, for CQDs prepared under certain conditions, multiphoton absorption has also been observed. This corresponds to an upconversion of the emitted photons [LI 10b], which is an essential property for biomedical imaging; excitation can be produced in the near infrared spectrum (800–900 nm), offering very high levels of penetration of biological tissues (several mm). This property was discovered by the Sun group [CAO 07] for CQDs with sizes of the order of 5 nm, passivated by a nitrogenous polymer (polypropionylimine-co-ethylenimine) (Figure 1.11).

One or two-photon absorption can be obtained through excitation in the visible (458 nm) or near infrared spectrum (800 nm), respectively. The photoluminescence generated in the two cases is essentially the same, centered around 500 nm. The intensity of "two photon" luminescence is comparable to that of inorganic QD semiconductors, and verifies the law of linearity between the intensity of the emission and the square of the excitation power (Figure 1.11b).

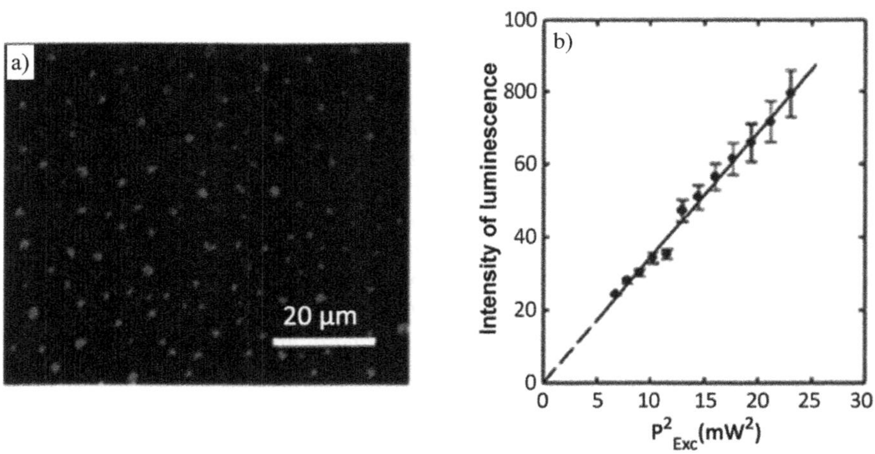

Figure 1.11. *Luminescence of CQDs deposited on a glass substrate*

COMMENT ON FIGURE 1.11.– *a) Image of CQDs excited by a femtosecond pulsed laser at 800 nm. b) Linear variation of the intensity of luminescence as a function of the square of the excitation power (P_{Exc}), characteristic of two-photon excitation. Adapted from [CAO 07].*

1.3.3. *CQD applications*

Given the ease with which CQDs can be produced and the numerous possibilities offered by surface functionalization, these have potential use in various multifunctional probes. For example, MRI and fluorescence imaging could be carried out simultaneously by combining magnetic and luminescent properties. This possibility has been illustrated by attaching nuclear resonance contrast agents (gadolinium complexes) to fluorescent CQDs [BOU 12].

Their solubility in aqueous media means that they are not only excellent diagnostic agents, but also very good vectors for transporting drugs to diseased biological cells; for instance, platinum-based anticancer drugs may be bound to CQDs [ZHE 14a].

Other applications have also been described [WAN 14b, HU 19] in domains as varied as analytical chemistry (sensor design), optoelectronics (photovoltaics and electroluminescence), and energy production and storage (supercapacitors and electrodes for lithium batteries).

1.4. Carbon nanotubes

Lijima [LIJ 91] was the first to demonstrate the formation of multi-walled carbon nanotubes (MWCNTs) by carrying out an electrical discharge between carbon electrodes in a reduced-pressure argon medium. Lijima thus obtained a mixture of MWCNTs with diameters ranging from 4 to 30 nm, around 1 μm long, and made up of two to seven concentric tubes. Single-walled carbon nanotubes (SWCNTs) were obtained shortly afterward, this time using cobalt-based catalysts [BET 93]. The structure of the walls of all of these nanotubes is identical to that of graphene: an assembly of a single layer of sp^2 hybrid carbon atoms in a regular hexagonal mesh.

Since CNTs were first produced and characterized, several teams have introduced different approaches to production and separation, described in the literature [GRO 07, HER 08, KOM 10, CHE 14c]. The main techniques used to produce CNTs include vaporizing carbon by laser ablation, electric discharge between graphite electrodes, catalytic chemical vapor deposition (CCVD), decomposition of CO under high pressure (HiPco process)[15] and the use of

15 The HiPco process was developed in the 1990s at Rice University. It allows high-purity SWCNT to be obtained by reacting carbon monoxide mixed with $Fe(CO)_5$ pentacarbonyl iron at high temperature (900–1,100°C) and under a pressure of around 30–50 atmospheres. The quantity of CNT produced using this process is of the order of 450 mg per hour, consisting of a mixture of SWCNT corresponding to about 10 distinct chiralities [BRO 01].

Co-doped silica nanoparticles (Co-MCM 41)[16]. The choice of catalyst and the technique used to fragment the carbon-based precursors play a determining role. The most widespread of these techniques is CCVD, which has the potential to produce large quantities of SWCNT and is thus best suited to production on a commercial scale (Figure 1.12).

Broadly speaking, the CVD technique consists of introducing a carbon source (ethylene, acetylene, toluene, CO, etc.) mixed with a carrier gas (H_2, argon) in vapor form into a reactor. The gas mixture, heated to a high temperature of at least 600°C, corresponding to the decomposition of the carbon-based precursor, comes into contact with the catalyst (in the form of metallic NPs) on which CNTs then develop.

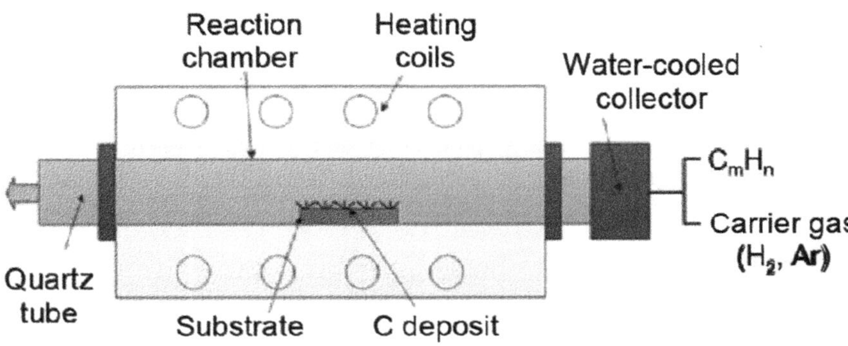

Figure 1.12. *Simplified diagram of the CVD apparatus used to produce CNTs. Adapted from Gore and Sane [GOR 11]*

The method is easily implemented on an industrial scale, as it only requires a tubular reactor, equipped with gas flow controllers, and an oven. However, the obtention of high CNT yields and well-defined chiralities remains problematic. The choice of a catalyst with a particular crystal geometry is crucial; a considerable number of different transition metal catalysts have been tested and described in the literature [WEN 10, TES 11, CHE 14c, WAN 15a, JAN 16].

The electrochemical route is also being investigated as a potential means of producing large quantities of carbon nanomaterials, such as graphene, carbon fibers, CDs and CNTs. One recent solution involves reducing CO_2 in the form of molten carbonate; this approach, developed by the Pint group for CNT production, has the

16 MCM-41 is a specific zeolite with pores that can be "tuned" between 2 and 10 nm. After depositing a thin layer of cobalt on the inner walls of the zeolite, it can be used as a nanoreactor for the production of SWCNT [LIM 03].

merit of converting CO_2 into valuable products [DOU 17]. Unlike in previous work, the authors showed that better selectivity with regard to CNT size could be achieved by using a thin layer of iron, deposited on a stainless steel cathode, as a catalyst. In contact with the molten electrolyte (Li_2CO_3), that is, at temperatures above 720°C, the thin metallic layer generates iron NPs on which CNTs will grow. The iron NPs grow as a function of electrolysis time and thus the diameter of the CNTs also increases; this makes it possible to control CNT size by adjusting electrolysis time. With an initial iron layer of 0.5 nm thickness, the average diameter of CNTs produced is 12.5 nm after 3 minutes and 23 nm after 1 hour of electrolysis using a current of around 100 mA/cm^2 [DOU 18] (Figure 1.13).

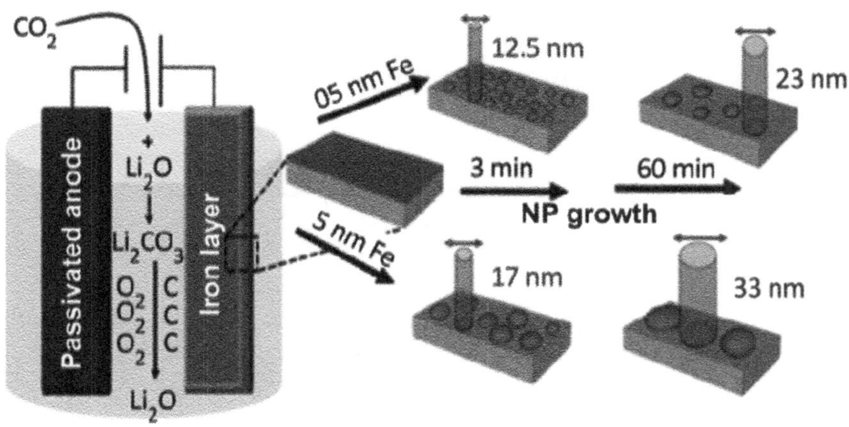

a) b)

Figure 1.13. *Schematic illustration of the electrolysis setup for CNT production*

COMMENT ON FIGURE 1.13.– *a) The cathode is covered with a thin layer of iron on which carbon is deposited. The anode is passivated by a layer of alumina on which O_2 is released. b) On contact with the electrolyte (Li_2CO_3), the iron layer generates iron NPs; the size of these NPs increases as a function of electrolysis time, as does the size of the CNTs. The diameter of the CNTs is calibrated by that of the iron NPs. Adapted from [DOU 18].*

According to the authors, the cost of producing CNT in this way is much lower than when using CVD techniques, and the yields are also higher. However, the process still needs to be refined in order to obtain CNTs of smaller diameter, and to increase selectivity.

Most of the techniques used result in CNT mixtures containing both MCVNTs and SWCNTs, with different architectures corresponding to different chiralities. Research into improving separation methods is still ongoing; researchers also aim to develop new catalysts which offer better selectivity in producing CNTs with a given configuration or chirality. Various theoretical models have been developed to further understanding of CNT formation mechanisms at an atomic level, notably in terms of the auto-assembly phenomena by which carbon atoms form CNTs and by which these CNTs grow [PAG 15].

1.4.1. *Chirality of carbon nanotubes*

The formation of an SWCNT may be seen as the result of "rolling up" a graphene fragment along a given path, defined by a C_h vector (Figure 1.14).

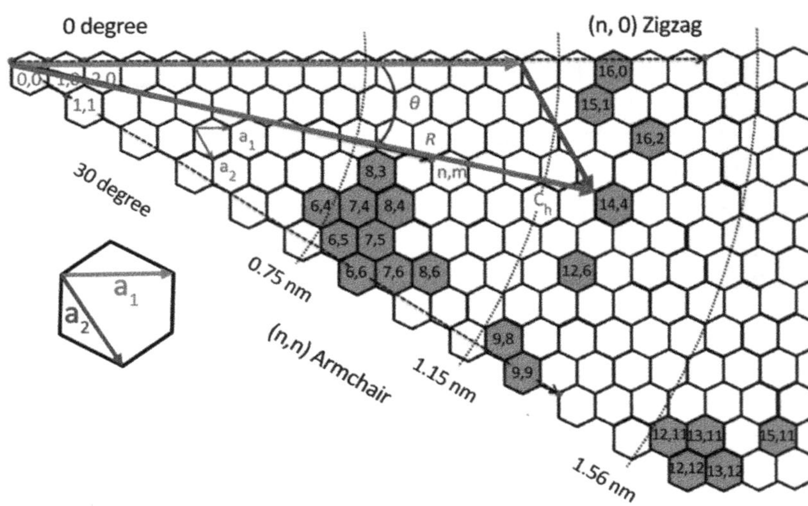

Figure 1.14. *Diagram showing the formation of an SWCNT defined by rolling up a graphene fragment along the chiral vector $C_h = na_1 + ma_2$*

COMMENT ON FIGURE 1.14.– *In the case shown in the figure, the C_h vector (red arrow) is equal to $14a_1 + 4a_2$ and corresponds to the formation of an SWCNT of chirality (14,4). The C_h vector is at an angle Θ with respect to the a_1 vector axis and defines an R (rectus)-type chirality. The diameter of the SWCNT (4,14) is the same as that of the SWCNT (14,4) with S (sinister) chirality. The hexagons shown in blue correspond to the most common chiralities of synthetically obtained CNTs with diameters between 0.7 and 2 nm. Adapted from [WAN 15a].*

The vector C_h, known as the chirality vector, is defined by the vector sum of two vectors $\mathbf{a}_1$ and $\mathbf{a}_2$ which characterize the network, such that C_h is equal to n $\mathbf{a}_1$ + m $\mathbf{a}_2$, where the angle between $\mathbf{a}_1$ and $\mathbf{a}_2$ is 60°. A CNT of chirality (n,m) is thus obtained by rolling up the graphene fragment along the direction of the vector C_h in such a way that it connects the two ends of the vector C_h. It follows, from this construction, that the circumference of the SWCNT is equal to the length of the vector C_h [BAC 02].

Chirality is particularly important because it has a decisive effect on the conductive properties of CNTs and, consequently, determines their use in fields such as nanoelectronics, photovoltaics and energy storage.

Within each family of SWCNTs with distinct chiralities (n,m), there are two limit forms: "armchair" (n,n) or "zigzag" (n,0); these correspond to winding directions at an angle Θ of 0° or 30°, respectively (Figure 1.15). Note that the armchair configuration offers metal-like conductivity. The "zigzag" (n,0) configurations, like SWCNTs of chirality (n,m), also possess metal-like conductivity when the difference between n and m is a multiple of 3 [MAU 05], and are semiconductors when this is not the case. The band gap widths are less than 1 eV and decrease as the diameter of the CNT increases [ODO 98].

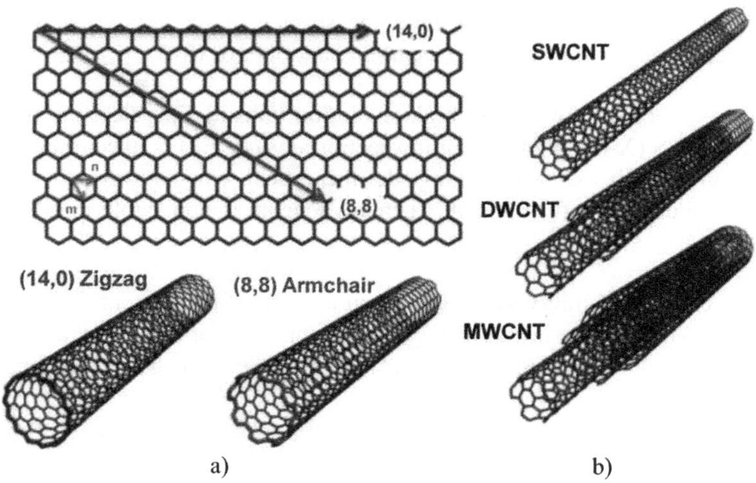

Figure 1.15. *Examples of different CNT configurations. a) Examples of two SWCNTs, (14,0) and (8,8), constructed from a single sheet of graphene and coiled along the vectors (14,0) (red) and (8,8) (green), corresponding to the zigzag and armchair forms. b) Images of single-, double- and multi-walled CNTs (SWCNT, DWCNT and MWCNT). Adapted from [SCH 11]*

Each of these configurations (n,m) has a distinct corresponding Raman spectrum and distinct CNT diameter, making it possible to identify them. The frequency of the Raman RBM (Radial Breathing Mode) band, corresponding to the radial mode of vibration of the nanotube (ω_{RBM}), is related to the diameter (d) by the relation $\omega_{RBM} = A/d + B$, where A and B are two experimentally determined constants [BAC 02, MAU 05].

Given the importance of chirality in terms of conductive properties, significant efforts have been made to develop separation methods [LIU 10a] along with preparation techniques for producing SWCNTs of defined chirality in the most selective way possible.

1.4.2. *Mechanistic models of CNT growth*

The growth of CNTs with a certain chirality is known to be closely correlated to the crystal structure of the catalyst and to the arrangement of atoms in the different planes of the crystal [TES 11]. The first models of CNT growth, VLS (Vapor Liquid Solid) followed by VSS (Vapor Solid), draw on the mechanisms at play in carbon fiber formation. According to these models, the carbon-based gaseous precursor, heated to a temperature of around 600°C, dissociates on contact with the catalyst, producing C atoms; these diffuse onto the surface of the catalyst and will then recombine, triggering the growth of a carbon nanofiber or CNT.

Numerous observations using high-resolution transmission electron microscopy (HRTEM), XPS (X-ray photo-electron spectroscopy) or EELS (electron energy loss spectroscopy) analysis, taken alongside theoretical calculations and molecular dynamic modeling, confirm that this general scheme is applicable to CNT growth (Figure 1.16).

However, CNT chirality still depends on a range of parameters such as the size of the NPs which make up the catalyst, the atomic composition of the alloy, the temperature at which the reaction occurs and the nature of the gas precursor.

For example, SWCNTs with a zigzag configuration (16.0) and a diameter of 1.25 nm are difficult to obtain in large quantities, but production techniques have been considerably improved in recent years by using W_6Co_7 crystals with specific cutting planes (116) [YAN 15]. According to the authors, this arrangement of the catalyst atoms results in a kind of "template" effect that directs the recombination of the carbon atoms toward the preferred CNT form (16,0), giving yields of around 80%.

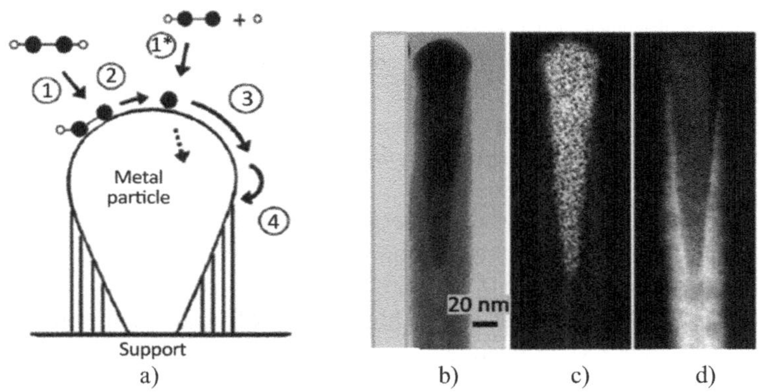

Figure 1.16. *Different steps in CNT growth using the VSS mechanism*

COMMENT ON FIGURE 1.16.– *a) (1) Adsorption; (2) decomposition of the carbon-based precursor, (3 and 4) diffusion of carbon atoms at the surface of the catalyst; (1*) partial dehydrogenation of the precursor. b) Bright field TEM. c) EELS image of Ni (absorption at 854 eV corresponding to the edge of the L band). d) EELS image of C (absorption at 284 eV corresponding to the edge of the K band). Adapted from [HOF 05].*

More recently, Wang *et al.* [WAN 18] have shown that high levels of selectivity can also be obtained by using nickel, cobalt or iron sulfates as catalysts; this amounts to introducing sulfur indirectly into the catalysts (Figure 1.17).

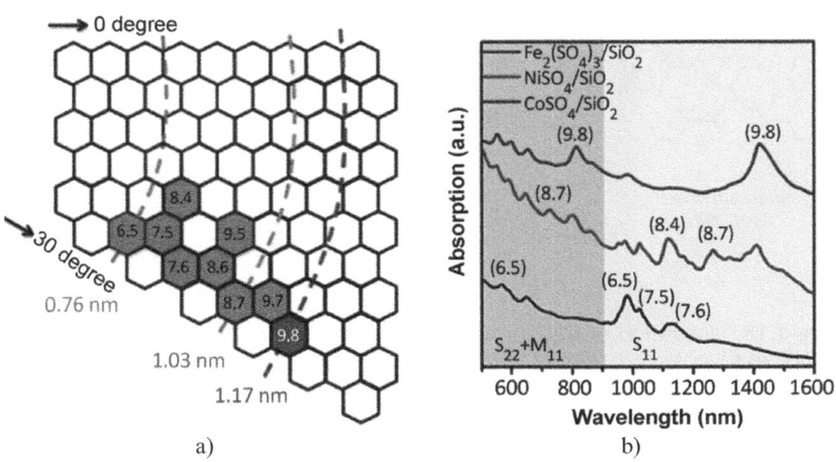

Figure 1.17. *Characterization of SWCNTs synthesized using different catalysts (mixtures of metal sulfates and SiO_2)*

COMMENT ON FIGURE 1.17.– *a) Distribution of the different SWCNTs (n,m) synthesized using ferric sulfate (green), cobalt sulfate (blue) and nickel sulfate (red) on a chiral "map". b) Near IR and IR absorption of different SWCNT mixtures obtained following separation by complexation using sodium dodecyl benzene sulfonate (SDBS) followed by ultracentrifugation. Adapted from [WAN 18].*

Thus, a dispersion of silica NPs impregnated with ferric sulfate ($Fe_2(SO_4)_3/SiO_2$) results, after heating, in a mixture of iron oxides and silicates, producing SWCNTs of chirality (6,5) and diameter 0.76 nm. Similarly, the replacement of ferric sulfate by nickel or cobalt sulfate results in high yields of larger diameter CNTs with chiralities (8,7) and (9,8), respectively.

These results confirm previous observations that the presence of a small amount of sulfur is necessary to stabilize the size of the metal catalyst particles, which are themselves critical for the growth of CNTs in SWCNTs or MWCNT form [JAN 16].

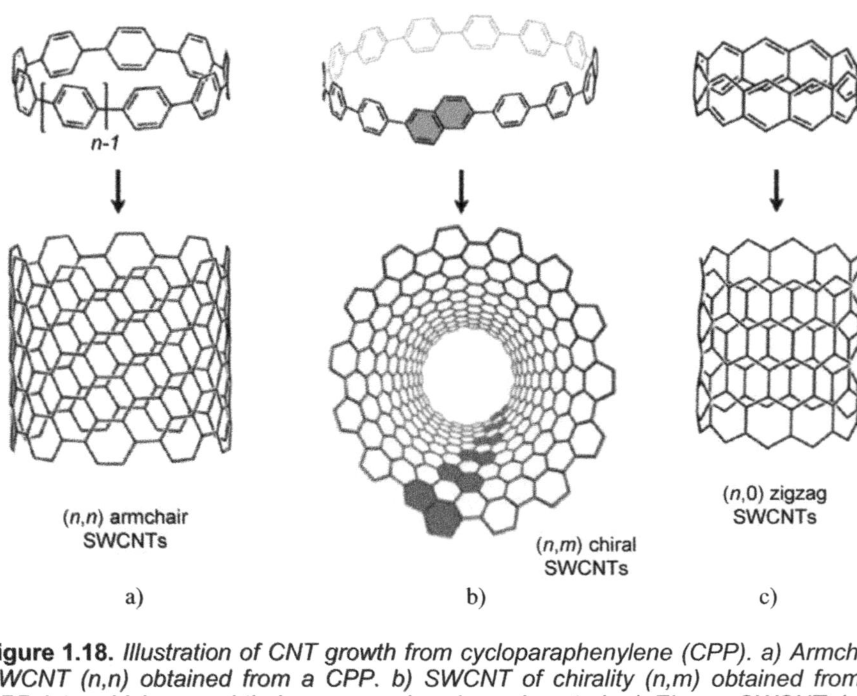

Figure 1.18. *Illustration of CNT growth from cycloparaphenylene (CPP). a) Armchair SWCNT (n,n) obtained from a CPP. b) SWCNT of chirality (n,m) obtained from a CPP into which a naphthalene group has been inserted. c) Zigzag SWCNT (n,0) which may be produced from n-acenes. Adapted from [PAG 15]*

This catalyst "template" effect has also been demonstrated using organic derivatives, such as cycloparaphenylene (CPP) carbon rings, which constitute the shortest segment of a CNT with an "armchair" configuration, in the place of metal crystals. In this case, the reaction is carried out at a temperature of 500°C, using ethanol as the carbon source. The growth of the CNT is thus calibrated to that of the carbon nanoring, meaning that the diameter of the CNTs can be adjusted by acting on the size of the CPP (Figure 1.18a) [OMA 13].

Note, however, that the level of selectivity obtained using this technique is not as high as those seen previously. For example, using a CPP made up of 12 phenylene nuclei ([12]-CPP) with a diameter of 1.7 nm results in a mixture of CNTs with diameters which mostly fall within a range of 1.7–1.3 nm; the most common diameter is 1.7 nm (over 20%).

A naphthalene group can be introduced into the ring [OMA 11] to break the symmetry and obtain chiral SWCNTs (Figure 1.18b). The use of a cyclic n-acene as a precursor is expected to result in SWCNTs with a zigzag configuration (Figure 1.18c), but at the time of writing, this hypothesis has yet to be tested.

1.4.2.1. *Classic separation methods for SWCNT mixtures*

Affinity or steric exclusion chromatography[17], electrophoresis and ultra-centrifugation are among the most common techniques used to separate materials of different shapes and sizes.

In the case of CNT, centrifugation techniques are most widespread due to their high efficiency, low cost of implementation and scalability.

The most popular method combines ultracentrifugation with an aqueous density gradient medium (DGU) which is widely used in biology. This technique has proven to be particularly effective for separating and sorting SWCNTs with determined chiralities.

This approach draws on small differences in the buoyancy density of particles dispersed in an aqueous solution with a density gradient. This density gradient is obtained by mixing a high-density compound with water in such a way that the density of the solution increases from the surface to the bottom of the solution. Under centripetal force, the analyzed species migrate toward the position at which the density of the species is equal to that of the solution, which has the effect of distributing the components of the mixture across different zones of the solution.

---

17 Steric exclusion chromatography is based on the difference in size of particles or macromolecules, whereas affinity chromatography is based on differences in physicochemical interactions between molecules and the chromatographic support.

As CNTs are hydrophobic, a surface modification is necessary to make them hydrophilic and facilitate their dispersion in an aqueous media. One approach involves the adsorption of polyaromatic derivatives, particular oligonucleotide sequences; a more common method involves the adsorption of amphiphilic surfactants such as sodium dodecyl sulfate (SDS) or sodium cholate (SC). These surfactants may coil and adsorb onto the CNT, transforming the CNT-surfactant mixture into a hydrophilic species compatible with the aqueous immersion medium. CNTs with very high purity levels can then be obtained following ultracentrifugation and washing (Figure 1.19).

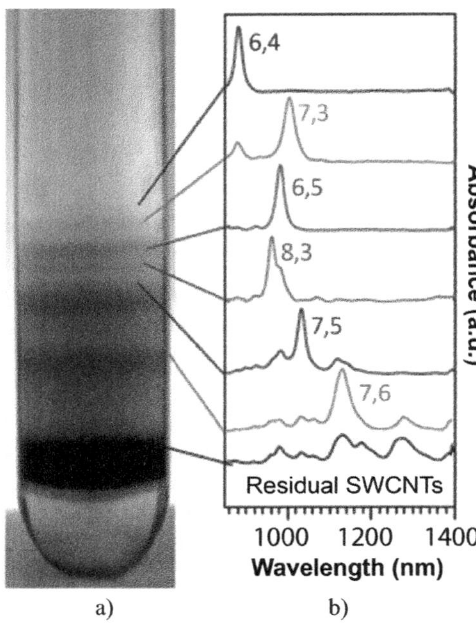

Figure 1.19. *Separation of SWCNTs of structure (n,m) using ultracentrifugation, applied to a mixture of SWCNTs obtained using the HiPCo method*

COMMENT ON FIGURE 1.19.– *a) The centrifuge tube initially contains seven distinct layers of water-iodixanol solutions with a concentration increasing from 15% to 30% by weight from the top to the bottom of the tube. After 18 h of centrifugation at 268000 g, different sedimentation regions are obtained, each containing SWCNT of defined chirality. b) Raman spectroscopy analysis of each colored slice confirms the presence of SWCNT which mostly correspond to well-defined (n,m) chiralities: (6,4), (7,3), (6,5), (8,3), (7,5) and (7,6). Adapted from [GHO 10].*

Various authors have obtained good results separating SWCNTs using this method [ARN 06, GHO 10, GRE 11]. Ghosh *et al.* [GHO 10] separated and isolated SWCNT from a mixture of CNT obtained using the HiPco method. Separation is carried out in a centrifuge tube in which several layers of liquid with distinct density gradients have been added, corresponding to solutions of increasing concentrations of iodixanol[18] in water, ranging from 15% (w/v) at the top of the tube to 30% (w/v) at the bottom of the tube.

1.4.3. *CNT arrays aligned horizontally or perpendicularly to a planar substrate*

CNTs which are horizontally or perpendicularly aligned with respect to a planar substrate present different properties and are thus suited to different applications. The excellent conductivity of vertically aligned CNTs, along with their ability to withstand very high electric currents, means that they can be used as vertical connectors between different layers of electronic components in 3D integrated circuits, using the TSV (Through silicon via interconnect) concept[19] [WAN 11]. This form is also ideal for use in designing cathodes which emit electrons under the effect of an electric field. Conversely, horizontal alignments are better for use in field effect transistors (FETs), which can be used in designing sensors, for example, for detecting traces of various chemical compounds. Substantial progress in this area has made it possible to obtain defect-free alignments, thanks to a particular set of experimental conditions, resulting in physical, mechanical and electronic properties which are close to the theoretical values [ZHA 17c].

1.4.3.1. *Vertically aligned CNT arrays*

Chhowalla *et al.* [CHH 01] were among the first to create vertically aligned CNT (VACNT) arrays on a silicon substrate using the Plasma Enhanced Chemical Vapor Deposition (PECVD) technique. They showed that the thickness of the catalyst layer, the deposition temperature and the proportions and flow rate of the acetylene/ammonia gas mixture play a decisive role in the vertical growth of the CNTs and in determining their structure [CHH 01]. Numerous variants have been tested, acting on the nature of the gaseous precursor and the catalyst, while maintaining the CVD technique; this has resulted in the development of VACNTs with various surface densities and lengths of up to several mm [HAT 04].

18 Iodixanol is a neutral iodine compound with formula $C_{35}H_{44}I_6N_6O_{15}$ which is used to create density gradients. It is available commercially in the form of an aqueous solution of 60% concentration (w/v), corresponding to a density of 1.32 g/ml.

19 The term VIA, used in nanoelectronics, stands for vertical interconnect access.

Zhang *et al.* [ZHA 08] suggested a particularly interesting use of VACNTS for producing electrochemical supercapacitors, formed by inserting redox NPs of manganese oxide (MnO_2) in contact with VACNTs deposited on a tantalum collector electrode (Figure 1.20). In this case, the VACNTs guarantee strong electrical contact between the metal electrode and the redox species inserted between the CNTs; furthermore, this results in a hierarchical porosity of the whole (macropores between the CNTs and micropores in the Mn oxides), which promotes the electrochemical diffusion of ions.

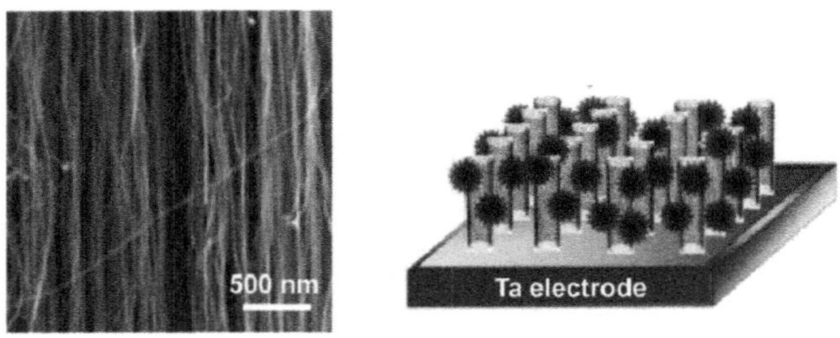

a) b)

Figure 1.20. *Diagram of a "composite" electrode, made up of a "forest" of SWCNTs (approximately 35 μm long) into which NP of MnO_2 have been inserted*

COMMENT ON FIGURE 1.20.– *a) SEM image of the VACNT array obtained by CVD (catalyst: iron phthalocyanine, carbon precursor: ethylene, CNT formation temperature: 550°C). b) Insertion scheme of MnO_2 NPs (electrodeposited from $MnSO_4$ solutions) inside the VACNT array. Adapted from [ZHA 08].*

Comparison with other composite electrodes, for example, with activated carbons (AC), clearly shows that the kinetic behavior of the VACNT solution is superior. While the specific capacities of the MnO_2/VACNT composite electrode and that of a MnO_2/AC electrode are comparable at low current densities (about 200 F/g each), the difference at high discharge intensities is considerable. Thus, for current densities of 10 A/g, the MnO_2/VACNT electrode maintains a capacity of around 130 F/g, whereas that of the MnO_2/CA electrode drops rapidly to around 50 F/g. For even higher intensities of 70 A/g, the capacity of the MnO_2/VACNT electrode is 101 F/g, making it particularly suitable for high-power applications [ZHA 08].

1.4.3.2. *Horizontally aligned CNT arrays*

Obtaining aligned CNT arrays on a substrate is essential for the production of various electronic components. Liu and coworkers were among the first to show that the orientation of CNTs could be controlled by CVD, by rapidly increasing the temperature of the gas stream, consisting of a CO/H_2 mixture, to 900°C [HUA 03]. In this way, it is possible to obtain 2D arrays of aligned SWCNTs, several mm in length, from monodisperse Fe/Mo NPs deposited on the surface of a Si/SiO_2 substrate.

Improvements to the process have since been made by Li and coworkers, resulting in the production of ultralong, horizontally aligned SWCNTs; these are highly useful for producing ultra-resistant fibers [JIN 07] (Figure 1.21).

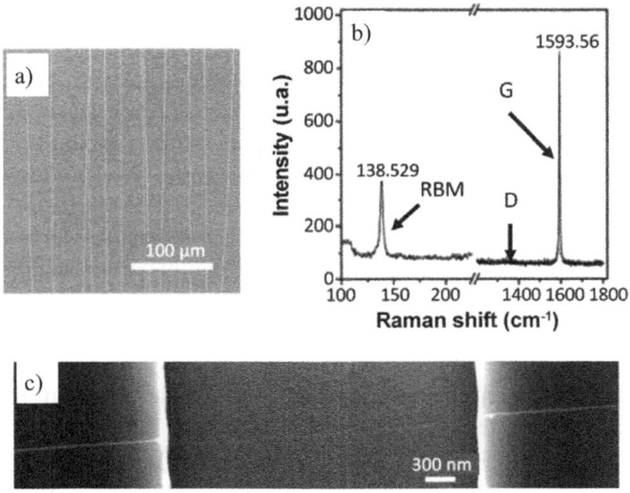

Figure 1.21. *Horizontally aligned SWCNT array on silicon substrate*

COMMENT ON FIGURE 1.21.– *a) SEM image showing the alignment and spacing of SWCNTs. b) Raman spectrum of a specific SWCNT. The RBM vibration corresponds to an SWCNT with a diameter 1.8 nm. The very low intensity of the D-band at 1350 cm^{-1} is an indication of the absence of defects in the CNT. c) SEM image of an SWCNT crossing a microtrench. Adapted from [JIN 07].*

They have shown that has shown that by slowly circulating the gaseous carbonaceous precursor (CH_4 + H_2) above the catalyst (iron and molybdenum NPs) at a very high temperature (970°C), CNTs grow steadily and can reach several cm in length [JIN 07].

One particularity of this growth mode is the floating character of the CNTs, which have the capacity to cross micro-trenches (~3 µm wide and 0.5 µm deep) or micro-obstacles without interrupting their progression; this is due to the fact that a catalyst NP remains attached to the end of the growing tube. Raman spectra confirm that the SWCNT obtained in this way present very few defects. High-quality horizontally aligned CNT (HACNT) arrays have also been obtained using iron-based catalysts. One advantage to this approach is that nanotubes can self-repair during formation; this eliminates most of the defects which may result in a loss of useful properties. The carbon source is a weakly humidified mixture of methane and hydrogen; the substrate is a silicon plate, and the gas mixture is maintained at a temperature of between 900°C and 1000°C [ZHA 17c].

1.4.4. *Key properties and applications of CNTs*

Space does not permit us to provide a full list of properties, current and future applications of CNTs here; an excellent summary has been published by De Volder *et al.* [DEV 13]. The diversification of preparation methods, notably the development of HACNT or VACNT assemblies and the production of CNTs of several cm length, has considerably increased the field of applications. Self-supporting films of agglomerated CNT with a paper-like form (buckypaper), with excellent properties from both a mechanical and electrical standpoint, are now available commercially and have a wide range of potential applications. Sheets of buckypaper (6 μm thick) can be manufactured by simple vacuum filtration of an MWCNT suspension; the film obtained has an elasticity modulus of 3 GPa and a conductivity of 2.10^4 S/cm depending on the state of deformation, meaning that it could be used in motion detection sensors [DEG 17].

CNTs have a wide range of uses in fields such as nanoelectronics, energy and biomedicine, where their exceptional electrical, mechanical and thermal properties are particularly helpful. These properties have contributed to the success of CNTs, to the point where many companies have developed large-scale production techniques. Over the space of a few years (2004–2011), world CNT production increased from 200 kg to 4500 tons per year, a 22500-fold increase [DEV 13]. However, it should be noted that a number of companies have recently stopped producing CNTs, presumably due to high production costs. The average cost of SWCNTs remains very high; prices depend on packaging methods and vary between manufacturers. In 2020, the cheapest SWCNTs fetched between $80 and $100/kg, while industrial grade MWCNTs cost between $200 and $400/kg.

1.4.4.1. *Electrical properties*

CNTs are characterized by high conduction due to their metallic nature. Wei *et al.* [WEI 01] have shown that MWCNTs can withstand current densities of more than 10^9 A/cm^2 over periods of two weeks, at temperatures of around 200–250°C, with no degradation of the CNT or variation in conductivity. These conductance properties are a result of ballistic-type conduction mechanisms, different from the classical electron displacement mechanisms found in metals. Semiconducting CNTs are also characterized by exceptionally high electron and hole mobility, far beyond that observed in silicon, reaching up to 79000 $cm^2\ V^{-1}\ s^{-1}$ [DÜR 04]. Bandwidth gaps are inversely proportional to the diameter of the CNT and are of the order of 1 eV for diameters of 1 nm, comparable to those of silicon [KAT 99].

1.4.4.2. *Thermal conduction*

CNTs are also excellent thermal conductors, but only along the axis. For a SWCNT 2.6 μm in length and 1.7 nm in diameter, thermal conductivity of 3500 $Wm^{-1}\ K^{-1}$ has been measured; that of copper, considered to be a very good thermal conductor, is only around 385 $Wm^{-1}\ K^{-1}$ [POP 06]. On the other hand, the thermal conductivity of CNTs in a radial direction is very low, only 1.52 $Wm^{-1}\ K^{-1}$, meaning that they can almost be considered thermal insulators in this direction [SIN 05]. This difference in thermal properties between the two orientations is the reason why CNT films, where assemblies are arranged randomly on a surface, have a thermal conductivity of the order of 1500 $Wm^{-1}\ K^{-1}$ [KOZ 17]. Note that these CNTs can withstand temperatures of up to 2800°C in a vacuum and up to 750°C in air.

1.4.4.3. *Mechanical properties*

Theoretical calculations performed on ideal CNT structures display exceptional mechanical characteristics, with a tensile strength of 100 GPa (10^{11} Pascal), a Young's modulus of 1 TPa (10^{12} Pascal) and a breaking strain of 18%. These values were obtained for defect-free CNTs of several cm in length. In terms of mass per kilogram, systems of this type have the potential to store a mechanical energy density equivalent to 1125 Wh/kg, and an immense (theoretical) power density of 144 MW/kg, far beyond that of all known devices – such as electrolytic capacitors and lithium batteries – in terms of performance. The power densities of these systems do not exceed 10 kW/kg and 1 kW/kg, respectively, and their respective energy densities of 3–4 Wh/kg and 100–200 Wh/kg are also much lower than the values obtained for CNT [ZHA 11].

Much work has also been carried out concerning the use of CNT as reinforcing fibers for polymers, resulting in the production of CNT/polymer composites. The techniques used to manufacture these composites, as well as their electrical and mechanical properties, have been described in detail in the literature [COL 06, SPI 10, SUN 13, CHE 17]. These properties not only depend on the nature of the polymer, the synthesis method and the dimensions of the CNTs, but also on the alignment of these CNTs and the degree of dispersion of the CNTs in the polymer matrix.

In practice, CNT/polymer composites are obtained by introducing CNTs into either a polymer solution or the polymer itself in a molten state. If the polymer is insoluble and infusible, the composite material can be synthesized by adding the corresponding monomer to the CNT suspension, then carrying out polymerization *in situ*. One further way of synthesizing a composite is to inject a polymer solution into a CNT array.

1.4.4.4. *Black body properties of CNTs*

Near-perfect black body properties have been observed using a vertical assembly of CNTs on a silicon substrate, created using the CVD process and with humidified ethylene as the carbonaceous derivative precursor [MIZ 09] (Figure 1.22).

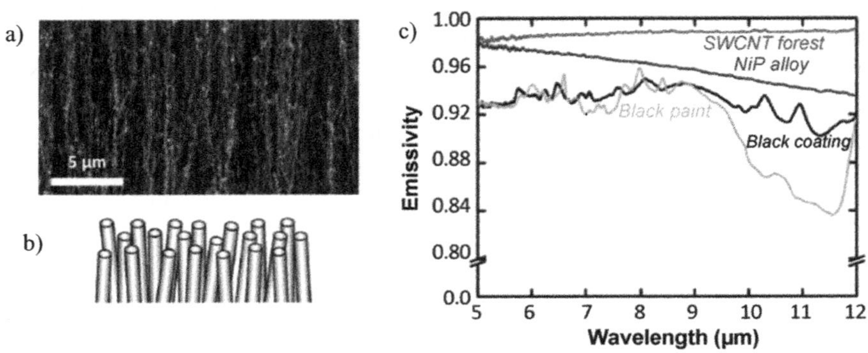

Figure 1.22. *SWCNT "forest" deposited vertically on a silicon substrate*

COMMENT ON FIGURE 1.22.– *a) Profile image of the CNT deposit. b) The slight defect in parallelism between SWCNTs is probably responsible for the very high emissivity. c) Emissivity of the SWCNT forest compared to that of NiP alloy, black paint and black coating. Note that beyond 9–10 µm, the emissivity of NiP, carbon black and black paint decreases significantly in comparison with SWCNTs. Adapted from [MIZ 09].*

Emissivity of 98%–99% is obtained across a whole spectral band from UV (200 nm) to far IR (200 µm)[20], much higher than values obtained for graphite or carbon black (between 80% and 85%); this level of performance is also very rarely obtained using other products, such as black paints or arrays of nickel-phosphorus alloy needles, which are particularly reputed for their capacity to absorb light radiation.

---

20 The emissivity of an object under irradiation is defined as the ratio between its emitted energy and that of the ideal black body.

1.4.5. *Conclusion*

The strategic importance of CNTs, as a family of nanomaterials, has become increasingly apparent since their first discovery by Lijima in Japan in 1991. Their exceptional electrical and mechanical properties have given rise to a significant number of applications, a development promoted by widespread commercialization. However, obtaining large quantities of CNTs with high purity levels and specific chiralities still presents a challenge. Current research is directed at identifying new catalysts to improve production on an industrial scale and to further broaden the field of applications, all while lowering production costs.

1.5. Graphene

Graphene was long considered as an ideal model, used in theoretical studies of 2D materials. A graphene fragment was first isolated by Geim and Novoselov [GEI 07] by mechanical exfoliation of a graphite sheet[21] (Figure 1.23).

Remarkably, the graphene film obtained in this way (with dimensions of around 10 μ) turned out to be extremely stable – contrary to previous predictions – and easy to subject to a whole series of electrical and mechanical measurements. Moreover, a monolayer of graphene can be easily identified using Raman spectroscopy; a characteristic line (G) appears at 1584 cm^{-1}, with a further line (2D or D′) at around 2700 cm^{-1}, the width of which increases as the number of graphene layers increases [FER 06, GRA 07].

Graphene presents exceptional physical, chemical and mechanical properties, leading many to consider it as the material of the 21st century. Its unprecedented electrical properties make it a material of choice for electronics, and have considerably improved the performance of various devices used in electronics and optoelectronics, in the energy sector and for biomedical analysis. The first graphene-based transistors were built in the 2010s. However, as Schwierz indicated [SCH 10], the fact that the band gap of graphene is zero means that it cannot be used as a FET for digital logic operations. A band gap opening is necessary for these applications; nevertheless, one solution is to use very fine graphene ribbons, for which band gaps of 200–300 meV have been observed. Graphene has been used in developing new

---

21 Although this technique may appear simple, the graphene sheet deposited on a Si substrate covered with a thin layer of SiO_2 only becomes visible if there is an optical contrast between the silica and the graphene. This only occurs if that the thickness of the silica layer is perfectly matched to that of the graphene sheet. Geim and Novoselov (who were awarded Nobel Prize for Physics in 2010) acknowledged that they were only able to identify the presence of a graphene monolayer through a fortunate combination of circumstances [GEI 07].

photonic and photoelectronic devices [BON 10], and has also been put forward as a potential component for batteries and capacitors (see Chapter 3) or photovoltaic cells (see Chapter 4) [BRO 11]. It also has promising applications in biosensors [PUM 11].

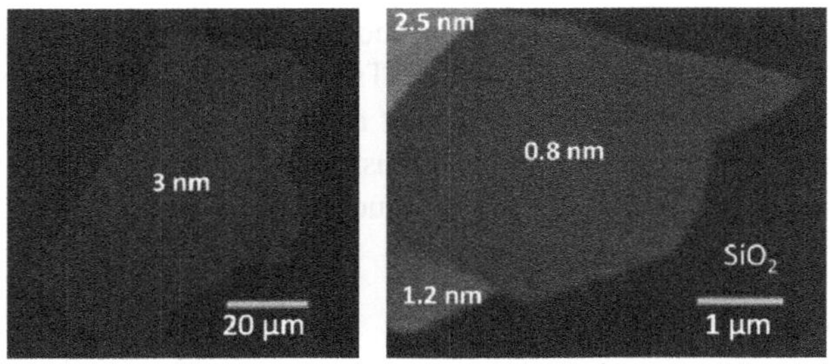

a) b)

Figure 1.23. *Graphene films deposited on a Si substrate covered with a thin layer of SiO_2*

COMMENT ON FIGURE 1.23.– *a) Photograph (in white light) of a multilayered graphene fragment, thickness 3 nm. b) AFM image of different thicknesses of graphene fragments. The central section, with a thickness of 0.8 nm, corresponds to a double layer of graphene. A thickness of 0.4 nm, corresponding to a single layer, is detected at the very bottom of the film. Adapted from [NOV 04].*

1.5.1. *Electrical properties of exfoliated graphene*

Graphene, which consists of an assembly of sp^2 carbons arranged in a perfect 2D hexagonal lattice, may be assimilated to a semiconductor material which has an energy band gap of zero (Figure 1.24).

The conductivity of graphene and the mobility of electrical charges are both extremely high, and these essential properties are what set graphene apart from other 2D materials, such as $NbSe_2$ or MoS_2. It can be doped with either positive or negative charges, by simply applying an electrical field. In contrast, the application of an electrical field in other 2D semiconductors only modulates the concentration of electrons, without reversing the passage from negative to positive charge carriers according to the polarity of the electrical field [NOV 05a]. The charge carriers in graphene do not obey the laws of non-relativistic classical quantum mechanics deduced from the Shrödinger equation; instead, they obey the laws of relativistic Dirac physics, in which low-energy electrons are assimilated to massless quasi-particles, with spin of 1/2, corresponding to 2D Dirac fermions. Their electrical properties have been extensively described in numerous articles by Geim and Novoselov [GEI 07] and Novoselov *et al.* [NOV 05b]. It is possible to interpret all of the observations relative to the electrical properties of graphene – notably those concerning the Hall effect – using this theory.

Dirac physics involves conical valence and conduction bands, known as Dirac cones. These intersect in the space of wave vectors k associated with these particles at a singular point of zero energy and zero charge (the Dirac point). One of the paradoxes inherent in this theory is the existence of non-zero conductivity at the Dirac point despite the fact that the corresponding charge density is zero (Figure 1.24a).

The differences in conductivity between graphene and 2D semiconductor materials such as $NbSe_2$ and MoS_2, each incorporated as a channel in a FET, are clearly shown in Figure 1.24b. In the case of graphene, the conductivity variation curve is perfectly symmetrical with respect to the gate voltage $V_g = 0$ V, while for bulk $NbSe_2$ and MoS_2, we observe a monotonic variation in conductivity as a function of V_g.

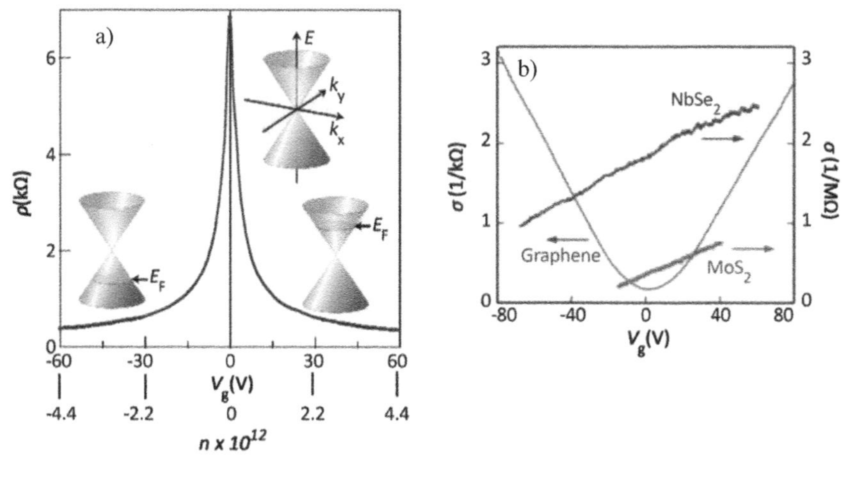

Figure 1.24. *Resistivity variations of graphene and two 2D materials as a function of gate potential (Vg) and charge densities*

COMMENT ON FIGURE 1.24.– *a) Resistivity ρ of a graphene sheet inserted as a channel in a field effect transistor subjected to a gate potential V_g. The perfect symmetry of the curve $\rho_{xx} = f(V_g)$ with respect to the axis $V_g = 0$ V corresponds to completely pure graphene and is obtained at very low temperatures. Note that for a gate voltage and a charge density of zero, graphene maintains a finite resistivity value of around 7 kΩ. The different conductivity regions are shown alongside energy band diagrams and the Fermi level (E_F) corresponding to each polarization;*

the negative and positive values of n correspond to holes and electrons, respectively. With the device used here, the density of charge carriers in electrons or holes varies linearly as a function of voltage, according to $n = \alpha V_g$ *where* $\alpha = 7.2 \times 10^{10}\ cm^{-2}V^{-1}$. *Adapted from Geim and Novoselov [GEI 07]. b) Comparative conductivity curves of graphene and two 2D semiconductors,* MoS_2 *and* $NbSe_2$. *Adapted from [NOV 05b].*

Using a FET and placing the graphene (FET channel) above a "trench" made in the gate dielectric (SiO_2), Bolotin *et al.* [BOL 08a] were able to show that the electrical properties of graphene are closely linked to the preparation method and are highly sensitive to the possible presence of impurities, which may simply result from contact with the support. The conduction and mobility curves (Figures 1.25a and 1.25b) show a significant increase in these two quantities following thermal treatment.

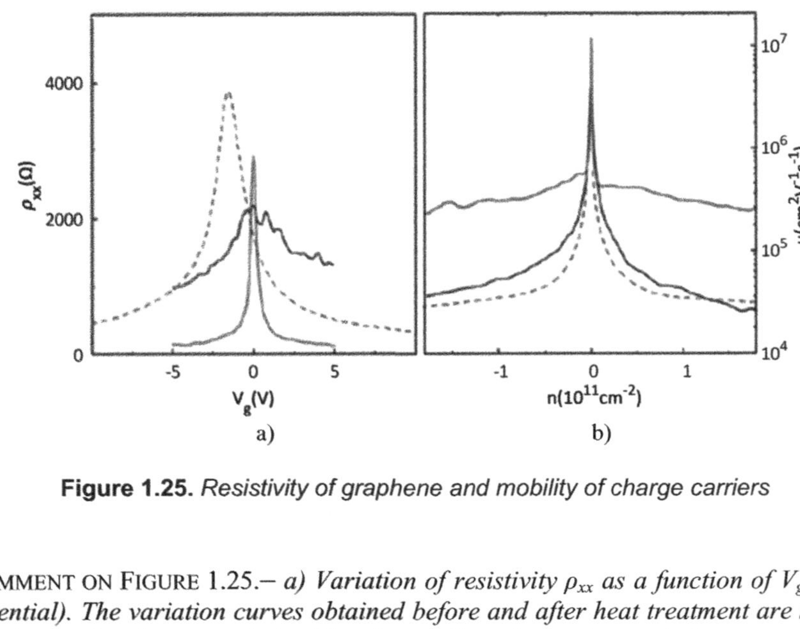

Figure 1.25. *Resistivity of graphene and mobility of charge carriers*

COMMENT ON FIGURE 1.25.– *a) Variation of resistivity* ρ_{xx} *as a function of* V_g *(gate potential). The variation curves obtained before and after heat treatment are shown in blue and red, respectively. The dotted line shows the variation curve for a standard graphene fragment. b) Variation of mobility* μ *as a function of charge densities, before (blue) and after (red) heat treatment. Adapted from [BOL 08b].*

This indicates that the very high mobility values of over 150000 cm^2/Vs cited in the literature can only be obtained for extremely pure graphene. These values correspond to ballistic-type charge transport mechanism, for which the mean free path is of the order of 1 µ. Mobility values for untreated graphene are much lower (around 15000 cm^2/Vs) and the transport mechanism which occurs in this case is partially dependent on diffusion; the mean free path of the charges is also smaller, at around 150 nm [BOL 08a][22].

1.5.2. *Graphene production techniques*

Several techniques have been developed for producing graphene on a commercial scale. Notable examples include the CVD method developed in 2008–2009 and the chemical and electrochemical exfoliation methods [PAT 14], which are less costly than CVD but which produce lower quality graphene.

One technique involves growing graphene epitaxially on a SiC crystal, under atmospheric argon pressure and at high temperature (1650°C). This particular technique eliminates graphene transfer operations and thus makes it possible to obtain a graphene film of the same size as the crystal, deposited directly on an insulating support; these properties are essential for the creation of FETs [EMT 09].

1.5.2.1. *CVD on a solid catalyst*

The first graphene deposits obtained using the CVD technique were made on mono- or poly-crystalline nickel and copper substrates. In both cases, a gaseous stream of methane and hydrogen heated to about 900°C–1000°C was used; this gaseous mixture decomposes on contact with nickel or copper, releasing carbon atoms. Variations of both processes have also been developed, essentially involving modifications to the gas flow rate and to the proportions used in the CH_4/H_2 mixture [ZHA 13]. Note that formation conditions using CVD are similar to those used for CNTs; the only differences relate to the nature of the catalyst and the higher temperatures used.

1.5.2.1.1. Graphene deposition on nickel

The fact that CVD can be used to form thick graphitic deposits on solid nickel substrates is a well-known process, which was adapted by Kim *et al.* [KIM 09] to obtain graphene monolayers. The authors replaced the solid nickel with a very thin film (300 nm), heated to about 1000°C, over which a mixture of methane and hydrogen flows. As the methane dissociates, carbon atoms are released, and these then solubilize inside the nickel; on cooling, the atoms re-diffuse on the surface and

22 For the majority of the most conductive materials, the mean free path, which depends on the applied electric field, is between a few nm and a few tens of nm. The mean free path of copper, for instance, is around 40 nm [GAL 16].

recombine to produce a highly homogeneous monolayer of graphene film (Figures 1.26a and 1.26b).

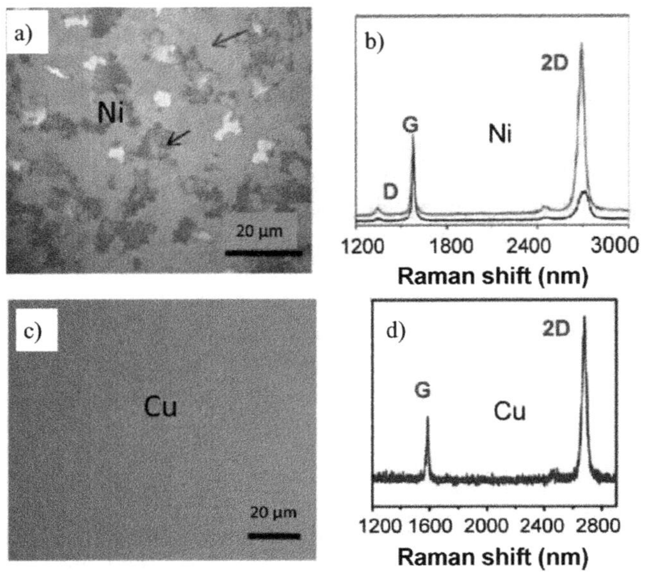

Figure 1.26. *Graphene deposits obtained on Ni and Cu substrates, transferred to a Si/SiO$_2$ substrate*

COMMENT ON FIGURE 1.26.– *a), c) Optical images of graphene fragments deposited on Ni and Cu, respectively. Note the homogeneity of the graphene film obtained on Cu in contrast with the deposits on a Ni substrate. b), d) Raman emission spectra of graphene for the two deposits. b) In the case of nickel, the heterogeneity of the deposit (fragments indicated by black and red arrows in image a) results in two distinct Raman spectra; the red spectrum corresponds to a monolayer of graphene with some defects (non-zero D band around 1300 cm^{-1}), while the spectrum shown in black corresponds to a multilayer deposit (characterized by a very low-intensity of 2D band). The perfect homogeneity of the graphene film obtained on copper is confirmed by the presence of the two characteristic 2D and G bands alone, with an intensity ratio I_{2D}/I_G of almost exactly 2; this is characteristic of a monolayer graphene film. Adapted from [ZHA 13].*

According to the authors, graphene formation is promoted by the small number of carbon atoms dissolved in the nickel film and by the fact that the lattice of nickel atoms (111) is, in geometric terms, highly compatible with the graphene lattice.

Single-crystal Ni must be used in order to obtain graphene of homogeneous thickness. Polycrystalline nickel features more grain boundaries, resulting in more homogeneity defects in the form of multiple layers along the grain boundaries in the metal; these can be seen in the respective Raman spectra[23].

1.5.2.1.2. Graphene deposition on copper

The experimental conditions (CH_4, H_2 and temperature) used for copper are similar to those used with nickel [LI 09b]. However, the use of a copper substrate offers considerably greater control over the formation of a monolayer of graphene with a larger surface area (dimensions of around 1 cm) (Figures 1.26c and 1.26d).

This difference is due to the fact that the carbon atoms coming from the dissociation of methane are only weakly soluble in copper; they thus adsorb directly onto the surface of the metal, producing graphene [ZHA 13]. The images in Figures 1.26c and 1.26d clearly show an improvement in the size and homogeneity of the graphene layers when using copper instead of nickel.

1.5.2.2. *CVD on a liquid catalyst*

One new trend in 2D material production methods involves the use of liquid metal catalysts in CVD. Applied to graphene, this approach might improve its quality. The technique described in detail by Zeng *et al.* [ZEN 14] is comparable to previous CVD techniques, except for the fact that the catalyst is in a liquid state. The carbon atoms resulting from dissociation of the carbon-based precursor (CH_4) are initially dissolved in the molten metal catalyst. The temperature is slowly reduced, leading to a phase change in the catalyst, from liquid to solid surface, and the dissolved carbon atoms then stop diffusing toward the surface. The solidification of the surface limits the concentration of carbon atoms at the surface, thus promoting the development of graphene monolayers; this result has been observed experimentally (Figure 1.27).

---

23 Different graphene films (G, 2D and D) have different Raman emission bands, meaning that it is possible to assess their thickness and the presence or absence of defects. A monolayer of graphene is characterized by an intensity ratio of approximately 2 between the I_{2D}/I_G bands. Thickness (in terms of the number n of layers) can be measured from the position of the G band, which verifies the wave number relation $\omega_G = 1581.6 + 11/(1 + n^{1.6})$ cm^{-1}. The D band appears only when there are defects present in the graphene, and is located around 1300 cm^{-1}. The 2D band, sometimes considered as a harmonic of the D band, is particularly intense for monolayer graphene and is located at 2750 cm^{-1}. The band decreases in intensity and widens as the number of graphene layers increases [MAL 09, WAL 11].

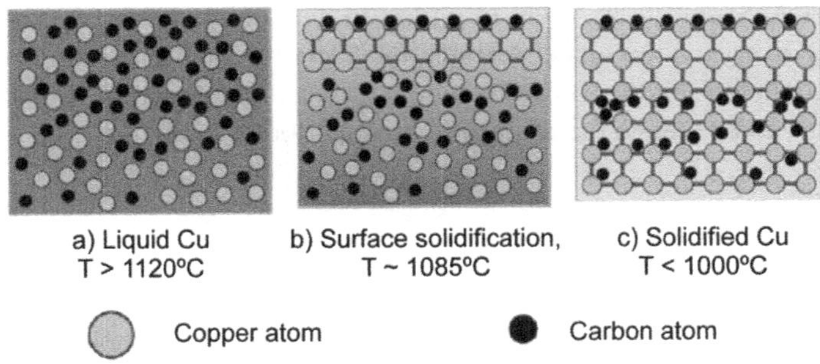

Figure 1.27. *Different steps in the catalytic process leading to a graphene monolayer*

COMMENT ON FIGURE 1.27.– *a) Copper in liquid state with carbon atoms in solution. b) Initial cooling causes surface crystallization of the copper, preventing C atoms in the core of the liquid from diffusing to the surface and thus maintaining a monolayer of carbon deposits on the copper surface, leading to graphene. c) Further cooling keeps the graphene monolayer on the copper surface. Adapted from [ZEN 14].*

In the case of copper, the CVD reaction lasts around 30 minutes at 1120°C; cooling is then progressively applied at the rate of 20°C per minute while maintaining a gas stream of a H_2/CH_4 mixture with a flow rate proportion of 300:6 sccm (standard cubic centimeters per minute). Other metals such as indium or gallium can also be used to synthesize graphene under similar conditions.

The technique can be used to produce homogeneous graphene monolayers of large dimensions. This results from the fact that isotropic graphene grains (IGG) form on the molten copper, and, due to their high mobility, have the ability to react together, with the addition of carbon atoms coming from the precursor, while still respecting the hexagonal geometry of graphene (Figure 1.28).

Remarkably, the charge carrier mobilities measured both intra- and inter-grain are approximately the same (around 3000 cm^2/Vs); this confirms the homogeneity of graphene after the coalescence of two IGGs. The coalescence of large numbers of grains results in dimensions of almost 1 cm^2; this constitutes significant progress toward obtaining large surfaces of high-quality graphene. This method is promising in terms of wider applicability, and may also be used for controlled synthesis of other 2D materials [ZEN 18].

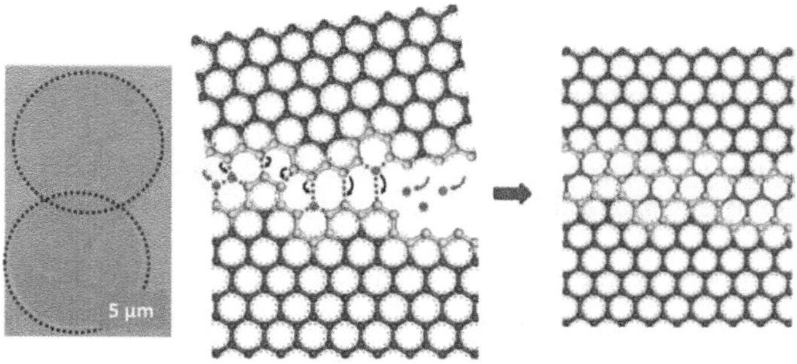

a) b) c)

Figure 1.28. *Formation of a single layer of graphene by "joining" two graphene grains (IGG). a) Optical image of two partially overlapping IGGs. b) Formation of –C–C– bonds between the two IGGs with the addition of carbon atoms (in red). c) A single layer of graphene is obtained when the IGGs interconnect. Adapted from [ZEN 16]*

1.5.2.3. *CVD for large-scale production*

Given the strategic importance of graphene, several improvements to the CVD process have been suggested over the last few years, aimed at increasing the scale and lowering the cost of graphene production. CVD is the only known process which is suitable for large-scale graphene production, insofar as the two successive operations involved in synthesizing graphene and transferring it onto a flexible support can be carried out in a continuous and automated manner.

One notable example is the adaptation of the roll-to-roll or "R2R" printing process for producing graphene in the form of films measuring 30 cm wide and several meters long. Research in this area has received significant financial support from industrial groups and is the driving force behind a number of university-industry consortiums[24].

Several variants of the CVD process used in R2R technology have been put forward in recent years, as described by Xin *et al.* [XIN 18]. This technique has been used to produce large graphene films on a flexible, transparent plastic support. Products of this type, which are essentially equivalent to transparent and mechanically flexible metallic electrodes, may be used in a wide range of applications; notable examples include photovoltaics, the production of large organic light-emitting diodes (OLEDs), opto-electronic sensors and, more generally, a whole range of portable and flexible electronic and photoelectronic devices [XIN 18].

24 The European Union has provided funding for a consortium of universities and industrial actors aimed at developing large-scale graphene production techniques.

Three major CVD processes which are compatible with the R2R technique have been developed for large-scale production of graphene on copper substrates. These include atmospheric pressure CVD (APCVD) [VLA 13], low pressure CVD (LPCVD) [KOB 13] and plasma-enhanced CVD (PECVD) [YAM 13] and mostly involve the use of a copper ribbon as catalyst. APCVD is the easiest of the three processes to implement and also produces the lowest quality of graphene [XIN 18].

The second step in the graphene synthesis process consists of detaching the graphene from the copper ribbon and transferring it to a transparent, flexible plastic support. Several different methods have been proposed for carrying out this operation in a continuous manner. One consists of hot-pressing the Cu/graphene film onto an adhesive plastic tape, then using chemical or electrochemical techniques to strip away the copper. The result is a composite graphene/plastic film with electrical conduction and optical transmission characteristics comparable to those of ITO (indium tin oxide) electrodes, with the added advantage of high mechanical flexibility in comparison with ITO.

Bae *et al.* [BAE 10], working in collaboration with Samsung, were among the first to produce hybrid rolls of graphene bonded to plastic, almost 1 m in length, using the R2R technique. Later, Kobayashi *et al.* [KOB 13], working for Sony, succeeded in producing graphene films of over 100 m in length, with a quality comparable to that of ITO plates, using this method (Figure 1.29).

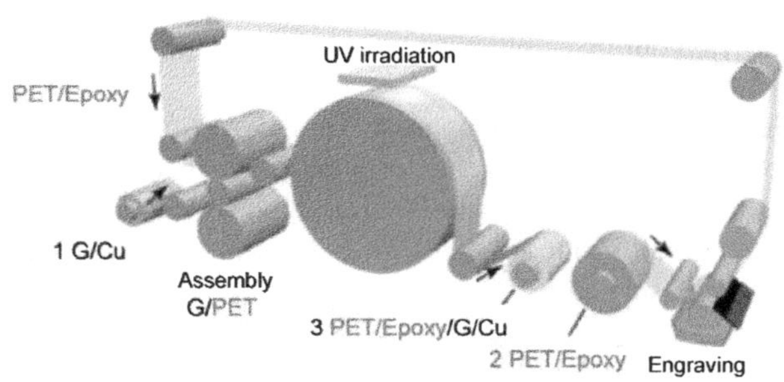

Figure 1.29. *Steps in continuous tape formation [PET/epoxy/G/Cu]*

COMMENT ON FIGURE 1.29.– *The first tape (PET + epoxy) is etched, then is brought into contact with the graphene-on-copper tape [G/Cu]. UV irradiation of the epoxy creates adhesion between the graphene and the PET, creating a [PET/epoxy/graphene/Cu] tape which is then wound onto a spool. The second step (not shown in the figure) is to unwind the composite film in a copper etching bath (acid solution), followed by various rinsing and drying operations needed to produce the final PET/G coil. The PET/G tape displacement speed is around 10 cm/minute. Adapted from [KOB 13].*

Kobayashi *et al.* [KOB 13] used copper tape, heated to around 1000°C using the Joule effect, which passes slowly through a chamber in which a gaseous mixture of methane and hydrogen was made to circulate at reduced pressure. Precise regulation of the gas flow and the copper tape feed speed was essential in order to obtain a high-quality monolayer graphene film.

Once the graphene film has been formed, a further continuous process is used to transfer it onto a plastic substrate. This is done by bringing the [graphene/Cu] tape into contact with a thermo-adhesive tape formed from a combination of PET (polyethylene terephthalate) and epoxy, which becomes adhesive under UV radiation; this initially produces a hybrid [Cu/graphene/epoxy/PET] tape. This tape is passed through an acid solution (to dissolve the copper), rinsed with water and then dried to produce a graphene film on a plastic substrate. Graphene prepared in this way (a tape of 100 m length and 20 cm wide, supplied in coil form) is characterized by a square resistance of between 150 and 250 ohms/sq and by an optical transmittance of around 85% [KOB 13].

This conductivity is significantly lower than that of extra-pure graphene, but can be improved by depositing metallic silver nanowires on the PET tape. In this case, the square resistance is less than 10 Ω; the optical transmittance value remains very high, of the same order of magnitude as before [DEN 15a].

1.5.2.4. *Chemical techniques for reduced graphene oxide (r-GO) production*

The starting material used in r-GO production is graphite oxide, which may be considered to be a multi-layered precursor for GO. A wide variety of graphite oxidation methods have been described. One of the oldest, which is still widely used, is the Hummers method, which consists of oxidizing graphite powder with potassium permanganate[25]. Exfoliation is then carried out using strong ultrasonic agitation in a liquid medium; the whole operation takes around an hour [EIG 13].

The chemical reaction described by Hummers results in significant degradation of the crystal lattice, as C–C bonds break down, following oxidation and the release of CO_2. The number of defects can be reduced by carrying out the operation at low

25 Hummers' method consists of gradually adding potassium permanganate to graphite powder suspended in a solution of silver nitrate and sulfuric acid. The reaction is exothermic and can become explosive if the temperature is not controlled [HUM 58]. Starting with 1 g of graphite powder, Eigler *et al.* [EIG 13] were able to obtain 500 mg of GO and 200 mg of graphene following chemical reduction.

temperatures, limiting the release of CO_2, bond breakage and the appearance of structural defects [EIG 13].

Once GO has been obtained, in the form of a suspension in a water-alcohol mixture (0.1 mg/mL), the GO is dispersed onto a solid support, then reduced using a mixture of hydroiodic acid (IH) and trifluoroacetic acid (CF_3CO_2H) in vapor form to produce reduced graphene, r-GO, in the form of flakes of between 500 nm and 1 μm in size[26]. Statistical Raman analysis of these r-GO flakes shows that after reduction, they can be classified into three distinct fractions based on the full width at half maximum (FWHM) of the 2D Raman band centered on 2700 cm^{-1}. These fractions are: (1) r-GOs with low structural defect densities (30 cm^{-1} < FWHM < 40 cm^{-1}); (2) r-GOs of "average" quality (40 cm^{-1} < FWHM < 50 cm^{-1}); and (3) a high proportion of r-GOs with large numbers of structural defects (FWHM > 50 cm^{-1}) (Figure 1.30).

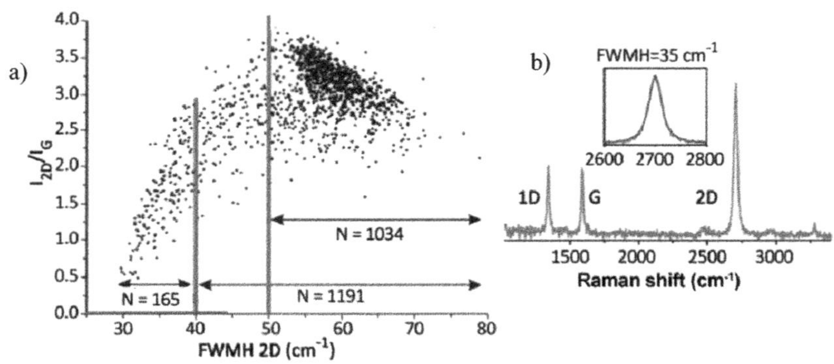

Figure 1.30. *Statistical evaluation of the quality of graphene obtained by GO reduction and adsorbed on Si/SiO₂*

COMMENT ON FIGURE 1.30.– *a) Statistical distribution of observed Raman spectra at different points on the surface and as a function of the width (FWHM) of the 2D scattering line centered at ~2700 cm^{-1} and of the I_{2D}/I_G ratio. The 165 spectra for which FWHM (at 2700 cm^{-1}) < 40 cm^{-1} correspond to graphene with very few defects; 1034 correspond to graphene with a high density of structural defects (FWHM > 50 cm^{-1}) and 157 are of medium quality (40 < FWHM < 50 cm^{-1}). b) Raman spectrum of a graphene flake with low defect density (FWHM ~ 35 cm^{-1}). Adapted from [EIG 13].*

26 Hydroiodic acid and trifluoroacetic acid offer non-toxic alternatives to hydrazine as GO reduction agents.

From these considerations, it appears that graphene obtained by GO reduction is not as pure as graphene obtained by mechanical exfoliation or by CVD. Nevertheless, although the conduction properties of this graphene are inferior, the fact that it can be produced at low cost with no need for heavy material investments represents a significant advantage. Furthermore, the mechanical and conductive properties of even "inferior" graphene are sufficient for many applications.

More recently, Cheng *et al.* [CHE 16] developed an original method for obtaining bulk (multilayer) graphene by laser irradiation of a GO aerogel[27], a reduction technique developed a few years earlier by Kaner and coworkers [STR 12]. The authors used this technique to obtain a variety of graphene which is exceptionally well suited for use in battery electrodes. The method is particularly striking because of the way laser irradiation (1 watt power) from a single point in the aerogel results in the transformation of the whole aerogel into bulk graphene (Figure 1.31).

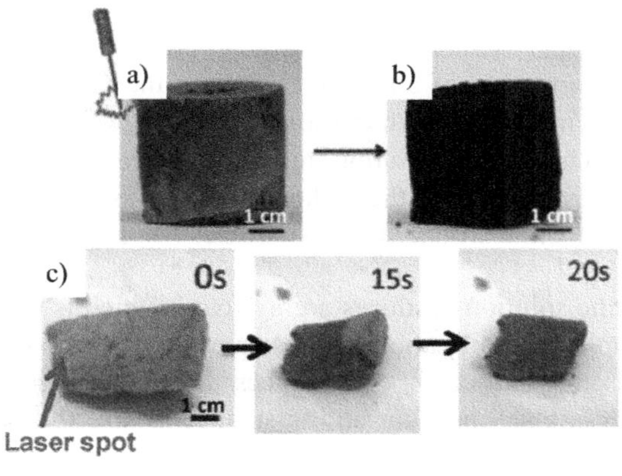

Figure 1.31. *Transformation of GO bulk aerogel into graphene by point laser irradiation. a) GO aerogel before transformation. b) Reduced graphene after transformation. c) Optical photographs showing the gradual and complete transformation of a GO aerogel sample (light gray) to graphene (dark gray) between 0 and 20 seconds. Adapted from [CHE 16]*

A transformation of this type, starting from a single point in the material, is only observed when the GO is in aerogel form; this differs from "conventional" GO in that pockets of air and water are trapped within the carbon network. The authors found that the impact of the laser beam results in localized GO reduction with the

27 GO aerogel is obtained by flash-cooling a suspension of GO in water (5 mg/ml) and drying it in a freeze dryer.

release of CO, CO_2 and H_2O; a variety of carbon-based fragments, such as $C_6H_4^+$ and $C_4H_2O_4^{2+}$, are also produced. These fragments (identified by mass spectrometry) react with the oxygen present in the aerogel, producing a significant amount of heat; this heat is sufficient to trigger local dissociation of new fragments and the removal of oxygen groups, thus triggering an uninterrupted propagation of the reduction reaction throughout the whole volume of the material. The result is a block of reduced graphene which retains the form and volume of the original material.

An added advantage of this technique is that it can be used to produce bulk graphene doped with various hetero-elements such as N, S, P and Pt. This is done by mixing chemical compounds containing the corresponding atoms with the GO suspension; during irradiation, the compounds are degraded and release these atoms, which are then implanted in the graphene [CHE 16].

The same authors have suggested similar strategies for producing highly porous graphene foams with innovative elastic properties [LV 16]. These foams are obtained using a suspension of GO, to which SDS (sodium dodecyl sulfate) then Nonidet P40 (4-nonylphenyl-polyethylene glycol) are added under strong agitation. SDS promotes the formation of air bubbles around the GO sheets, while Nonidet P40 acts as a solidifying agent when the solution is freeze-dried[28]. With the addition of these two surfactants, the solution turns from black to gray in color, and significantly increases in volume; the authors explain this due to the presence of air bubbles which agglomerate around the GO particles (Figure 1.32b).

Freeze-drying the solution produces a gray foam, which is then heated to 200°C for 1 hour to complete solidification; a subsequent heating to 1000°C in argon for 2 hours eliminates the surfactants and reduces the oxide to a solidified graphene foam (SGF) which is black in color and is smaller in size than the foam obtained after freeze-drying (Figure 1.32).

The resulting foam is extremely elastic, and these properties are perfectly conserved over repeated cycles (1000 cycles) of compression and relaxation of 80% in the axial direction. The authors state that their foam performs much better in this respect than the vast majority of known foam materials. The mechanical properties of the foam are closely linked to the ratio of the two surfactants used, along with the annealing temperature. The elasticity modulus E and the conductivity σ are optimized using an annealing temperature of 1000°C, where E = 292 kPa and $\sigma = 55.2$ S m^{-1} [LV 16].

28 The graphene oxide solution is freeze-dried by freezing the solution in liquid nitrogen, then vacuum drying by ice sublimation for 72 hours.

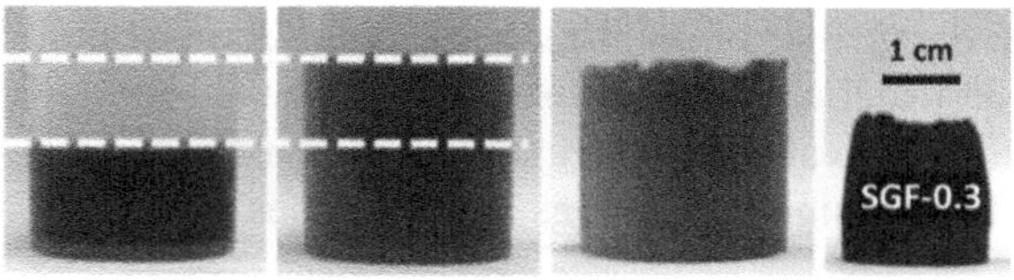

a) b) c) d)

Figure 1.32. *Different stages in the production of solidified graphene foam, SGF*

COMMENT ON FIGURE 1.32.– *a) GO suspension before addition of surfactants. b) Volume increase of GO suspension after the addition of surfactants (under strong agitation). c) Solid GO foam obtained after freeze-drying. d) Solidified graphene foam, SGF-0.3, after heat treatment. 0.3 refers to the amount of Nonidet P40 introduced into the GO solution (0.3 ml). Adapted from the Supporting Information for [LV 16].*

1.5.3. *Applications of graphene and graphene derivatives*

The exceptional mechanical, electrical and thermal properties of a graphene monolayer were discovered not long after the material itself. In addition to its low mass density (2.26 g cm^{-3}), its mechanical characteristics include a tensile strength of 130 GPa and an elasticity modulus E of 1100 GPa; in comparison, stainless steel has an elasticity modulus E of around 200 GPa and a tensile strength of approximately 0.5 GPa. The electrical conductivity of a graphene monolayer (10^8 S m^{-1}) is higher than that of metals such as copper (6×10^7 S m^{-1}), and a graphene monolayer is able to withstand current densities of up to 10^{10}–10^{11} A m^{-2} without damage. Its thermal conduction properties are also excellent, with a maximum recorded value of 5300 Wm^{-1} K^{-1}, compared to 400 Wm^{-1} K^{-1} for copper. Evidently, these values only hold for extremely pure, defect-free graphene. The graphene produced for large-scale operations does not perform at this level; however, the properties of this "average" graphene are sufficient in many cases. It is helpful to establish clear requirements in terms of the quality, uniformity and reliability of graphene required for particular applications, in order to create products and devices which are able to compete with existing systems in terms of both cost and performance[29].

29 The Manchester GEIC (Graphene Engineering and Innovation Centre) website, along with that of the Graphene Council, was created in 2013 for the development and exploitation of graphene and provides a general idea of the many present and future applications of graphene material in various forms. For example, the interview with T. Barkan (Graphene Council) on "The future of graphene", published in February 2019, is particularly informative (AZOnano, https://www.azonano.com/article.aspx?ArticleID=5135).

For instance, the quality of graphene used in the form of multilayered sheets to produce electrode materials, used in solar energy conversion devices or for energy storage in the form of various battery systems, is not of critical importance. The same is true of fibers designed for use as electric cables.

Graphene has many actual and possible applications in the fields of energy, electronics and medicine. A small selection of products derived from graphene and GO, available in various forms (paper, fibers, tape, etc.), will be described below, along with some specific applications relating to polymer-graphene or polymer-GO composites. Graphene nanofragments, known as graphene quantum dots (GQD), will be discussed separately as their specific properties are very different from those of graphene and of CDs.

1.5.3.1. *Graphene oxide*

Graphene oxide, obtained by oxidizing graphite using the Hummers method, takes the form of a lamellar product consisting of a stack of graphene sheets, functionalized in the plane and periphery by oxygenated groups (carboxylic on the edges, phenolic and epoxydic in the plane). Sheets are exfoliated by ultrasound stirring in an aqueous medium, creating a colloidal suspension of GO nanosheets in water. These nanosheets are around 1 nm thick, with lateral dimensions varying between a few nm and several hundred microns (Figure 1.33).

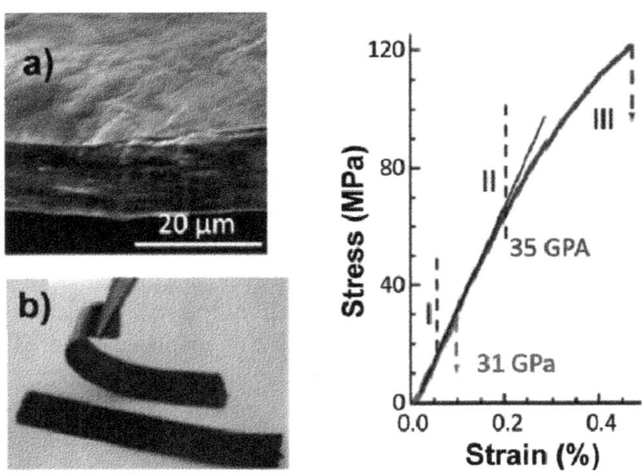

Figure 1.33. *Mechanical characteristics of a film of graphene oxide (GO)*

COMMENT ON FIGURE 1.33.– *a) SEM image of a GO film approximately 10 µm thick. b) Folding a GO tape approximately 25 µm thick. c) Stress/strain curve of a*

GO film 5.2 µm thick. The regions labeled I, II and III correspond to strictly elastic, quasi-elastic and plastic variations, respectively. The red curve shows a first variation in strain (with elasticity modulus E of 31 GPa) while the blue curve corresponds to the second deformation cycle (with E = 35 GPa). Adapted from [DIK 07].

The suspension is filtered very slowly through a porous disk, producing a thick, compact film after about 12 hours. The thickness of this film can vary from 1 to 30 µm, depending on the quantity of solution used. The resulting paper-like product has very good mechanical properties due to strong interactions between hydroxyl groups. For films of a few µm thickness, the elasticity modulus (E) is around 30 GPa [DIK 07].

According to the authors, the fact that the elasticity modulus of the GO increases with the application of several stress/strain cycles is evidence of an axial realignment of GO fragments, leading to increased rigidity. Although the elasticity modulus of GO varies in a non-negligible manner according to the method used to produce it, it remains high, much higher than the elasticity modulus of similar papers resulting from the agglomeration of CNT (buckypaper: ~7 GPa), or of inorganic lamellar compounds such as vermulite (E ~12 GPa). Its tensile strength, σ, is also high (between 38 and 125 MPa); this value is higher than that of buckypaper (between 10 and 75 MPa) and comparable to that of a vermulite paper (between 25 and 150 MPa). These mechanical properties outperform those of many materials; however, the mechanical characteristics of a graphene monolayer are still higher (E ~1 TPa and σ ~130 GPa) [DIK 07].

The presence of ionizable carboxyl groups (hydrophilic) on the periphery of the NPs and of large graphitic domains (hydrophobic), along with some epoxy and phenolic groups, in the plane of the GO sheets means that the material is amphiphilic and possesses tensioactive properties. For this reason, it can be used as a dispersing agent in water for highly hydrophobic products, such as CNTs. This amphiphilic property is also the reason why GO sheets adhere to gas bubbles (N_2, CO_2) and can migrate to the liquid-air separation surface, forming a compact layer, which can then be transferred to a solid support using the Langmuir-Blodgett technique. The hydrophilic properties of the material may also be increased by increasing the alkalinity of the medium, promoting the ionization of the carboxyl groups into carboxylates; in this case, the amphiphilic character decreases [KIM 12] (Figure 1.34).

The amphiphilic properties of GO sheets make them suitable for a wide range of applications. A GO sheet may be seen as a 2D di-block copolymer with, on one side, a strongly hydrophobic graphitic block and, on the other side, a strongly hydrophilic carboxylic block. The amphiphilicity of the sheet can be used to improve the crystal order of another polymer interacting with the GO.

A result of this type occurs when a conductive polymer such as poly(3,4-ethylenedioxythiophene): poly(styrene sulfonate) (PEDOT-PSS) is mixed with non-conductive GO sheets. Instead of the expected decrease in conductivity, this value increases by a factor of 10, and a highly viscous, gel-like adhesive product is obtained. This seemingly paradoxical effect results from improved organization of the PEDOT chains, induced by contact with the GO sheets [TUN 11].

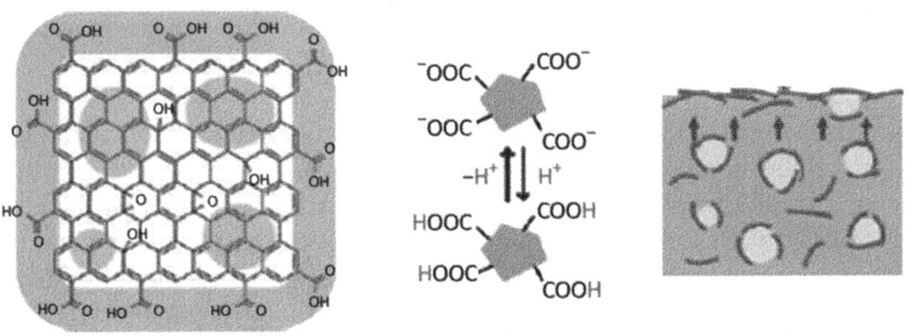

Figure 1.34. *Graphene oxide (GO) sheets*

COMMENT ON FIGURE 1.34.– *a) Distribution of oxygenated functional groups on the surface of a GO sheet (green: hydrophobic graphitic regions, yellow: hydrophilic carboxylic regions). b) Ionization of COOH groups and increase in the hydrophilicity of GO as pH increases. c) Adsorption of GO sheets on the surface of gas bubbles (yellow) and migration to the water/air interface. Adapted from [KIM 12].*

Finally, an important property of GO suspensions is their ability to organize and form liquid crystals. This property has been demonstrated by several authors [KIM 11, XU 11], and is useful for many applications, including large-scale fabrication of graphene fibers (GFs) (see section 1.5.3.3).

1.5.3.2. *Graphene paper: a semi-metallic r-GO sheet obtained after reduction of graphene oxide*

The technique used to create GO films from suspensions in water using filtration has also been adapted for the production of graphene films. This approach, developed by Chen *et al.* [CHE 08], consists of reducing a suspension of GO using hydrazine; following vacuum filtration, this produces metallic-like films of controllable thickness (from a few tens of nm to several microns) (Figure 1.35).

These "paper" sheets obtained through GO reduction may be assimilated to a variety of graphene noted r-GO. This material exhibits high mechanical resistance and good electrical conductivity, and this latter characteristic differentiates it from GO, which is an insulator.

The mechanical and electrical properties of graphene paper are strongly dependent on the strength of the cohesion established between graphene sheets. This cohesion may be weakened by the presence of residual gases and liquids left over from previous chemical treatments. Located between the sheets, these residues are difficult to remove due to the impermeability of the material. An American team has recently managed to solve this problem by creating holes on the surface of the graphene sheets (making "holey graphene" , noted h-graphene) in order to evacuate impurities, then using a heat treatment to rebuild the hexagonal structure of the graphene mesh.

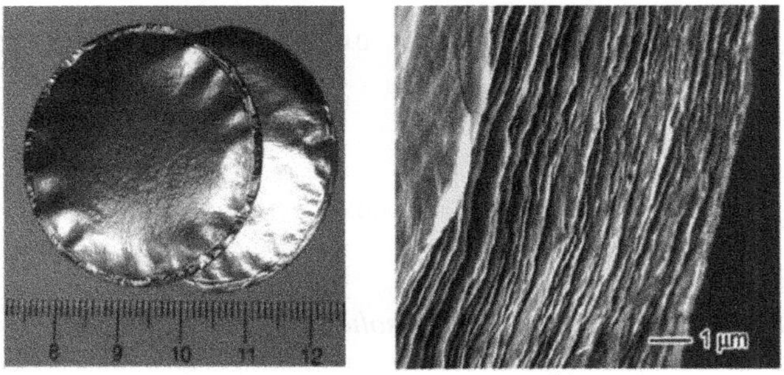

a) b)

Figure 1.35. *Graphene paper (r-GO) obtained after reduction of graphene oxide (GO) sheets. a) Photograph of two metallic graphene sheets (front and back views). b) Scanning electron microscopy (SEM) image of a graphene paper sheet showing stacked sheets of graphene. Adapted from Chen et al. [CHE 08]*

This method, described by Chen *et al.* [CHE 19], consists of heating graphene sheets in air over a long period (10 hours at 700 K); this results in a porous structure, with holes in the graphene sheet of approximately 15 nm in diameter (Figure 1.36a).

Treating the h-graphene obtained in this way in an inert atmosphere and at very high temperatures, using the Joule effect (2700–3000 K), the crystalline structure of the graphene is restored; this is confirmed by the Raman spectra (Figure 1.36b).

Reprocessed h-graphene exhibits very good electrical and thermal properties. The conductivity of the graphene sheets increases from 128 S cm^{-1} before treatment to 2209 S cm^{-1} after treatment at 2700 K; similarly, the mobility of the charge carriers increases from 26 $cm^2 V^{-1}s^{-1}$ to 673 $cm^2 V^{-1}s^{-1}$. Paradoxically, the density of charge carriers after reconditioning is lower than that observed in h-graphene (2.18×10^{19} cm^{-3} compared to 3.3×10^{19} cm^{-3} before restoration of the graphene structure); this signifies that treatment at 2700 K has the effect of removing the defects and impurities present in h-graphene which are responsible for weakened conductivity [CHE 19].

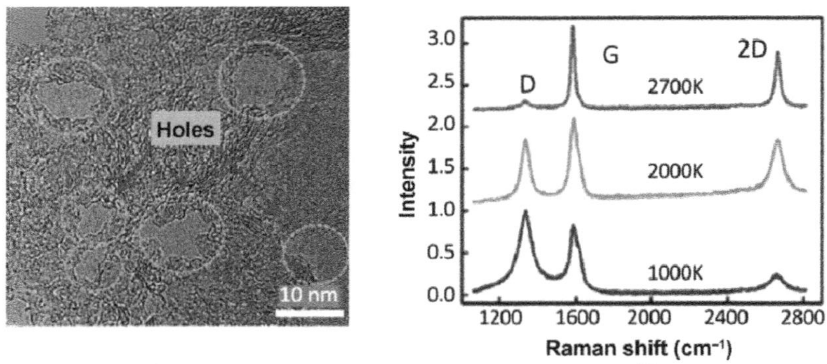

a) b)

Figure 1.36. *Reprocessed h-graphene*

COMMENT ON FIGURE 1.36.– *a) High-resolution transmission electron microscopy (HRTEM) image of a graphene sheet with multiple perforations (circled in green) created by air heating (h-graphene). b) Raman spectra of h-graphene sheets reconditioned at different temperatures. Only heat treatment at very high temperature (2,700 K) produces a defect-free graphene, as we see from the very low intensity of the D-band. Adapted from [CHE 19].*

Thermal conductivity is also greatly improved, increasing by a factor of 7 up to 863 $Wm^{-1}K^{-1}$ after treatment at 2700 K; this corresponds to a thermal conductivity more than twice that of copper. These thermal and electrical properties far exceed the known characteristics of graphene sheets, suggesting new potential applications for this type of material in the field of electronics and energy, including the development of new electrode materials.

Another feature of r-GO films is their impermeability to gases, liquids and many aggressive chemical species, such as hydrofluoric acid. This property is only effective if the r-GO films are free from structural defects (holes); one of the methods that seems to be most effective in producing non-porous r-GO is to reduce the GO using hydroiodic acid or ascorbic acid (vitamin C) [SU 14].

Impermeability is a particularly useful property for protecting organic electronic devices against humidity; these devices are usually encapsulated in PET (polyethylene terephthalate) films covered with a thin layer of aluminum. Su *et al.* [SU 14] have shown that an optically transparent 30 nm r-GO film, obtained by reducing GO with hydroiodic acid (HI), is almost completely impermeable to hydrogen and water vapor; the water permeation rate through a membrane of this type is around 10 mg m^{-2} day^{-1}, 100 times lower than the value for "standard" protection (aluminum-coated PET membrane).

In this, r-GO film differs from GO film, which exhibits abnormally high water permeability, which can also be controlled electrically [ZHO 18]. A further advantage of r-GO is that it is biocompatible, meaning that it can be used in the biomedical field for designing biosensors.

1.5.3.3. *Graphene fibers*

Qu and coworkers were among the first to propose a method for large-scale GF production using an easy and inexpensive approach, based on a suspension of GO in water [DON 12]. This method is very simple, consisting of heat treating a suspension of GO (8 mg GO/ml) in a glass tube (inner diameter 0.4 mm), closed at both ends, at 230°C for 2 hours. This leads to a fiber with a diameter of 150 μm when wet, reduced to 35 μm after drying.

From a mechanical perspective, the stress/strain variation curves improve significantly following treatment at 800°C for 2 hours in argon, increasing the tensile strength from 120 to 420 MPa. The electrical conductivity of these fibers is around 10 S/cm, a value which is comparable to that of fibers made from CNT; it remains stable after over 1000 repetitive bending/relaxation cycles. These fibers can also be knotted or twisted, showing their exceptional flexibility (Figure 1.37).

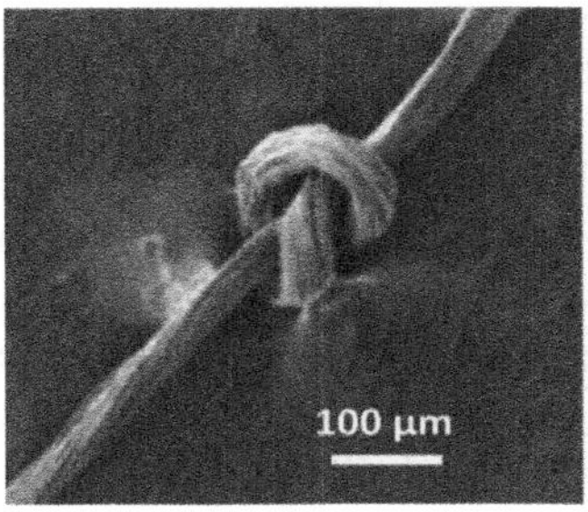

a)

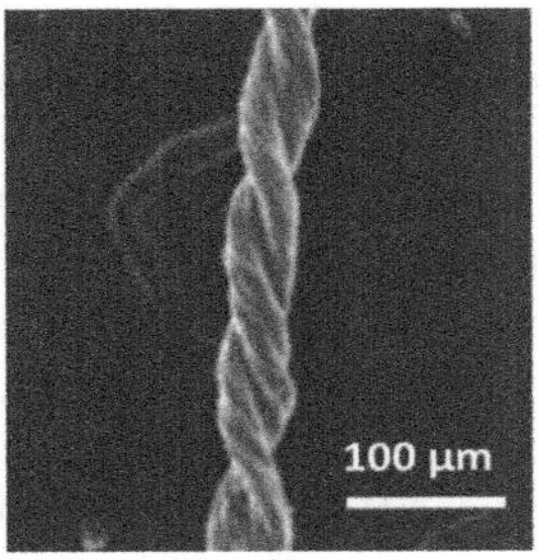

b)

Figure 1.37. *SEM images of graphene fibers. a) Knotted fiber. b) Twisted two-strand graphene fiber. Adapted from [DON 12]*

These fibers can be given new properties by doping with inorganic compounds. For example, incorporating a semiconductor such as titanium oxide (TiO_2) into the initial GO solution results in the creation of a photoelectronic device in which the GF generates an electrical response when subjected to a photon pulse. The addition of ferric oxide, Fe_3O_4, to the GO suspension produces a "magnetic" GF, which moves or bends elastically when held near a magnet [DON 12].

Given the technological importance of these fibers, they are a subject of active research, with many teams attempting to produce GFs with properties as close as possible to those of pure single-layer graphene films. We recall a graphene film which is mechanically exfoliated from a graphite crystal has an elasticity modulus of ~ 1 TPa, tensile strength of 130 GPa, electrical conductivity of 10^8 S/m with an ampacity of 10^{12}–10^{13} A/m^2 and thermal conductivity of up to 5300 W/mK.

The Gao group was recently able to demonstrate large-scale, continuous production of very high-quality GFs [XU 16]. Their improvements largely concern an improved orientation of the GO sheets, which are always, initially, obtained as a dispersion of GO monolayers in water (from 1 to 10 mg GO/ml solution). This dispersion is then transferred into a solution of N,N dimethyl formamide (DMF), which has the effect of creating a liquid crystal structure, with a parallel arrangement of the GO sheets. This substance is then extruded through a small diameter die (60 µm), then the fiber is unwound by rotating rollers which maintain a slight stress in the GO thread. The resulting GO fiber (GOF) is wound directly into a coil. Heat treatment under nitrogen at 3000°C is used to transform the GOFs into GFs (Figure 1.38). These GFs have very good mechanical properties along with excellent electrical and thermal conductivity. The Young's elasticity modulus is between 385 and 400 GPa (measured using a 5 mm length of wire), and the tensile strength is in the range of 1.8–2.2 GPa. Conductivity at room temperature is up to 0.8×10^6 S/m and is independent of the tested wire length; the ampacity is extremely high and corresponds to current densities of 2.3×10^{10} A/m^2. These values are comparable to, and sometimes better than, those measured with wires formed from CNTs.

The possibility of producing extremely flexible, mechanically strong GFs with electrical properties comparable to those of metals such as copper and very low mass densities on a large scale opens up a wide range of possible applications. The authors showed that wound GFs may be used to replace copper wires in electric motors, something which represents a significant advance in terms of mass, considering the differences in density between GF (0.23 g/cm^3) and copper wires (8.96 g/cm^3). Similarly, the high ampacity of GFs could be exploited in producing incandescent lamps which, all other things being equal, would emit light with a higher intensity than that obtained using a tungsten filament [XU 16].

Alongside the work cited above, the same group showed that the conductivity of GFs can be greatly increased by doping with certain elements. Using GFs with a high purity level, prepared according to the method described above, doping agents in the vapor state can be inserted into the fibers: the use of ferric chloride ($FeCl_3$), bromine (Br_2) or potassium (K) results in GFs with conductivities of 0.77×10^7 S/m (GF-$FeCl_3$), 1.5×10^7 S/m (GF-Br_2) and 2.24×10^7 S/m (GF-K), respectively [LIU 16b]. These values come very close to those of copper (~ 6.10^7 S/m at 20°C).

Finally, the same team has recently shown that GFs can become superconductors when doped with calcium, using the same process as before [LIU 17c]. This superconducting property, which is not specific to GFs, has already been observed in different carbon allotropes (diamond, graphite, C_{60} fullerenes, CNT and graphene), but always in very small quantities. In the present case, the authors show that this superconducting property can be obtained in a continuous manner and without limitation along the whole length of a graphene wire, opening up more promising perspectives in terms of applications. The transition to the superconducting state is observed at 11.3 K and becomes complete at 4 K[30]. Remarkably, after transformation, the GF retains its mechanical flexibility when subjected to multiple bending/relaxation cycles.

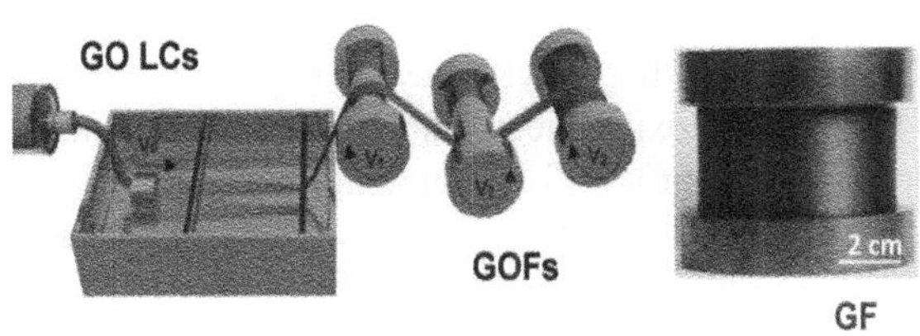

a) b)

Figure 1.38. *Diagram of the apparatus used in large-scale graphene wire production*

COMMENT ON FIGURE 1.38.– *a) Extrusion of GO in liquid crystal form (GO LC) at speed V_0. The extruded product is subjected to a first heat treatment intended to produce strong agglomeration of the GO particles while maintaining the axial orientation of these particles. Rollers V_1 and V_2 rotate at speeds slightly higher than the extrusion speed V_0, maintaining a slight tension in the extruded fiber. b) A coil of GF obtained after heating the GOF fiber at 3000°C under argon. Note the reflective metallic appearance of the GFs, very different from the yellowish color of GOF yarns. Adapted from [XU 16].*

30 For comparison, commercial NbTi superconducting wire has a transition temperature of 9.7 K.

1.5.3.4. *Graphene-polymer nanocomposites*

The idea of using small amounts of graphene to significantly improve the mechanical, electrical and thermal properties of compositive polymers is attractive. The appeal of graphene as a polymer reinforcement agent, in comparison with other sub-micrometer additives, is its extremely high aspect ratio: this suggests that strong graphene-polymer interactions could be obtained using only small amounts of graphene. Ruoff and his group were the first to exploit this concept, using a suspension of NP-GO in water to develop large-scale composite polymeric materials, consisting of a mixture of polymer and graphene, with significantly improved electrical and mechanical properties [STA 06].

The main difficulty with this process lies in the fact that the graphene nanoparticles (NP-Gs) which must be inserted into the polymer matrix are generally obtained by reduction of NP-GOs which are suspended in water; after reduction, the NP-Gs aggregate immediately, given the hydrophobic character of the NP core. This makes it difficult to insert individual graphene NPs into a polymer solution, particularly considering the fact that most polymers are insoluble in water.

A solution is to make is to make the NP-GO less hydrophilic by chemically treating the graphite oxide with phenyl isocyanate, thus transforming the OH and COOH groups of the graphite oxide sheets into amide and carbamate functions, which are weakly hydrophilic [STA 06]. Exfoliation is not possible in aqueous medium, but is easily achieved in an aprotic solvent such as N,N dimethylformamide (DMF). The authors were thus able to obtain a suspension of functionalized GO NPs (GOf) in an organic medium, one in which many polymers can be solubilized (Figure 1.39).

It is notable that the hydrazine reduction of the NP-GOfs in the presence of a polymer such as polystyrene only results in an aggregation of the graphene NPs with the polymer, and not an aggregation of the NPs themselves; this permits the formation of a graphene/polymer composite (G/P). The transfer of the suspension into alcohol leads to coagulation of the NPs in the form of a grayish powder (Figure 1.39c), which after drying and hot molding under pressure produces black pellets of G/P composites (Figure 1.39d).

The electrical conductivity of the composite is a function of the concentration of graphene NPs by volume, and results from a percolation mechanism that becomes effective at a very low concentration threshold of 0.1%. The conductivity of the composite increases rapidly from 10^{-5} to 0.1 S/m as the volume concentration of the graphene is increased from the percolation threshold up to a concentration of around of 1%. This conductivity value is comparable to that obtained for CNT/polymer composites, but the material is much more cost effective.

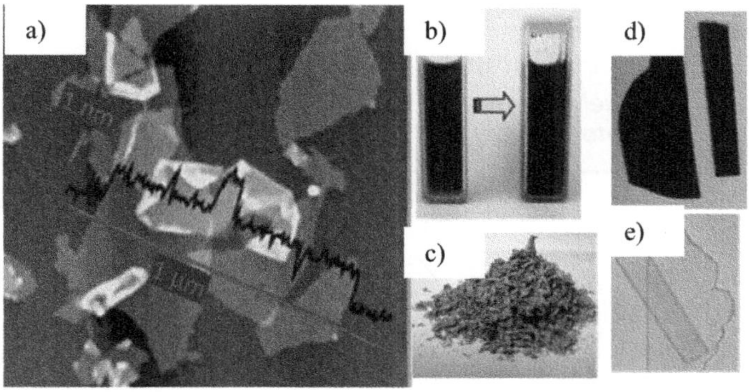

Figure 1.39. *Successive steps in the formation of G/PS composite*

COMMENT ON FIGURE 1.39.– *a) AFM image and profile of functionalized GO nanoparticles (NP-GOf) after reaction with phenyl isocyanate. The NP-GOfs deposited on mica have lateral dimensions of between 1 and a few μm; their thickness, of the order of 1 nm, confirms that monolayer sheets have been obtained. b) Color change (from brown to black) of the NP-GOf suspension in the presence of polystyrene (PS) after reduction with hydrazine and obtention of NP-Gf/PS. (c) NP-Gf/PS powder obtained following coagulation of the suspension in alcohol. (d) and (e) NP-Gf/PS and PS pellets obtained under the same thermal molding conditions. Adapted from [STA 06].*

Since the publication of these first works, which have been abundantly cited in the literature, numerous teams have invested in this research, proposing various improvements and variations to the graphene/polymer composite fabrication process. Several reviews have been published on this subject [POT 11, DAS 13, HU 14, BER 17, MAL 17], which detailing the composites obtained using a wide range of polymers of technological interest, along with variations in production methods, notably techniques for functionalizing graphene NPs; these have led to significant improvements on both mechanical and electronic levels (Figure 1.40).

These new properties have contributed to a vast expansion of the field of application of composite polymers in four essential areas: energy (storage and conversion), biomedical (medical imaging and drug delivery), electronic devices (sensors, transistors, electrodes and electromagnetic shielding) and mechanical development (artificial muscles).

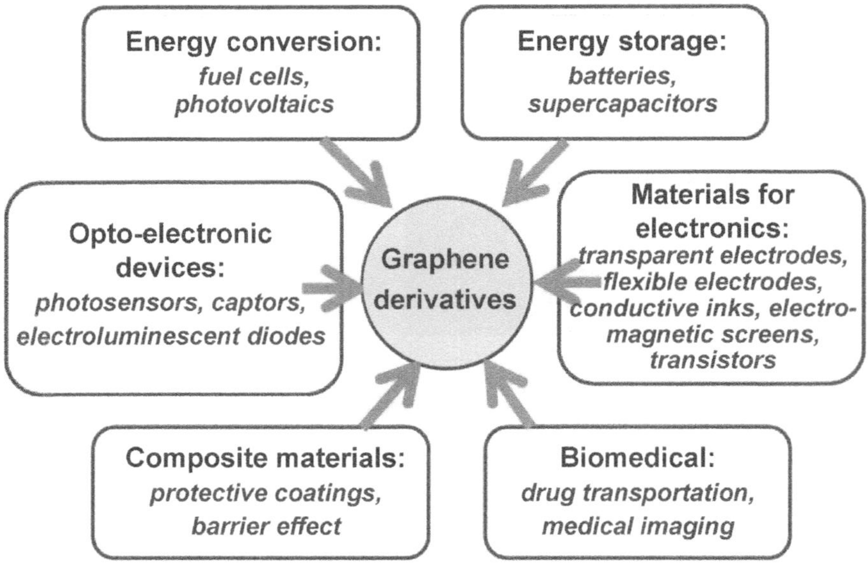

Figure 1.40. *Uses of graphene derivatives in different fields*

1.5.4. *Conclusion*

Graphene is one of the very few materials which combine exceptional electrical and thermal conductivity, the ability to withstand extremely high temperatures and electron current densities higher than those of the most conductive metals (such as copper), mechanical properties (elasticity and tensile strength) which far exceed those of steel and a low mass density (2.3 g/cm^3). However, these properties are only found in almost pure, mechanically exfoliated graphene sheets, which can only be obtained in very small quantities.

CVD techniques have undergone considerable developments, making it possible to produce continuous graphene ribbons of very high quality. However, this technique requires significant material investment and can only be envisaged in an industrial context. The chemical methods used to transform GO into graphene, on the other hand, are more readily available and have also matured in recent years. GO, which can be obtained in large quantities by a simple oxidation treatment of graphite, is the precursor *par excellence* for obtaining large quantities of graphene in various forms. It can take the form of very long fibers of only a few microns in diameter and of very high homogeneity; it may also be used in manufacturing composite materials. The quality of the graphene produced in this way can be considerably improved by a thermal treatment at very high temperatures and in an inert atmosphere; this has the effect of restoring the graphitic structure of the graphene.

1.6. Graphene quantum dots

GQDs are essentially graphene fragments of nanometric size, and can be assimilated to point NPs (0D) which share the physicochemical characteristics of both graphene and carbon dots (CDs). The quantum confinement effects and edge effects encountered in CDs are expected to be observed in this case, alongside certain properties which are characteristic of the structure of graphene.

GQDs are less than 10 nm in size, and their thickness is less than 10 layers of graphene. They are fluorescent, due to the confinement effect; unlike graphene, GQDs also have a non-zero band gap, meaning that they are semiconductors. This opens up a range of possible applications, for example, in the field of optoelectronics.

1.6.1. *GQD production methods*

Since GQDs were first obtained in 2008 by Ponomarenko *et al.* [PON 08], by fragmenting CNTs, a large number of other techniques have been proposed. These may be grouped into two categories. "Top-down" techniques consist of cutting fullerenes, CNTs, carbon fibers or graphene into fragments, while "bottom-up" techniques aim to synthetize GQDs via a chemical transformation of carbonaceous molecules used as precursors. Citric acid, glucose and a range of other compounds with benzene cores have been used as precursors in GQD production. A detailed description of recent advances may be found in articles such as [TIA 18] and [VAL 17]. For our purposes, we have chosen to present two examples of fabrication methods from the categories cited above.

1.6.1.1 *Top-Down GQD production methods*

One frequently used technique consists of fragmenting a carbon-based material, preferably made up of aromatic nuclei. Graphite is the preferred option, but other materials, such as cotton fibers, have also been processed into GQD using this method [ZUO 17a].

The combined use of microwaves and a chemical treatment of graphite in an acid medium permits the creation of large quantities of GQD and, moreover, does not require significant material investments. This method was used by Luo *et al.* [LUO 16] to obtain emissive GQD films for use in white light diodes (Figure 1.41).

In step 1, a suspension of powdered graphite in a mixture of sulfuric and nitric acid (in proportions of 3:1) is subjected to sonication for 2 hours, then to a microwave treatment at 80°C for 4 hours, leading to the formation of GQD. With average lateral sizes of about 2.5 nm and thicknesses of about 2 nm, corresponding to three to five layers of graphene, these GQDs are characterized by luminescence spectra in the yellow-green range.

In step 2, the GQDs are subjected to further microwave radiation for 12 hours, this time in a strongly basic medium (pH 13) and at a higher temperature (175°C), which has the effect of creating numerous oxygenated and carbonylated groups and removing part of the graphitic structure. These new materials are characterized by a wider fluorescence spectrum that can result in white light-emitting layers. They present a significant advantage over other white light OLEDs (organic light-emitting diodes) in that no multiphase combination of emissive materials is required. However, their luminance value, around 200 candelas/m^2, and their quantitative efficiency of around 0.2% are still modest compared to the performance of current LED lamps.

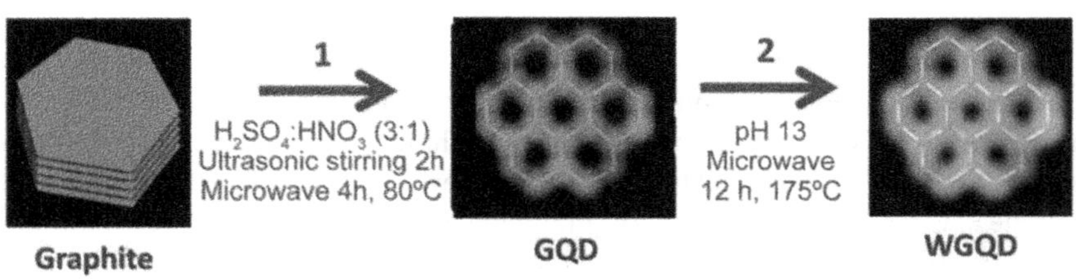

Figure 1.41. *Two-step transformation of graphite into GQDs then into WGQD white luminophores. GQDs emit greenish yellow light, while WGQDs emit white light. Adapted from [LUO 16]*

1.6.1.2. *Bottom-up GQD production methods*

Unlike the previous method, which involved degrading graphitic structural materials, the bottom-up method for GQD production is based on growth reactions using carbon compounds, which, by thermolysis in aqueous medium, are likely to produce condensed aromatic structures. The compounds used as a carbon source may be polycyclic aromatic compounds (pyrene, dinitro 1.5-naphtalene), but non-aromatic compounds, such as citric acid or glucose, may also be used for GQD production.

This technique presents a clear advantage of simplicity in comparison with top-down techniques for obtaining GQDs doped with elements other than carbon. Thus, nitrogen-doped GQDs (N-GQDs) can be obtained in a single step, by directly treating a mixture made up of a nitrogen derivative and the carbon source compound. N-doped nanomaterials have significantly improved photoluminescence compared to undoped GQDs, and are therefore particularly interesting for medical imaging: in addition to their luminescence properties, they are highly stable in aqueous media and are non-toxic. They also have applications in the energy field and in sensor design [KAU 18].

Gu *et al.* [GU 16] performed thermolysis of citric acid in the presence of ethylene diamine to obtain N-GQDs, with sizes between 5 and 10 nm and thicknesses between 0.8 and 2.5 nm, equivalent to two to five layers. These N-GQDs exhibit a remarkably high internal quantum yield of blue photoluminescence, of the order of 75%, far exceeding the quantum yields usually observed with this type of material.

Another means of obtaining N-GQDs is to start with a single compound, such as 1,3,5-triamino-2,4,6-trinitrobenzene (TATB), including both the carbon and nitrogen atoms needed to form N-GQDs [LI 16c]. A pyrolysis reaction is carried out at high temperatures (750°C for around 20 minutes) in a nitrogen atmosphere. In this case, a high yield of N-GQDs is synthesized, with an atomic nitrogen doping rate of around 10% with respect to the carbon atoms (Figure 1.42).

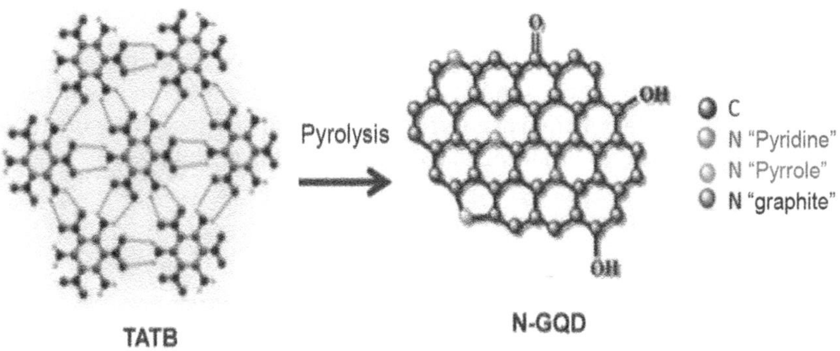

Figure 1.42. *N-GQD production by TATB pyrolysis*

COMMENT ON FIGURE 1.42.– *The TATB precursor is made up of a supramolecular planar network of TATB molecules linked by hydrogen bonds between the H in the NH_2 and the O in the NO_2 (the red, blue and gray spheres represent the oxygen, nitrogen and carbon atoms, respectively; the gray points in TATB are H atoms). Note that in N-GQD, the nitrogen atoms are incorporated into the graphitic network and occupy distinct positions corresponding to functionalities of different types (pyridine, pyrrole or graphite). Adapted from [LI 16c].*

The TATB precursor compound is already a bidimensional planar network with hexagonal symmetry. The neighboring TATB molecules are linked by intermolecular hydrogen bonds between NH_2 and NO_2. This supramolecular organization prefigures a graphene-like structure, likely facilitating the formation of N-GQDs.

This heteroatom doping technique has been widely applied to other atoms (boron, sulfur, phosphorus, halogens, alkali metals, etc.) using a wide variety of precursor systems, documented in an article by Tian *et al.* [TIA 18]. Simultaneous doping with two or more heteroatoms is also possible, and in this case, it is possible to produce GQDs with multiple emissions [QU 15].

1.6.2. *Properties and applications of GQDs*

Unlike graphene, GQDs have a non-zero energy band gap, the width of which depends on the size, shape and edge effects of the NPs. Considering the wide variety of methods used to produce GQDs and the wide range of possible sizes and shapes, it is not surprising that their optoelectronic and photoemission properties also vary widely. Photoemission is a key property in GQDs, and is used in various detection systems, with important applications in medical imaging.

1.6.2.1. *Photoluminescence properties of GQDs*

The photoluminescence of GQDs is determined both by the core of the nano-material, which may be assimilated to a graphene fragment, and by the surface states corresponding to functional groups positioned at the periphery of the NPs, along with the heteroatoms produced in the case of doping (Figure 1.43).

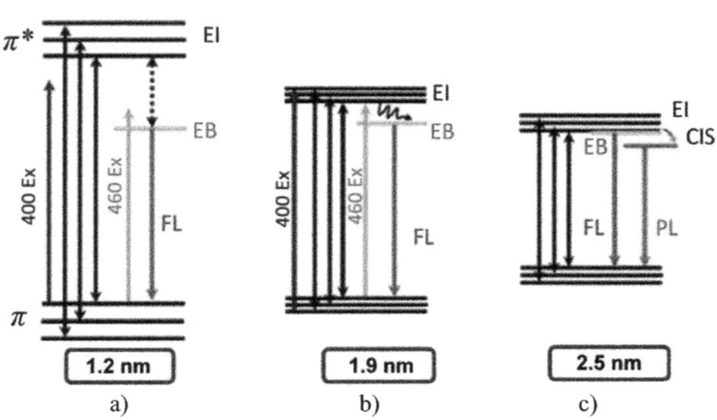

Figure 1.43. *Photoluminescence of GQDs, interpreted as a function of size and light excitation energy*

COMMENT ON FIGURE 1.43.– *a) GQD ($C_{42}H_{18}$, size 1.2 nm): fluorescent emissions (FL) are only obtained with excitation at 460 nm. b) GQD ($C_{96}H_{30}$, size 1.9 nm): a-intensity fluorescence emission is observed following excitations at 400 and 460 nm, resulting from an energy transfer from the intrinsic state EI (π*) to the edge state EB (trap level, due to the functional groups located on the periphery of the GQDs). c) GQD ($C_{132}H_{34}$, size 2.5 nm): the EB and EI states have the same energy, and there is a possibility of an intersystem crossing (ISC) between the intrinsic state and triplet state, resulting in simultaneous emissions of fluorescence (FL) and phosphorescence (PL). Adapted from [ZHU 14b].*

In a highly detailed spectroscopic study (both static and dynamic), Zhu *et al.* [ZHU 14b] showed that the photoemission properties of GQDs are determined by the energy level offset between the intrinsic state and the edge state in the GQD. The intrinsic state energy (EI) corresponds to the π* orbitals of graphene fragments, and the edge state energy (EB) corresponds to chemical edge groups, grafted onto the periphery of the GQD, which can be assimilated to energy traps.

Using GQDs of increasing size ($C_{42}H_{18}$-1.2 nm, $C_{96}H_{30}$-1.9 nm, $C_{132}H_{34}$-2.5 nm), the authors clearly show that the energy of the intrinsic $\pi \to \pi^*$ transition decreases with the size of the GQD, along with the energy difference ΔE between the intrinsic states (π*) and edge states (EB). It is this energy difference ΔE that determines the nature and intensity of the luminescence originating from the deactivation of the edge states. The emission is essentially of the fluorescent type for non-zero ΔE, but this may be combined with phosphorescence when ΔE is zero. The fluorescence intensity may be very low, or even non-existent, when the energy level of the intrinsic state is close to that of the edge state; this is the case for GQDs of formula $C_{132}H_{34}$ and $C_{222}H_{42}$ [ZHU 14b].

A similar approach may also be used to describe the photoluminescence properties of GQDs which have been doped using different O and N heteroatoms in various proportions. In this case, the Jablonski diagram includes a $n \to \pi^*$ transition in addition to the $\pi \to \pi^*$ transition.

Qu *et al.* [QU 15] observed that nitrogen-doped GQDs of increasing sizes can produce blue (B), green (G) or yellow (Y) fluorescence, determined by the relative position of the n, π* and edge state energy levels[31]. The energy levels of these GQDs, named N-GQD-B, N-GQD-G and N-GQD-Y, respectively, are shown in Figure 1.44, and clearly illustrate the changes in fluorescence colors, which correlate with slight increases in GQD size corresponding to redshifts of the $\pi \to \pi^*$

31 The different N-GQDs are obtained by treating a mixture of citric acid and diethylene triamine (DETA) in an aqueous medium or N,N dimethyl formamide (DMF), resulting in different C/N and C/O ratios, which determine the fluorescence.

transitions. Compared to undoped GQDs of identical sizes, we note the presence of a red shift in fluorescence emissions. This is due to the fact that the electronic transition responsible for fluorescence occurs between the edge state and the non-bonding n-orbitals of the heteroatoms (O, N), rather than between the edge state and the lower energy π-orbital.

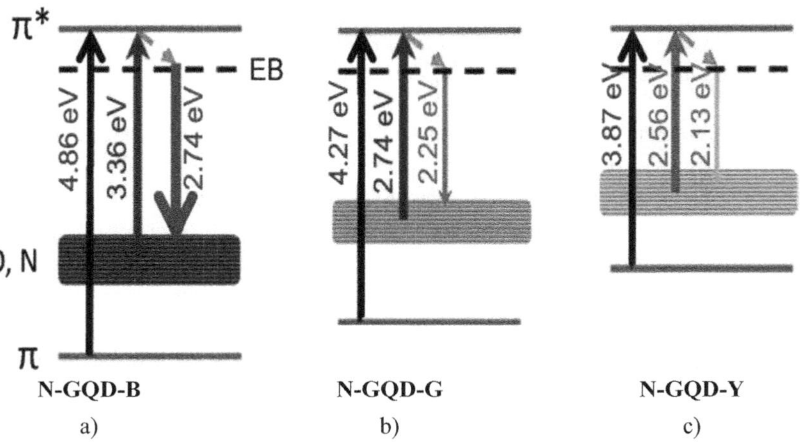

Figure 1.44. *Perrin-Jablonski diagrams of nitrogen and oxygen doped GQDs characterized by blue (B), green (V) and yellow (J) fluorescence*

COMMENT ON FIGURE 1.44.– *Energy diagrams a, b and c are deduced from the excitation and fluorescence spectra of different GQDs. Each of the compounds is characterized by two excitation bands π→π* and n→π* accompanied by a fluorescence emission band corresponding to the EB→n transition, resulting from a prior transition from π* to the EB state. The different energy values are deduced from the maxima of the excitation and emission bands of the corresponding spectra of the different GQDs. The values of 2.74 eV, 2.25 eV and 2.13 eV correspond, respectively, to blue (450 nm), green (550 nm) and yellow (580 nm) light emissions. Adapted from [QU 15].*

1.6.2.2. *Applications of GQDs*

A significant body of work has been produced over the last few years focusing on potential uses of GQDs for energy and biomedical applications. In the energy field, GQDs offer the potential for important improvements in battery electrode performance and in supercapacitor performance. Their photoluminescence properties have also been used to improve the operation of OLEDs and photovoltaic devices; furthermore, GQDs also have a positive contribution to make to photocatalytic processes. The largest number of potential applications, however, is found in the biomedical field; GQDs may be used in applications as varied as imaging and drug delivery, various forms of phototherapy, and in the production of antibacterial products. We have chosen to limit our discussion to a few examples of applications in photodetection devices, used in trace analysis or to detect radiation; in these contexts, GQDs are often used in the form of composite materials.

1.6.2.2.1. Use of GQD composites for detecting volatile organic compounds

Certain devices used to detect gases or volatile organic compounds (VOCs) at low concentrations make use of the variations in the conductivity of a solid film which occur in contact with a gas. Conductive polymers such as polypyrrole and PEDOT (poly[3,4-ethylenedioxythiophene]) exhibit observable variations in conductivity when interacting with volatile compounds. Graphene, and more recently GQDs, has been suggested as additives for these films, allowing an increase in sensitivity and in the specificity of VOC detection.

A device of this type has been made using a nanocomposite consisting of N-GQD and poly(3,4-ethylenedioxythiophene)-poly(styrenesulfonate) (PEDOT-PSS). The N-GQD/PEDOT-PSS mixture took the form of a suspension in a water-dimethyl sulfoxide medium (10 wt% DMSO); the classic hydrothermal method of transforming of a mixture of citric acid and urea was used to create the N-GQDs (diameter between 2 and 7 nm, atomic ratio N/C equal to 0.058). One drop of the previous mixture was deposited on a flat surface with interdigitated gold electrodes, arranged according to conventional lithography techniques, then the whole assembly was annealed at 80°C. This had the effect of covering the interdigitated electrode area with a N-GQD/ PEDOT-PSS film. The contacts are two conductive wires connected to the gold electrodes with silver solder (Figure 1.45).

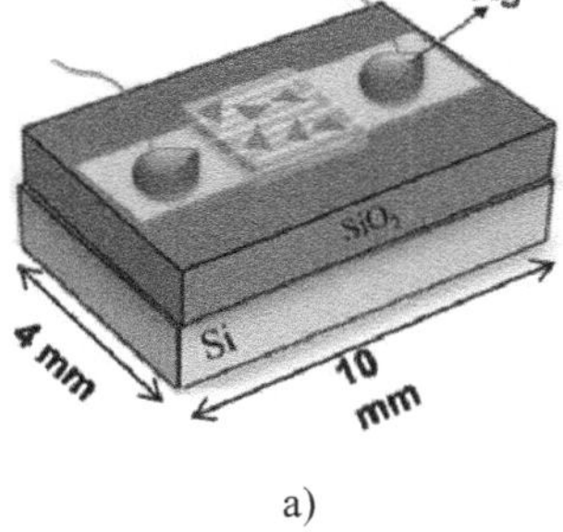

a)

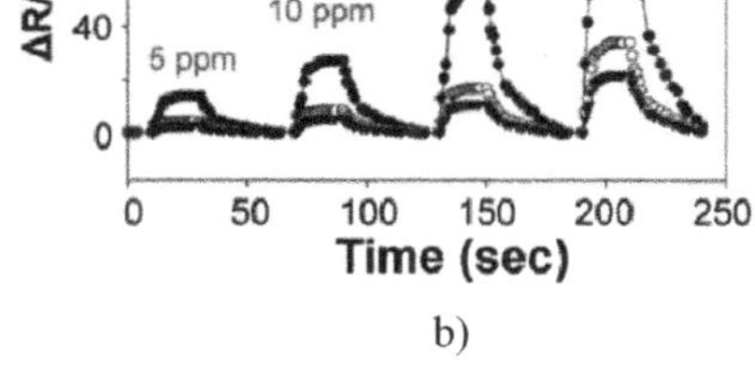

b)

Figure 1.45. *VOC detection device, using variations in the resistance of a N-GQD/PEDOT-PSS film*

COMMENT ON FIGURE 1.45.– *a) View of the device showing the area of interdigitated gold electrodes covered by nanocomposite film. The interdigitated gold electrodes are 100 nm thick; their width is 200 μm and their spacing is 400 μm. b) Relative changes in the resistance of the composite film subjected to different concentrations in air (5, 10, 30 and 50 ppm) of methanol, ethanol and acetone vapor. Adapted from [GAV 15].*

When in contact with methanol, ethanol and acetone vapors, the resistance of the film, measured between the two contacts, increases as a function of VOC concentration. The relative change in resistance (ΔR/R) is nearly linear between 0 and 50 ppm concentration, after which the system is saturated and the resistance no longer depends on concentration. Note that the response time is of the order of 10 seconds; the return to equilibrium, while slightly slower, is of the same order of magnitude.

The presence of N-GQDs in the composite is necessary for a strong increase in resistance to be observed in the presence of different VOCs. The resistance of a film made up of N-GQDs alone hardly varies in contact with these vapors, while in the case of a film composed only of PEDOT-PSS, the relative variation in resistance does not exceed 10%. With N-GQDs as additives, the relative variation in R is 130%. Moreover, the response is noticeably greater in the presence of methanol than with ethanol and acetone: this must be attributed to the stronger reducing character of methanol, which partially reduces the PEDOT and thus increases its resistivity. The return to equilibrium occurs due to the oxygen in the air, which re-establishes the initial oxidation state of the PEDOT and thus its original conductivity.

Paradoxically, methanol only has a weak effect on a PEDOT:PSS film alone and almost none on a film composed only of N-GQDs, but when the two are combined, a considerable variation in resistance is observed. The authors attribute this difference in behavior to the fact that (i) the initial σ-conductivity of PEDOT:PSS is greatly increased by the presence of N-GQDs ($\sigma_{PEDOT:PSS}$ = 635 S/cm and $\sigma_{N\text{-}GQD/PEDOT:PSS}$ = 1365 S/cm) and (ii) the initially compact PEDOT:PSS film becomes porous in the presence of N-GQDs, allowing increased diffusion of VOCs inside the film and thus a greater reduction of the PEDOT, resulting in increased resistance.

1.6.2.2.2. Photodetectors for UV radiation

Conventional UV detectors are fabricated from wide band gap inorganic semiconductor films, such as ZnO, GaN, AlGaN or Si, but these may be replaced by GQDs; this latter option presents the advantage of being easy to obtain in solution, which results in clear benefits in terms of both design and cost. Unlike graphene, which has a zero energy band gap, the energy band gap of GQDs can be easily tuned to be compatible with UV absorption energies: this is done by reducing their size to that of 0 dimensional (0D) NPs.

Zhang *et al.* [ZHA 15] have designed a UV photodetector in which the sensor layer of the device (80 nm thick) is made of GQDs sized between 2.5 and 6 nm with a thickness equivalent to 1 and 4 graphene layers (from 0.4 to 2.5 nm)[32]. The band gap, measured by electrochemistry, is equal to 3.6 eV.

The principle used to detect luminous radiation is a classic one, and consists of measuring, for a voltage applied between two electrodes, the electric current resulting from the creation of an electron-hole pair in the semiconductor under the effect of the irradiation (metal/GQD/metal junction). The authors showed that the highest sensitivity was obtained in the case of an Au/GQD/Ag junction in which a voltage of 5 volts is applied between the electrodes and where the gold electrode is positively polarized. This configuration has the effect of making the Au/GQD (+) and Ag/GQD (–) contacts ohmic: one for the transfer of electrons, the other for the transfer of holes. The device is sensitive to UV radiation of wavelengths below 300 nm and insensitive to white light (Figure 1.46).

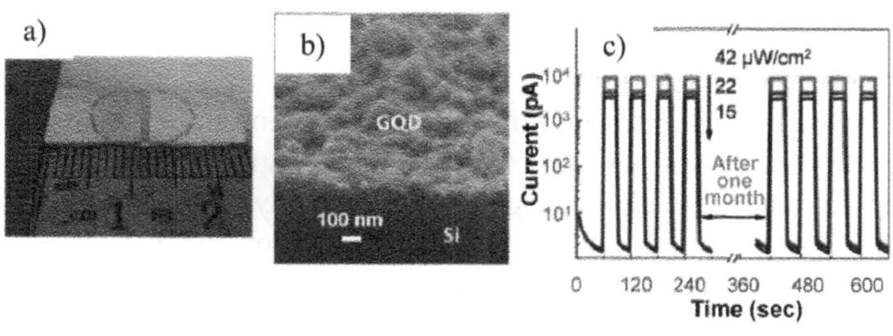

Figure 1.46. *UV detection device. a) Photograph of the Au/GQD/Ag junction. b) SEM image of the GQD film covering the channel between the two electrodes. c) Variations of the photocurrent subjected to UV light pulses at 342 nm of different intensities (42, 22 and 15 μW/cm^2). Note the high stability of the device after one month. Adapted from [ZHA 15]*

32 The GQDs are synthesized from graphene sheets (GS) obtained by deoxygenation (300°C under nitrogen atmosphere) of graphene oxide (GO). The GS are then oxidized in a mixture of sulfuric acid and nitric acid under ultrasonic agitation, diluted in water, and then filtered to obtain a first purified suspension of oxidized GS. A final autoclave treatment of the suspension at 200°C (10 h) and filtration results in the obtention of GQDs.

1.6.3. *Graphdiyne: a new alternative to graphene*

A new family of 2D carbon-based compounds, known as graphynes, was first identified in the 1980s by Baughman and Eckhardt [BAU 87]. Graphynes are graphene-derived compound planes, in which the benzene rings are covalently linked by n conjugated –C≡C– ethyne groups instead of being fused together (Figure 1.47).

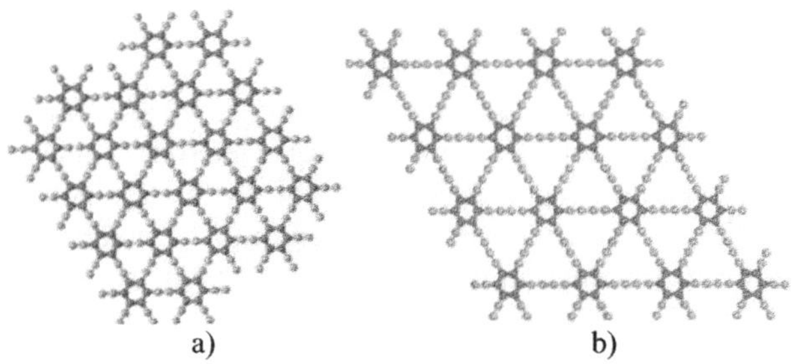

Figure 1.47. *Diagrammatic representation of graphynes. a) Simple graphyne (n = 1) in which the benzene rings (shown in blue) are linked, at their vertices, by acetylenic groups –C≡C– (in pink). b) Structure of graphdiyne: this differs from the previous example in that the simple –C≡C– bonds have been replaced by diethynyl groups –C≡C–C≡C– (n = 2). Adapted from [LIU 17a]*

These ethyne groups form rigid linear links between benzene rings, resulting in the formation of planar and porous networks of sp and sp^2 hybridized carbons. These compounds are named graphyne-1, graphyne-2, graphyne-3, etc., according to the number (n) of ethyne groups linking the vertices of the benzene rings. Like graphene, these compounds are mechanically robust; however, their calculated rigidity characteristics are lower than those of graphene, and decrease when the number of ethynyl groups increases.

Graphyne-2 (or graphdiyne, GDY) is the most stable and, at present, the only graphyne to have been synthesized in appreciable quantities. This synthesis was carried out by Yuliang Li [LI 10a] and colleagues in 2010; since then, many works have been published on the subject of GDY from both theoretical and experimental perspectives. This interest is justified by the remarkable structural, electronic, mechanical and thermal properties of the material, which suggest new potential applications, notably in the fields of energy, catalysis and in the design of systems based on the photoelectronic conversion principle [GE 19, KAN 19a].

The formation reaction described by Li *et al.*, used to obtain GDY in large quantities and with good yields, consists of reacting hexaethynylbenzene (HEB) on a copper substrate in the presence of pyridine and in a nitrogen atmosphere at a temperature of 60°C [LI 10a]. This produces a large-area homogeneous film, consisting of multiple layers of GDY with a total thickness of about 1 μm. Since then, other variants of the method have been proposed, notably the deposition of GDY films on different solid substrates (Si, Ag and Au) as well as on graphene sheets.

Generally speaking, most of these methods use the same HEB precursor; the details of these reactions are described in detail by Gao *et al.* [GAO 19]. As for other carbon-based materials, GDY can also be doped with heteroatoms; this has been demonstrated using nitrogen, and also with simultaneous insertion of two elements such as F and N, S and N or B and N. These may substitute or link to the sp carbons in the ethynyl bonds, resulting in GDY with new catalytic properties. The material obtained by doping GDY with F and N, for example, is an excellent catalyst for the electrochemical reduction of oxygen in a basic medium [ZHA 16a].

It is significant that GDY can be produced without the use of high temperatures, unlike graphene and its derivatives; this is an additional advantage when considering possible large-scale applications.

1.6.3.1. *Properties of graphdiyne*

The presence of diethynyl bonds between benzene rings results in new properties which are not found in the condensed structure of graphene, paving the way for new applications.

In terms of electronic properties, there are notable similarities to graphene, particularly with regard to the mobility of charge carriers. Their calculated mobility is very high, between 10^4 $cm^2/Vsec$ for holes and 10^5 $cm^2/Vsec$ for electrons; this latter value is of the same order of magnitude as that observed for graphene, for which the mobilities of electrons and holes have been estimated at 3×10^5 $cm^2/Vsec$ (theoretical) and measured at between 2 and 25×10^4 $cm^2/Vsec$ [CHE 13a].

However, GDY differs from metal graphene in terms of its low-gap semiconductor properties, evaluated according to different methods at between 0.46 and 1.10 eV for single-layer GDY. Unlike graphene, GDY thus has the potential to be used in creating various electronic devices in which organic transistors are used. The conductivity of a multilayer GDY film of 1 μm thickness, synthesized on a copper substrate, is around 2.5×10^{-2} S/cm; this value is comparable to that of silicon [LI 10a]. In the case of GDY wires produced using a different technique, much higher conductivities of 1.9×10^5 S/cm have been measured, but in this case the mobility value is lower, around 7×10^2 $cm^2/Vsec$. The authors attribute this difference in conductivity between films and wires to the absence of defects in the process used to produce GDY wires [QIA 12b].

The presence of calibrated triangular pores (with an approximate size of 2.5 Å) [HAL 08] is another notable difference between GDY and graphene, offering perspectives for further applications, notably in selective membranes for gas separation. Another important property of GDY is the possibility of inserting metals in the atomic state, which can be used, for example, to produce electrodes. Furthermore, this property can be used in fuel cells to improve the electrochemical reactions of oxygen reduction or proton reduction in an acidic medium (see Chapter 5, "Electro and photo-electrocatalysis").

These characteristics of GDY make it suitable for use in a wide variety of applications, a comprehensive list of which can be found in Gao *et al.* [GAO 19] and Yu *et al.* [YU 19]. Notable potential applications for GDY include use as a membrane for the separation of gases (H_2, O_2 and N_2), a functional role as a catalyst support in the atomic state (see Chapter 5, section 5.3.1), and significant improvements to electrode function in lithium-ion batteries (LIBs).

1.6.3.2. *Some specific applications of graphdiyne*

The porosity and electronic conductivity GDY mean that it has great potential for use in electrode manufacture. These electrodes may be used in batteries, or as catalysts for the reduction of oxygen and protons in fuel cells. Two examples of GDY use in these fields are described below.

1.6.3.2.1. Catalysts for oxygen reduction reaction

One important objective in the development of fuel cells with an oxygen cathode is to replace the excellent, but very expensive, platinum or platinum/carbon electrodes with catalytic systems based on low-cost non-noble metals.

GDY has been doped with heteroelements with the aim of developing efficient electrodes, likely to improve the kinetics of the oxygen reduction reaction (ORR). Doping with nitrogen and fluorine produces an NF-GDY material with a high catalytic power for the ORR, comparable to that of C/Pt electrodes (20 wt% Pt) [ZHA 16a], and much higher than that of electrodes consisting of GDY alone or GDY doped with N and S (NS-GDY), or N and B (NB-GDY) (Figure 1.48).

Doping is carried out by reacting GDY powder first with NH_4F and then with NH_3 at high temperatures (700°C). This reaction increases the number of structural defects in the initial GDY, resulting in greater porosity. The increase in porosity is due to the fact that fluorine binds preferentially to sp carbons, leading to the bonds between a benzene ring and an ethynyl bond breaking down. Nitrogen atoms, on the other hand, can either take the place of sp^2 carbons in the benzene rings or bind to sp carbons[33].

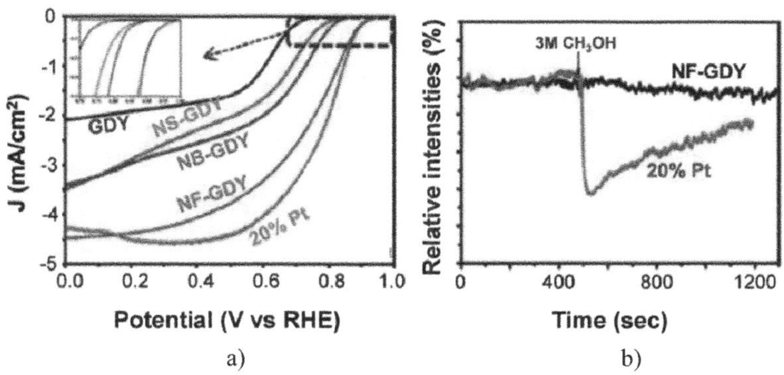

Figure 1.48. *Electrochemical reduction of O_2 on GDY electrodes doped with different elements in an alkaline medium (0.1 M KOH solution saturated with O_2) in comparison with a Pt/C electrode*

COMMENT ON FIGURE 1.48.– *a) J/V curves obtained using a linear voltage sweep (10 mV/sec). b) An addition of 3 M CH_3OH to the oxygen-saturated solution has no effect on the behavior of the NF-GDY electrode, in contrast to the Pt electrode, which exhibits a significant drop in the oxygen reduction rate. Adapted from [ZHA 16a].*

The electrocatalytic properties of NF-GDY electrodes, evaluated by electrochemical reduction of an O_2-saturated solution in an alkaline medium, are comparable to those of a Pt composite surface made up of a classic C/Pt mixture. Moreover, contrary to Pt, the behavior of the NF-GDY electrode is little affected by the presence of CO or methanol, often indicated as the reason for a loss of efficiency in Pt electrodes.

In short, with regard to oxygen reduction, N and F-doped GDY has electrochemical characteristics that are comparable to those of commercial Pt/C composite electrodes. The presence of N and F appears to be decisive in this respect. Comparable effects have also been demonstrated with mesoporous graphene electrodes doped with N and F heteroatoms [ZHA 16a].

33 This multiplication of the number of defects in the GDY structure is illustrated by Raman spectroscopy, and can be seen in the ratio of the I_D/I_G intensities of the two characteristic bands D and G located at 1350 cm^{-1} and 1620 cm^{-1}, respectively. The number of defects increases as the I_D/I_G ratio increases.

1.6.3.3. *Use of a modified GDY structure as a super-cathode in Lithium-ion batteries*

The presence of channels running perpendicular to the GDY plane constitutes an improvement in comparison with graphite, allowing ions such as Li^+ to penetrate more easily into the planes. The fact that the interior of the vacancies may be modified with chemical groups which have the capacity to stabilize these ions is a further improvement, developed recently by Li and coworkers [WAN 17].

They aimed at synthesizing a chlorinated hexadiethynyl benzene (HEB) monomer, such as 1,3,5-(triethynyl-2,4,6-trichlorobenzene), in which chlorine atoms alternate with ethynyl groups (Figure 1.49a) [WAN 17]. This monomer can be polymerized under the same conditions as HEB, resulting in a chlorinated GDY compound (Cl-GDY) (Figure 1.49b). The resulting product displays larger pores (1.6 nm), which are hexagonal in shape, rather than triangular as in the case of GDY, for which vacancies are around 0.25 nm in size [HAL 08]. This larger pore size enables the insertion of a greater quantity of alkaline atoms and facilitates their diffusion inside Cl-GDY; the moderate electronegative character of the chlorine atoms which are uniformly distributed throughout the Cl-GDY contributes to stabilizing the implantation of these alkaline ions.

This material, deposited as a thick film of about 2.8 µm on a metallic copper substrate (Cu/Cl-GDY), was used by the authors to build a cell using a Li film as the anode and the Cu/Cl-GDY electrode as the cathode[34].

The charge/discharge curves between 0.005 and 3 V show Li^+ ion storage capacities which are much higher than those obtained using GDY alone. For current densities of 50 mA/g, the Li^+ storage capacities are 1150 mAh/g; a value of 500 mAh/g was obtained for charge/discharge rates (C/D) of 2 A/g (Figure 1.49c). In the case of a GDY electrode, the storage capacity is only 450 mAh/g for C/D rates of 100 mA/g[35] .

These results confirm the stabilizing role of chlorine atoms with respect to Li^+ ions. In addition, the larger pore size of Cl-GDY compared to GDY is likely to result in the storage of larger amounts of Li^+.

---

34 The conductivity of the Cl-GDY film, which is higher than that of GDY ($2.5.10^{-4}$ S/m), is around 2.10^{-3} S/m. The two electrodes are separated by a polypropylene film. The electrolytic solution is a classic mixture of organic solvents (dimethyl carbonate and ethylene carbonate in equal volumes) with the addition of $LiPF_6$ (1M).

35 For most current lithium-ion batteries, the specific capacities of the lithium insertion cathodes do not exceed 250 mAh/g [GUO 17b].

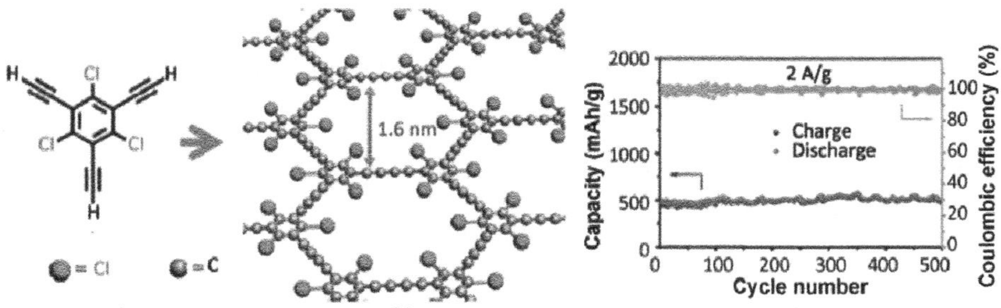

Figure 1.49. *Production of Cl-GDY and lithium-ion storage capacities*

COMMENT ON FIGURE 1.49.– *a) Structure of the monomer used to obtain Cl-GDY. b) Structure of the chlorinated derivative: Cl-GDY. c) Charge-discharge curves for Li^+ ions at high current intensity (2 A/g) with a Cl-GDY electrode. Note that the charge and discharge capacities (of the order of 500 mAh/g) are nearly equal to one another (corresponding to 100% coulombic efficiency), with no noticeable change after 500 charge/discharge cycles. Adapted from [WAN 17].*

1.7. Conclusions and perspectives of carbon-based nanomaterials

Over the last 20 years, new families of carbon-based nanomaterials have emerged. These materials present striking new properties and have the potential to be used in a wide range of applications, solving problems encountered in the fields of energy, electronics and biology. Given two materials with similar properties, preference will be given to that which is easiest to produce on a large scale and at reasonable cost: essential considerations for any potential industrial or commercial application.

Fullerenes and CNTs (some of the oldest materials discussed here), graphene and their derivatives take center stage; surprisingly, given a basic structure based solely on carbon chains in sp^2 hybridization, these materials present very clearly differentiated properties according to their dimensional characteristics: 0D for fullerenes, 1D for CNTs and 2D for graphene and GDY. The first three of these materials are now commercially available, a fact which is explained by their range of uses in crucial areas of application. Certain fullerenes, such as a C_{70} derivative, have been used to improve the energy yields of photovoltaic cells; CNT, graphene and their functionalized derivatives have been particularly successful because of their exceptional electrical and mechanical properties, which allow them to be used in new hybrid plastic materials with improved mechanical and electrical properties.

As we shall see later, these materials have also been used in the production of composite electrodes for electrocatalysis, energy storage and conversion applications.

Currently, the largest number of actual and potential applications has been found for graphene and its derivatives (GO); these take the form of plasticized tapes, foams, papers or fibers, which combine both excellent electrical and mechanical properties. Graphene fragments of nanometric dimensions – GQDs – present remarkable fluorescence properties and are characterized by non-zero energy band gaps, allowing their use in constructing various optoelectronic devices.

Most of these nanomaterials compete with activated carbons for applications in the field of electrochemistry, notably in the design of capacitive devices. In many cases, their cost is significantly higher than that of activated carbons, meaning that their use is only justified if a small quantity of product can be used to offer a significant improvement in terms of properties.

Nanodiamonds (ND), carbon NPs in the form of CQD, constitute a further family of these new materials. NDs are distinguished by their small size – less than 10 nm – and by different types of carbon-carbon bonds involving variable proportions of sp^2 and sp^3 hybridizations. NDs feature a majority of bonds between sp^3 carbons; in CQDs, the core of the particle is essentially made up of graphitic-type sp^2 carbons. These materials are commercially available, and are much easier to obtain than fullerenes, CNTs and graphene, which also makes them cheaper. For some electrochemical applications (energy storage and electrolytic capacitors), CQDs are now in a position to compete with activated carbons. They are also characterized by fluorescent properties which clearly differentiate them from fullerenes, CNT and graphenes, which are intrinsically non-luminescent. Moreover, the presence of hydrophilic functional groups on the surface means that these particles are soluble in aqueous media and, due to their non-toxicity, they are excellent candidates for use in medical imaging.

Research into new and improved methods for producing carbon nanomaterials, and on developing uses for these materials in a wide variety of fields, continues to progress. New carbon nanomaterials, such as GDY – which is a semiconductor – are also emerging. GDY possesses remarkable properties including controlled porosity and the ability to adsorb elements in the atomic state, properties which open new perspectives for uses in catalysis, for example. Described as a “rising star”, GDY looks set to take its place as an important member in the family of 2D semiconductors in the near future.

2

Inorganic Nanomaterials

Among the major families of inorganic nanomaterials currently being studied, Portehault *et al.* [POR 18] note that the number of studies focusing on Au, Ag and their alloys is by far the most important, followed by oxides (TiO_2, MnO_2, Fe_3O_4, etc.) and, to a lesser extent, chalcogenides (CdS, CdSe, Bi_2Te_3, etc.); the latter are mostly used as quantum dots (QDs) due to their semiconducting properties (Figure 2.1).

In the first case, mono- or bimetallic particles may be synthesized in various forms, such as spheres, cubes, wires, rods, tubes, platelets or combined core/shell structures. These vary in size from a few nm to around 100 nm. This criterion is somewhat subjective, but is of crucial importance for the very smallest nanoparticles (NPs), in which a quantum confinement effect is observed, which profoundly modifies the energy structure of the particles. Thus, when their size – initially of the order of 10 nm or more – decreases to less than 2 nm, the conduction and valence energy bands change, from those observed in bulk metal to a series of discrete energy levels, comparable to those found in isolated atoms or molecules. This structural change results in new photo-electronic properties, specific to these NPs. The particles may then be classified as "atomic aggregates", "nanocrystals" or "nanoclusters". The crystalline structure of nanoclusters (NCs) is an additional parameter which strongly affects their catalytic properties. This is a relatively new theme in fundamental research, and highlights the interest and importance of controlling the crystallinity of the faces that characterize an NP.

All of these nanomaterials can be easily functionalized using organic molecules or polymers, making them soluble in aqueous media in colloid form; the same approach can be used to obtain transport and vectorization properties, particularly helpful for biomedical applications. They may also be arranged into networks with

---

For a color version of all the figures in this chapter, see: www.iste.co.uk/lacaze/nanotechnology.zip.

new properties; one key example is the cooperative plasmonic resonance phenomena in the UV-visible spectral range, mostly observed in the case of gold or silver.

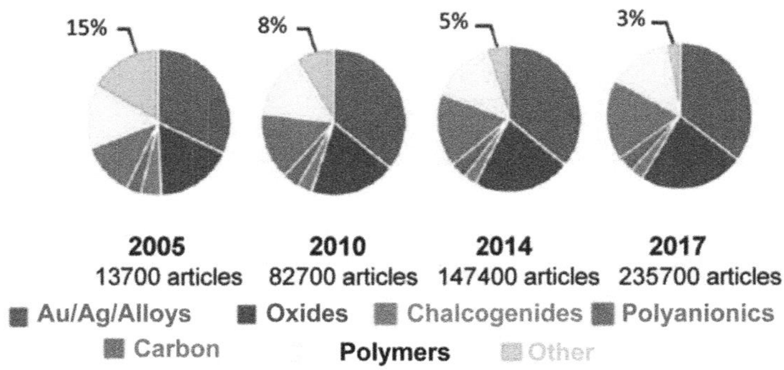

Figure 2.1. *Cumulative growth of the number of publications concerning different types of NPs between 2005 and 2017*

COMMENT ON FIGURE 2.1.– *Note that the number of publications devoted to noble metals and oxides has grown continuously from 2005, when it represented approximately half of the total publications, to 2017, when it represented two-thirds of total publications. Interest in materials other than carbon-based derivatives, polymers and inorganics appears to be falling away, with a reduction from 15% to 3% of total publications in the same period. The number of works on chalcogenides and polyanionics has remained virtually unchanged, accounting for only a small part of total publications. Adapted from [POR 18].*

2.1. Metallic nanoparticles

Research in the field of metallic NPs mainly concerns noble metals, particularly gold, which has been the subject of a very large number of publications. Active research on gold is still ongoing because of the importance of potential applications in analytical chemistry, catalysis, the development of optoelectronic devices and, notably, the biomedical field, due to its non-toxicity. The first publications concerning production methods and the properties of these NPs date back to the mid-2000s, and include notable contributions from the Astruc [DAN 04a] and El-Sayed [EUS 06] groups. More recent journal articles have focused on specific fields, notably involving NCs [JIN 16] and bimetallic nanocrystals [GIL 16].

The development of methods for the production of monodispersed NPs of controlled shape, with a precise definition of the physicochemical criteria necessary to obtain good reproducibility, is a major focus of current research. Two distinct methods are currently used. The most common and most accessible approach is chemical-based; this consists of producing colloidal solutions of NPs, based on the reduction of metal salts in an aqueous medium in the presence of surfactants. Other methods, which require greater material investment and rely on a more physical approach, involve laser ablation of solids in solution [DOL 02], thermal plasma synthesis of NCs [KOR 16] and sputtering of metallic materials [ISH 17].

2.1.1. *Gold nanoparticles (Au NPs)*

Gold NPs have received the most attention from researchers, and a large proportion of publications on the subject concern the definition of production methods and growth mechanisms; the main challenge in this area relates to the selection of physicochemical parameters which make it possible to control NP size and morphology.

2.1.1.1. *Chemical synthesis methods in the liquid phase*

In the 1950s, Turkevich proposed a precise, innovative protocol for obtaining "citrated" spherical gold NPs of controlled size; this method consisted of reducing Au^{3+} ions in an aqueous solution using trisodium citrate [TUR 51][1]. Au^{3+} ions are solubilized in water in the form of a tetrachloroauric acid complex ($HAuCl_4$). Citrate is then progressively added to the Au^{3+} ion solution, which has a double effect: first, the Au^{3+} ions are reduced to the Au° metallic state, and, second, the NPs formed in this way are stabilized as the citrate adsorbs on their surface. This results in the formation of a stable colloidal solution of Au NPs in water. The size of the spherical NPs produced in this way is determined by the concentrations of reagents. The reaction mechanism involves several reduction and complexation steps; these have been interpretated in various ways [KUM 06], and are still the subject of debate.

A number of different techniques have been developed with the aim of improving the monodispersity of metallic NPs. The use of reverse micelles in an organic medium, a technique developed by Pileni [PIL 03], results in good monodispersity and produces nanocrystals with a size correlated to that of the micelles. The seed-mediated growth technique developed by the El Sayed [HUA 09] and Murphy [EUS 06] groups has been widely adopted, and has been shown to result in a considerable improvement in NP monodispersity. The use of nucleation seeds also makes it possible to produce NPs with a variety of different shapes (rods, cubes, discs, rectangular or triangular plates, etc.).

---

1 The use of gold in divided form has been known since antiquity, when it was used as a color pigment. However, it has almost exclusively been used and mixed, in nanosphere form, with different materials such as glasses and polymers. Michael Faraday (1850) was the first to recognize that the coloration of these particles was due to their extreme division.

Spherical Au NPs of controlled sizes, between 5, 10, 30, 50, 100 and 150 nm, can also be synthesized using this method. Increasingly large gold seeds are injected each time, and a solution of tetrachloroauric acid ($HAuCl_4$) with ascorbic acid is used as a reducing agent in the place of citrates. A surfactant such as cetyltrimethylammonium chloride (CTAC) is also added. In this way, Au NPs with a diameter of around 30 nm can be produced from 10 nm Au NP seeds; these 30 nm NPs can then be used as seeds to obtain larger NPs, and so on up to a size of around 150 nm (Figure 2.2).

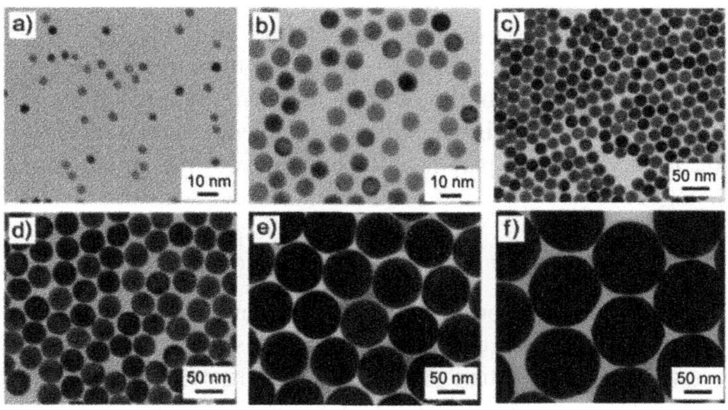

Figure 2.2. *Transmission electron microscopy (TEM) images of Au NPs of different sizes. a) 5 nm, b) 10 nm, c) 23 nm, d) 46 nm, e) 100 nm and f) 150 nm. Adapted from [ZHE 14b]*

Since the early 2000s, several works have also been published on the production of NPs in rod form known as nanorods (NRs), including contributions by the El Sayed [HUA 09] and Murphy [SAU 04] groups. These can be obtained relatively easily using a technique of anisotropic growth from nucleation seeds, different to that used to produce nanospheres.

One of the proposed approaches involves the use of seeds in nanopolyhedral form, characterized by faces with different crystallinities and different reactivities toward nucleation. A particular surfactant and a non-gold metal ion are used, in such a way as to block growth on certain faces, promoting anisotropic growth, which results in the formation of NRs[2].

2 The addition of Ag^+ ions to the reaction medium, alongside a polymer such as poly(vinylpyrrolidone) (PVP), is known to have this effect, blocking certain crystal faces of the gold nanopolyhedra and thus promoting the anisotropic growth of NRs [SEO 08].

In Korea, the Song group demonstrated this growth mechanism using decahedral gold seeds, which were synthesized in the presence of high concentrations of poly(vinylpyrrolidone) compared to that of the $HAuCl_4$ used [SEO 08, SEO 09] (Figure 2.3).

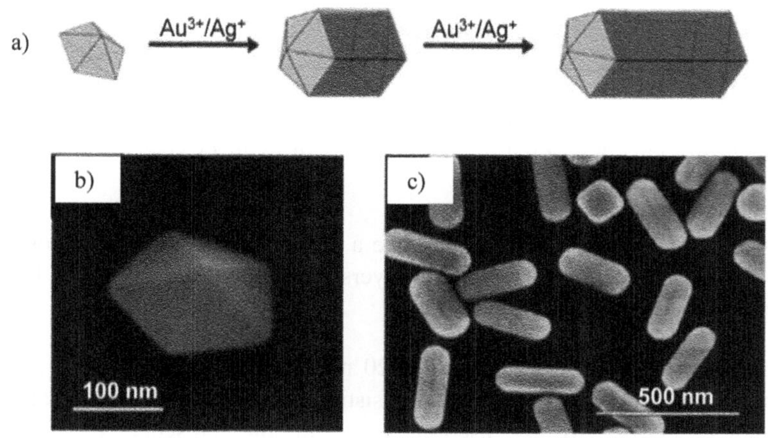

Figure 2.3. *Scanning electron microscopy (SEM) images of Au NR obtained from decahedral seeds. a) Schematic diagram of Au NR growth from decahedral seeds in the presence of silver salts. b) SEM image of a gold decahedron of large dimensions (80 nm). c) SEM images of Au NR obtained from gold decahedra, with average edge length of 60 nm. a) and c) Adapted from [SEO 09]. b) Adapted from [SEO 08]*

This method of growth from metal seeds may be used to produce a large variety of NP shapes, and can be applied to many materials in addition to gold. Several variations on the method have also been proposed and different shapes of gold NPs have been obtained; these are described in detail by Xia *et al.* [XIA 17a].

The optical and physical properties of NRs are highly dependent on their dimensions, particularly their aspect ratio, defined as the length-to-width ratio. The higher the aspect ratio, the more light absorption shifts toward the red, corresponding to a shift in the plasmon resonance of these NPs toward the lowest frequencies [EUS 06].

2.1.2. *Core-shell type bimetallic nanoparticles*

Fuel cells based on oxygen reduction and hydrogen oxidation use electrodes containing platinum, chosen for its catalytic properties, which are far superior to those of all other materials. Although these electrodes are designed to use a minimal amount of Pt[3], the expense of the material constitutes one of the main hindrances to commercial production of oxygen/hydrogen cells. The development of core-shell type bimetallic NPs, in which the Pt is deposited in the form of atomic layers on a lower cost metal, makes it possible to reduce the quantity of platinum required, and may offer a solution for lowering the cost of these fuel cells.

Palladium, which is more common than platinum and costs around half as much, is used as a metallic support: its crystalline properties are very similar to those of Pt, a fact which promotes high-quality epitaxial growth, with Pt layers that are firmly fixed to the Pd.

Xia *et al.* [XIE 14] were able to produce a system of this type by depositing a small number of homogeneous Pt atomic layers (between 1 and 6), on the faces of Pd cubes.

Pd nanocubes have an average size of 20 nm, and are obtained in an aqueous medium via a relatively simple protocol, consisting of the reduction of the palladium salt Na_2PdCl_4 6×10^{-2} M by ascorbic acid in the presence of KBr and poly(vinylpyrrolidone)[4]. The nanocubes are then retrieved by ultracentrifugation and dispersed in ethylene glycol (Figure 2.4a).

The second step consists of slowly injecting a solution of Na_2PtCl_6-$6H_2O$ in ethylene glycol (EG) into another solution of EG containing Pd nanocubes, ascorbic acid, KBr and PVP. The solution is heated to 200°C, increasing the reducing power of the EG and promoting the diffusion of Pt atoms onto the faces of the Pd seeds. The number of Pt atomic layers (nL) can be controlled by adjusting the amount of platinum salt as a function of the number of layers to be deposited on the faces of the nanocubes. This method results in the creation of core-shell NPs noted $Pd@Pt_{nL}$ (with n between 1 and 6) (Figures 2.4b, 2.4c and 2.4d).

This proposed mechanism relies on competition between the surface diffusion (V_{diff}) and deposition (V_{dep}) velocities of Pt atoms, which have a tendency to adsorb on the corners of the nanocube, corresponding to {111} faces.

---

3 Commercially available platinum electrodes consist of divided Pt mixed with carbon particles and a polymer binder in proportions of around 20 wt% Pt.

4 The reaction is carried out in aqueous medium at a temperature of 80°C. A 6.10^{-2} M Na_2PdCl_4 solution is added to a solution containing a mixture of ascorbic acid (4.10^{-2} M), KBr (0.6 M) and an amount of poly(vinylpyrrolidone) which corresponds to a concentration of $2.4.10^{-4}$ M for a molecular mass of 55,000.

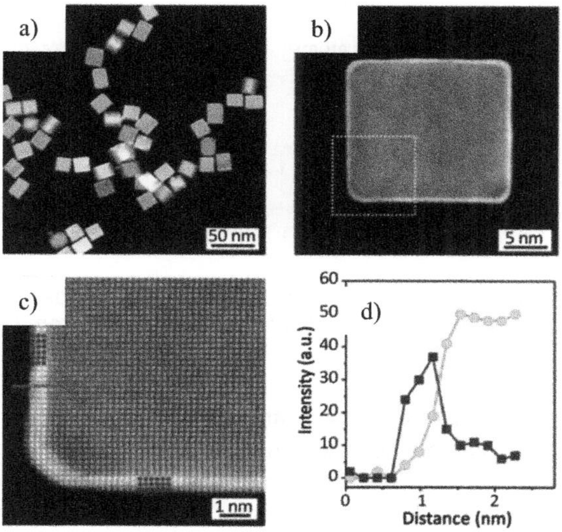

Figure 2.4. *Controlled deposition of Pt atomic layers on Pd nanocubes*

COMMENT ON FIGURE 2.4.– *a) Scanning Transmission Electron Microscopy (STEM) images of Pd nanocubes with side 18 nm. b) STEM image of a nanocube, showing the Pt deposit at its periphery. c) Enlarged portion of image b), showing three distinct atomic layers of Pt deposited on the surface of the Pd nanocubes. d) Concentration profile of Pt and Pd elements along the red arrow (image c) obtained by EDX (energy dispersive X-ray spectroscopy). These high-resolution electron microscopy images are obtained using HAADF-STEM (High-Angle Annular Dark-Field STEM). Adapted from [XIE 14].*

When V_{dep} is lower than V_{diff} (Figure 2.5, (i)), the Pt is uniformly deposited on the {100} faces of the Pd nanocubes, developing layer by layer. However, when V_{dep} is higher than V_{diff} (Figure 2.5, (ii)), as in the case of high Pt salt concentrations, the Pt atoms are mainly located at the periphery of the {111} faces of the cube, resulting in nanocubes with concave faces.

By mixing the nanocubes with carbon powder, new Pd@Pt/C particles are obtained. The active Pt surface area is 104.1 m^2/g_{Pt} for Pd nanocubes coated with a single monoatomic layer of Pt; for commercial C/Pt catalysts (20 wt% Pt), the active Pt surface area is only 51.4 m^2/g_{Pt}, despite the fact that Pt NPs are very small in size (around 3.2 nm) compared to Pd@Pt nanocubes (18–21 nm).

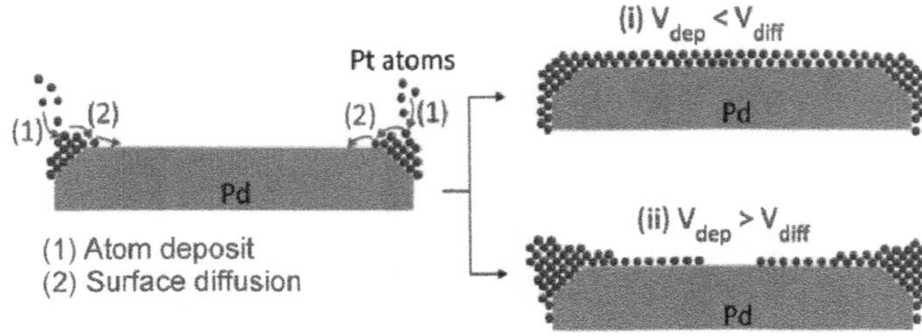

Figure 2.5. *Deposition mechanisms of Pt on Pd nanocubes as a function of relative deposition and diffusion rates. Adapted from [XIE 14]*

When used for oxygen reduction in an acidic medium, catalysts composed of Pd@$Pt_{1\text{-}2L}$/C particles exhibit, for the same Pt specific active areas, significantly higher activity than other catalysts, and perform far better than commercial C/Pt 20% electrocatalysts. The difference diminishes as the number of Pt layers increases progressively from 2 to 6 (Pt_{2L} to Pt_{6L}) (Figure 2.6).

The comparison of current densities calculated for 1 mg of Pt on electrodes made from different electrocatalysts such as C/Pt (20 wt%), Pt nanocubes (9 nm side) and Pd@$Pt_{1\text{-}2L}$/C shows higher current densities for the Pd@$Pt_{1\text{-}2L}$/C electrodes; these also exhibit lower overpotential compared to the redox potential of the O_2/O_2^- couple (not shown). The same group also obtained better results using Pd octahedra instead of nanocubes as a support for the Pt layers [PAR 15].

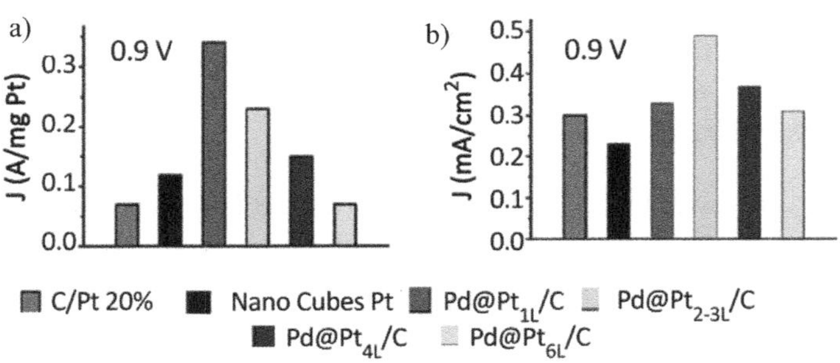

Figure 2.6. *Comparative performance of different electrocatalysts used for oxygen reduction in an acidic medium (0.1 M $HClO_4$)*

COMMENT ON FIGURE 2.6.– *a) Current intensities expressed for 1 mg of Pt in the case of electrodes made from mixture of carbon and Pd nanocubes covered with one or more Pt monoatomic layers (Pd@Pt_{1L}/C). b) Current densities (mA/cm^2) for the same electrodes. The highest current densities were obtained for electrodes made of Pd nanocubes coated with 2–3 Pt (Pd@$Pt_{2\text{-}3L}$/C) monoatomic layers. The applied* *potential of 0.9 V was measured with respect to the RHE (Redox Hydrogen Electrode) reference electrode. Adapted from [XIE 14].*

2.2. Metal nanoclusters

Metal nanoclusters (MNC) are extremely small NPs, corresponding to aggregates of atoms (between 10 and 300 atoms) with diameters from a fraction of a nm up to a maximum of 2.2 nm. In comparison to larger NPs (above 10 nm), whose optical properties are unaffected by small size fluctuations of only 1 nm, those of MNCs vary rapidly for similar size variations. Size fluctuations of only 1 Å are sufficient to affect MNC properties. The reason for this difference lies in the quantum confinement effect which occurs below the 2–3 nm threshold, causing the band structure (which characterizes bulk metals) to be replaced by discrete energy levels, similar to those observed in individual atoms or molecules. Light absorption in larger NPs corresponds to a plasmonic resonance of the electrons located in the conduction band, whereas for very small NPs, light absorption corresponds to electronic transitions between occupied (HOMO) and vacant (LUMO) energy levels. Jin [JIN 10] explains this transition between NPs and NCs using the free electron model for metals, showing that the change in behavior between the solid metal and the NC occurs when the size of the NP oscillates between 200 and 300 gold atoms, that is, at dimensions of around 2 nm (Figure 2.7). A rapid transition from the non-metallic to the metallic state has been shown to occur as the number of gold atoms increases from 246 to 279 [HIG 18]; the emergence of plasmonic resonance is noted in the 279-atom NC.

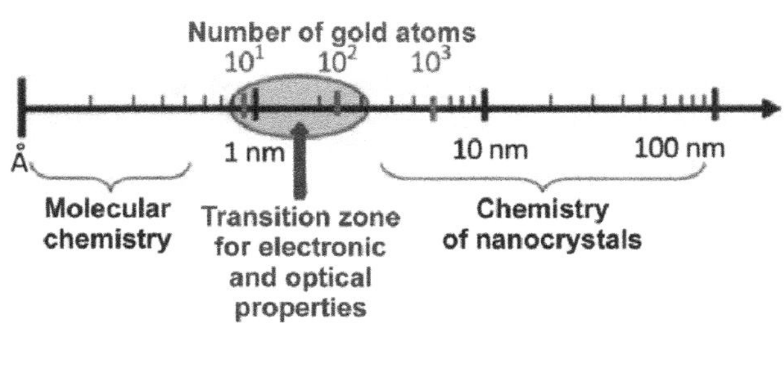

Figure 2.7. *Defining dimensional range of nanoclusters corresponding to a transition zone (between 3 and 0.8 nm) leading to electronic and optical properties which are different from those of metallic molecules and NPs. Adapted from [JIN 10]*

The MNC field as a whole is vast, but much of the work in the area – notably in terms of practical applications – focuses on gold and silver nanoclusters (Au NC and Ag NC). Many of these applications draw on the fluorescent properties of the NCs and, in the case of the smallest MNCs, their nonlinear optical properties [RUS 14]. These properties are particularly useful for biomedical imaging applications; the ease of functionalization and small size also make these MNCs an effective means of drug delivery.

In what follows, we have chosen to focus on gold and silver NCs, for which large amounts of data are available. Work in this area is still ongoing; key goals include the production of NCs with a precise and predefined number of atoms, which also exhibit specific luminescence properties.

2.2.1. *Production methods for gold nanoclusters*

Since Faraday first obtained colloidal solutions of gold, by reducing tetrachloroauric acid using phosphorus, in 1857, many other methods have been developed to produce and stabilize gold NPs. These methods are well-documented [CHE 08, CHE 15a, JIN 16, KHA 17, ZHE 17, HAL 18a]. Significant work has also been carried out in the same area with regard to silver and other metals [ZHA 14]. Given the very high variability of properties which result from very small variations in size, the ability to control this size precisely and to produce chemically stable MNCs is particularly crucial.

The most common method used to obtain highly stable Au NCs involves reducing gold salts in the presence of thiol derivatives with aliphatic or aromatic chains. The process generally results in mixtures of Au NCs, which are then separated using exclusion chromatography or gel electrophoresis techniques; these are tricky to implement, and yields are low. A certain number of thiolate-stabilized Au NCs, with formulas $Au_n(SR)_m$, were identified as early as the 1990s by Whetten *et al.* [WHE 96]; several different Au NCs were subsequently separated, isolated and characterized by Negishi *et al.* [NEG 05][5] .

One-pot, one-size techniques, which we will introduce later, avoid the need for long and tedious separation methods with low yields. Jin's group developed a "survival principle" which states that, in a mixture of Au NCs subjected to a particular chemical treatment, the most robust NCs will survive while the others will not. A similar method was also developed by Heaven *et al.* [HEA 08].

5 The structures isolated by Negishi *et al.* were obtained using thiolate glutathione (SG, a tripeptide resulting from the condensation of three amino acids: glutamic acid, cysteine and glycine) as a complexing agent for Au NCs with the general formula $Au_n(SG)_m$. The compositions, in terms of the number of gold atoms, ranged from 10 to 39. The most stable structures isolated by the authors were $Au_{10}(SG)_{10}$, $Au_{15}(SG)_{13}$, $Au_{18}(SG)_{14}$, $Au_{22}(SG)_{16}$, $Au_{22}(SG)_{17}$, $Au_{25}(SG)_{18}$, $Au_{29}(SG)_{20}$, $Au_{33}(SG)_{22}$ and $Au_{39}(SG)_{24}$.

Zhu *et al.* [ZHU 08] obtained a monodisperse Au_{25} nanocluster, stabilized by 2-phenylethanethiolates ($-SC_2H_4Ph$) and of global formula $Au_{25}(SR)_{18}$. Their highly rigorous method consists of a first reduction of Au(III) from chloroauric acid to Au(I) using thiol at low temperature (0°C). The gold obtained in this way, in the form of Au(I), is then complexed using the SR^- thiolate, resulting in the formation of gold complexes (Au(I)-SR). These complexes form aggregates of about 400 nm, and are then further reduced by $NaBH_4$, resulting in the monodisperse formation of $Au_{25}(SR)_{18}$ with a high yield of about 40% relative to the mass of gold (Figure 2.8).

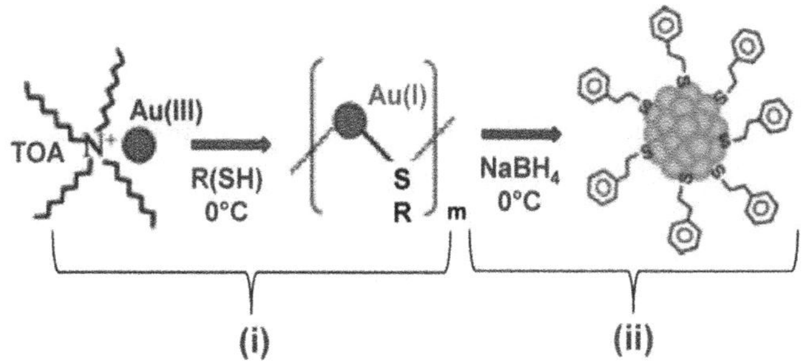

Figure 2.8. *Stages in $Au_{25}(SR)_{18}$ formation*

COMMENT ON FIGURE 2.8.– *(i) First step: partial reduction of Au(III) to Au(I) by thiol R(SH) in the presence of tetraoctyl ammonium bromide (TOAB), resulting in the formation of the Au(I)SR complex, which assembles into aggregates of about 400 nm. The chloroauric acid ($HAuCl_4$), initially in an aqueous medium, is transformed into ammonium salt (TOA-$AuCl_4$) and solubilized in toluene. (ii) Second step: the Au(I)SR aggregates are subjected to the action of a stronger reductant ($NaBH_4$), creating $Au_{25}(SR)_{18}$ clusters in the form of TOA^+ $Au_{25}(SR)_{18-}$ salt. Low temperature conditions, controlled stirring of the solutions and precise time intervals between the different steps are essential in order to obtain high yields of monodisperse clusters. Adapted from [ZHU 08].*

The reaction used to produce $Au_{25}(SR)_{18}$, which appears to depend on slow kinetics in the formation of Au(I)SR complexes and on slow aggregation, has been applied to the formation of other Au NCs, with appropriate modifications to the experimental protocol. These modifications can concern the nature of the thiol derivatives, the choice of surfactants associated with the kinetics of gold thiolate aggregation and/or variations in the concentration of $NaBH_4$ used in step 2 to reduce Au(I) to the metallic state. Details of procedures used to create large numbers of Au NCs, corresponding to precise and thermodynamically stable atomic structures, have been described in recent articles [JIN 15, JIN 16]. Similar methods have also been proposed for producing NCs of metals such as silver, platinum, palladium and copper [CHA 17].

The 25-atom gold cluster is one of the most widely studied Au NCs, due to its high stability and the relative ease with which it can be obtained in useful quantities. The opportunities offered by functionalization, via ligand exchange or by doping with other metals, pave the way for a wide variety of applications, both in catalysis and in imaging. Although this form of Au NC was first identified more than 20 years ago, practical applications have only begun to emerge in the last decade, and research on the subject is still ongoing [KAN 18].

2.2.2. *Structure and stability criteria of Au NC*

NCs are generally characterized by a core composed of metal atoms alone, and a "shell" consisting of organometallic ligands that contribute to the stability of the structure. In the case of the $Au_{25}(SR)_{18}$ cluster, the ionic form $[Au_{25}(SR)_{18}]^{-}TOA^{+}$ exhibits exceptional stability.

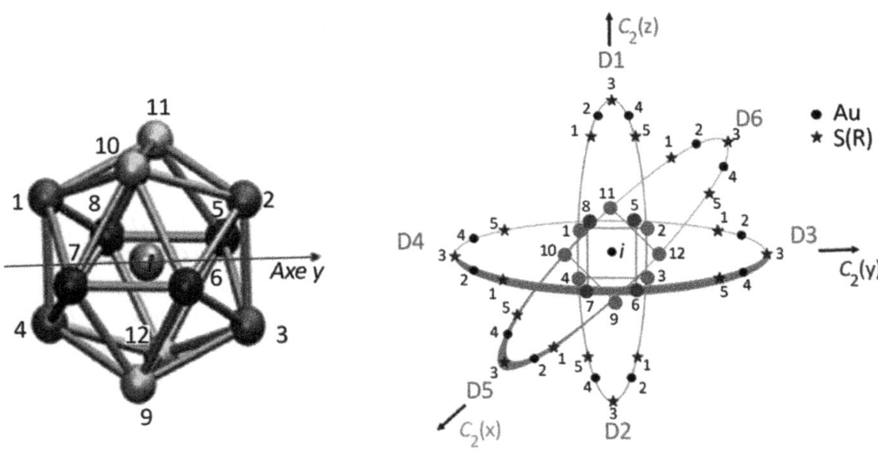

Figure 2.9. *Crystal structure of $Au_{25}(SR)_{18}$*

COMMENT ON FIGURE 2.9.– *a) Core (icosahedron) of the cluster showing 12 gold atoms distributed at the vertices of three rectangles (1, 2, 3, 4 red), (5, 6, 7, 8 blue) and (9, 10, 11, 12 green) located on three orthogonal planes, and one gold atom (i) in the center of the icosahedron. b) Arrangement of the -S(R)-Au-S(R)-Au-S(R)- "staples" with respect to the gold atoms of the central nucleus (represented in green, red and blue). "Staples" D1 and D2 are linked to the gold atoms 1, 2, 3, 4 (in red), D3 and D4 are linked to atoms 5, 6, 7, 8 (in blue) and D5 and D6 are linked to atoms 9, 10, 11, 12 (green). The other gold atoms in the "staples" are shown in black. Adapted from [NAT 15].*

The crystal structure of this molecule comprises a first nucleus, with 13 gold atoms distributed at the vertices and in the center of an icosahedron (Figure 2.9a), which is itself encapsulated by six [-S(R)-Au-S(R)-Au-S(R)-] chains. These links, which Natarajan *et al.* [NAT 15] describe as "staples", take the form of three interlocking (Borromean) rings, each encompassing four gold atoms of the central nucleus, arranged at the vertices of rectangles numbered 1, 2, 3 and 4 (red), 5, 6, 7 and 8 (blue), 9, 10, 11 and 12 (green). Each ring is of elliptical shape and includes two -S(R)-Au-S(R)-Au-S(R)- "staples", connected by two sulfurs to the gold atoms of the nucleus; the three ellipses are located on planes which are orthogonal to one another (Figure 2.9b).

Most gold NCs are synthesized with -SR thiol ligands, with no apparent restrictions on the nature of the R functional groups. A list of these groups, as well as the growth and symmetry rules for these Au NCs, can be found in recent reviews [CHE 15a, HIG 19, PEI 19, WHE 19].

The high stability of some Au NCs, such as $Au_{25}(SR)_{18}$, $Au_{38}(SR)_{24}$, $Au_{102}(SR)_{44}$ and others, may be explained by the "superatom" theory. Essentially, when the global number of electrons of the superstructure $Au_x(SR)_y$ is equal to one of the "magic" numbers, 2, 8, 18, 34, 58, 68, etc., corresponding to the complete filling of the S, P, D and F orbitals of a superatom, it takes on a stable structure [JIA 09, LUO 14].

Thus, for a NC of formula $[M_x(SR)_y]_z$, assimilated to a sphere, the global number of electrons n* of the superstructure is given by the formula: $n^* = x\,V_M - y - z$, where V_M is the valence of the metal atom and z the electric charge of the cluster. In the case of $[Au_{25}(SR)_{18}]$, the number of electrons $n^* = (25 \times 1) - 18 - (-1) = 8$, which is equivalent to a complete electronic layer of a rare gas, $1S^2 1P^6$. Similarly, for the $Au_{102}(SR)_{44}$ gold cluster, isolated by Jadzinski *et al.* [JAD 07] in 2007, the number n* of electrons is equal to 58, one of the aforementioned "magic" numbers. Nevertheless, the structure of Au NCs and their mode of formation are still the focus of active research; alternative theories, based on geometrical criteria, have also been put forward to account for their stability, as described in a recent article by Pei *et al.* [PEI 19].

2.2.3. *Luminescence properties of Au NCs*

As we saw in the case of $Au_{25}(SR)_{18}$, the structure of Au NCs involves two distinct parts: the central core, an assembly of 13 gold atoms (Au_{13}) with various crystal structures (essentially fcc, bcc and hcp)[6], and an organometallic "cap" covering the central core ($[Au_2(SR)_3]_6$), corresponding to "staples" attached to the atoms situated on the periphery of the core. This structure is common to most Au NCs, whatever the number of metallic atoms. It thus seems evident that the properties of these NCs will depend on both the metallic core and the organometallic cap, with possible synergistic effects between the two, particularly in the case of optical properties.

One characteristic in which MNCs differ from gold NPs of sizes larger than 2 nm concerns their energetic structure, similar to that of a molecule; notably, a band gap E_g is found between the higher energy HOMO and the lower energy LUMO, the width of which is a function of the size of the NC. Jin [JIN 15] established a correlation between the number of gold atoms (n) and the amplitude of E_g, showing a zigzag decay curve. When the points representing the maxima and minima are considered separately, this gives rise to two linear variations of E_g versus $n^{-1/3}$. This means that E_g decreases linearly as a function of the volume of the NC, from 2.5 eV for n = 13, to a value of around 0.1 eV once the number of gold atoms exceeds 150 (Figure 2.10).

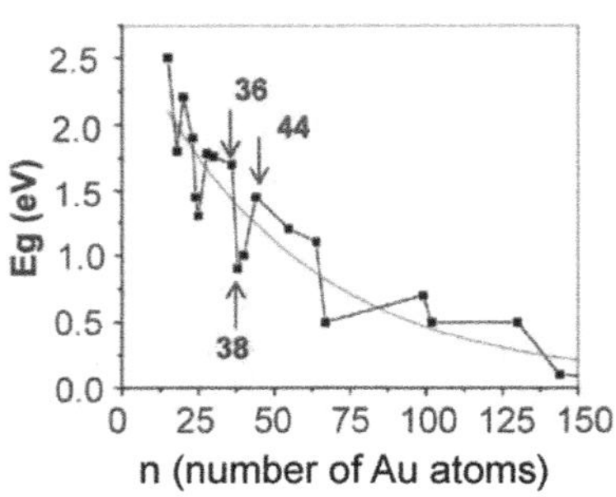

a)

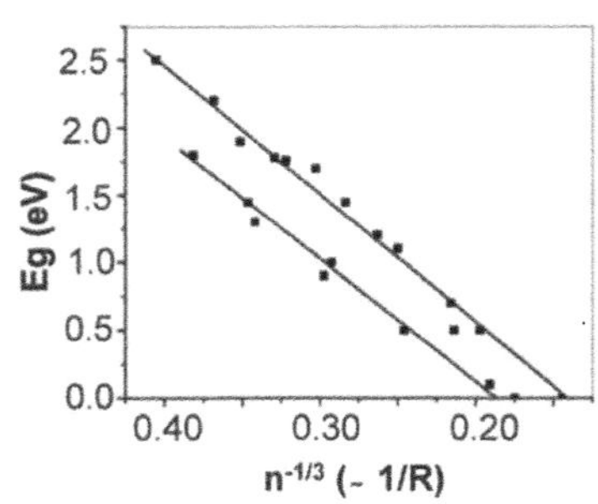

b)

Figure 2.10. *Variation of Eg as a function of Au NC size*

COMMENT ON FIGURE 2.10.– *a) Zigzag curve showing E_g decay as a function of the number of gold atoms in NCs. b) Linear variation of E_g as a function of $n^{-1/3}$, indicating that E_g is inversely proportional to NC size. The two lines correspond to the upper and lower points of the zigzag curve, and appear to correspond to two distinct families of symmetries. Adapted from [JIN 15].*

6 fcc, bcc and hcp: face-centered cubic, body-centered cubic and hexagonal-closed packed crystal structures, respectively. In the bulk state, gold mainly adopts the fcc structure; the gold atoms in nanoclusters may also take on bcc or hcp structures.

The luminescence of gold NPs was demonstrated in the early 2000s for Au NCs between 1 and 2 nm in size. Their quantum yields (QYs) are very low (between 10^{-5} and 10^{-3}), but still much higher than those observed for gold films (of the order of 10^{-10}) [ZHE 12]. El Sayed and Whetten interpreted the presence of two observable emission bands in the case of the 28-atom gold cluster stabilized by 16 glutathione molecules ($Au_{28}SG_{16}$) using a solid-state energy model with two transitions. These included a low-energy intra-band HOMO-LUMO transition within the sp band and a higher energy, high-intensity inter-band sp-d transition, characterized by a very low quantum efficiency of 3.5×10^{-3} [LIN 02]. These low QYs appear to be present for most Au NCs stabilized by thiol ligands.

The origin of the fluorescence of Au NCs is puzzling, at first glance, since neither the thiol ligands nor the metal core fluoresce [JIN 15, JIN 16] Jin *et al.* were the first to suggest that a charge transfer mechanism between the ligands and the metal core may be responsible for an increase in fluorescence. There are two effects to consider: a charge transfer from the ligands to the metal core of the NP (ligand to metal nanocluster charge transfer, or LMNCT) and an electron donor effect from the electron-rich ligand atoms to the metal core [WU 10]. High-intensity fluorescence [JIN 15, JIN 16] has been observed in the case of Au NCs with 25 gold atoms (Au_{25} NCs) with electron-donating atom-rich ligands and QYs of a few percent.

Shortly after Jin's publication [WU 10], drawing on the development of new preparation methods, Luo *et al.* [LUO 12] were able to observe luminescence QYs of over 10% for core-shell structured Au_{25} NCs, complexed with glutathione (GS) ligands. The difference between these NCs and conventional Au_{25} NC, where the metal core is complexed with thiolated groups -S(R)-Au-S(R)-Au-S(R)-, lies in the fact that the metal core is completely covered by a shell of thiolate oligomers. The authors showed that this structure, denoted Au(0)@Au(I)-thiolate NC, gives rise to intense luminescence, with a QY of 15%. These nanomaterials can be synthesized in a single step, following a very precise protocol with respect to timings, temperature and relative concentrations of gold and thiolates. According to the authors, this increase in QYs corresponds to a new luminescence phenomenon they call aggregation-induced emission or AIE[7], resulting from the aggregation of emissive units transformed into oligomers around the metal core (Figure 2.11).

---

7 This AIE effect was highlighted in the 2000s by Tang's group, who compared the luminescence of 1-methyl-1,2,3,4,5-pentaphenylsilole in solution and in the solid state [LUO 01]. The luminescence quantum yield for a solution in alcohol is very low (0.6×10^{-3}), while a yield of 0.2 (over 300 times higher) is obtained for the material in solid film form. This increase in luminescence is due to increased flatness, a result of the formation of aggregates, which leads to a stronger conjugation of electrons. It should be noted that this phenomenon, whereby luminescence increases as a result of aggregation of the emissive

Au NC synthesis using glutathione (GSH) results in a mixture of 5 Au NCs, with structures $Au_{29}SG_{27}$, $Au_{30}SG_{28}$, $Au_{36}SG_{32}$, $Au_{39}SG_{35}$ and $Au_{43}SG_{37}$; the luminescence characteristics of these NCs are essentially the same. Luo *et al.* [LUO 12] consider that this increase in luminescence, compared to conventional Au NCs, results from a "stiffening" of the NC that weakens all non-radiative electronic deactivation processes. According to the authors, the large energy shift between excitation (365 nm) and luminescence emissions at 605 nm may promote phosphorescent emissions, resulting from a ligand to metal-metal charge transfer (LMMCT). This corresponds to a triplet state of Au...Au bonds resulting from strong aurophilic interactions between the gold atoms.

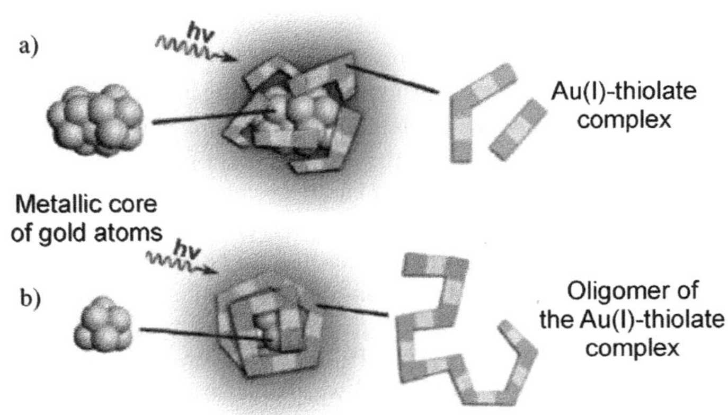

Figure 2.11. *Comparative luminescence of Au NC chelated by single Au(I)-Thiolate complexes or by oligomers of the Au(I)-Thiolate complex. a) UV excitation (360 nm) of the Au(I)-Thiolate complex results in no luminescence. b) Excitation of the Au(0)@oligomer-Au(I)- thiolate NC complex under the same conditions as before results in strong luminescence around 600 nm (orange). Adapted from [LUO 12]*

These results apply to a whole series of ligands possessing an atom (sulfur, nitrogen, etc.) with the capacity to bind onto a metallic core of atoms. The QYs reported in the literature for different gold NCs vary from 1% to several tens of %. Very high QYs (10%–70%) have been obtained by complexing Au NCs with poly(amidoamine) dendrimers. Zheng *et al.* [ZHE 04] observed luminescence emissions ranging from UV-visible, for the smallest sizes, to near-IR for the largest sizes (Figure 2.12).

molecules, is exceptional and runs contrary to the accepted general law whereby a condensed state results in a loss of fluorescence.

Doping Au NCs with Ag also results in high QYs. Thus, by replacing 13 Au atoms by Ag atoms, the NC $Au_{25}(PPh_3)_{10}(SC_2H_4Ph)_5Cl_2]^{2+}$ (characterized by a QY of around 0.1%) becomes highly fluorescent and reaches QYs of around 40%–60% [WAN 14a]. The rigidification effect of the NC structure, combined with the effect of protecting the NC core from the external environment via a careful choice of ligands, often results in high QYs. Based on this principle, a wide range of syntheses using a variety of different ligands have been proposed [JIN 16].

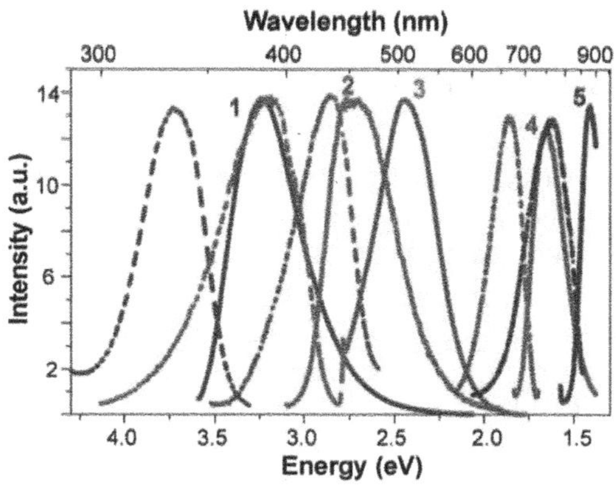

Figure 2.12. *Luminescence of Au NCs complexed with poly(amidoamine) dendrimers and of increasing size, from 5 to 31 gold atoms*

COMMENT ON FIGURE 2.12.– *The dashed and solid curves marked 1, 2, 3, 4 and 5 correspond, respectively, to the excitation and emission spectra of different NCs: Au_5 (1), Au_8 (2), Au_{13} (3), Au_{23} (4) and Au_{31} (5). The respective quantum yields of 70%, 42%, 25%, 15% and 10% are seen to decrease as NC size increases. Adapted from [ZHE 04].*

2.2.4. *Applications using the luminescent properties of Au NCs*

The luminescence of Au NCs is one property which makes them exceptionally well suited for use in the analysis and detection of metal ions, molecules of biological interest, proteins and nucleic acids, and for fluorescent labeling of biological molecules used in imaging. Au NCs may be preferred to organic fluorophores or semiconductor QDs due to a combination of favorable physicochemical characteristics: these include ease of production and functionalization; stable and intense luminescence; ease of modulating color according to NC size, high biocompatibility and non-toxicity. Their extreme thinness results in a high surface/volume ratio, presenting an additional advantage for catalytic purposes. This size criterion is also important when NCs are used for drug delivery; their diffusion through living tissues is facilitated, and accumulation in organs such as the kidneys and liver is also limited.

The number of applications of Au NCs or other MNCs in the detection and analysis of metal ions or biological molecules is particularly high, and these are well documented in the literature [HAL 18a]. For example, the luminescence properties of MNCs have been used to perform bioassays of molecules such as dopamine (DA), a crucial neurotransmitter, for which many different detection methods have been proposed (*ex situ* and *in situ*); these differ in terms of selectivity and detection thresholds.

The selectivity criterion is particularly important. The possibility of grafting a functional group with the ability to complex a particular molecule to the exclusion of all others onto fluorescent Au NCs constitutes a significant advancement. Cyclodextrins (α, β and γ), for instance, are well known for their specific complexing properties with certain biological molecules. This effect has been exploited by researchers, who have grafted β-cyclodextrin (β-CD) onto (non-luminescent) gold NPs. They then used the solution fluorescence properties of certain organic luminophores, which are quenched when the luminophore is complexed with β-CD and regenerated when the luminophore is displaced from the β-CD by a biological molecule of higher affinity, for detection purposes. This system has been used to detect cholesterol levels, using rhodamine-B as a fluorophore. Rhodamine complexed with β-CD@AuNC is not fluorescent. When displaced from β-CD@AuNC by cholesterol, it regains its fluorescent property, with an intensity directly proportional to the amount of cholesterol complexed by β-CD [ZHA 16c].

A different method draws on the fact that the fluorescence of Au NCs can be quenched by different biological molecules. Xu and coworkers used this technique for DA assays in solution [HAL 18a,b]. As before, one-pot synthesis of a fluorescent Au NCs (with emissions in the green spectrum) coated with β-CD molecules results in the production of core/shell NCs (β-CD@Au NC) (Figure 2.13).

When DA in solution is complexed by β-CDs, quenching of β-CD(@)Au NC fluorescence occurs, due to an electron transfer from Au NC to DA. The fluorescence intensity decreases linearly with the DA concentration in the range from 100 nM to 80 μM; the detection threshold is around 20 nM. This method also offers high selectivity, with no observed interference in cases where the DA is mixed with different biological molecules.

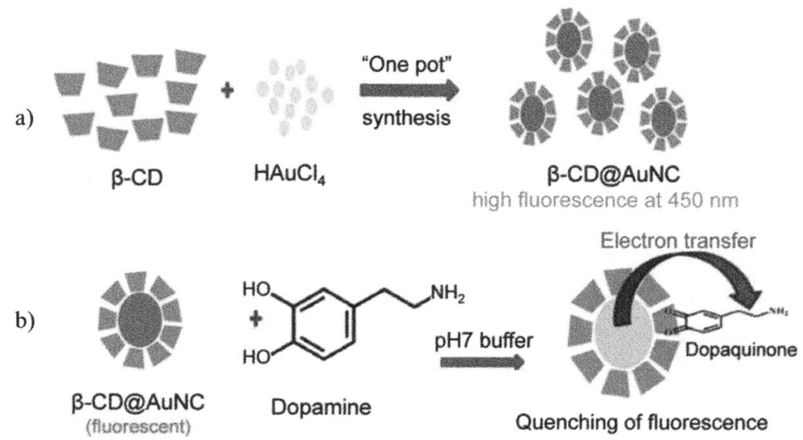

Figure 2.13. *Principle of dopamine (DA) assay by β-CD-modified Au NCs. a) One-pot synthesis of β-CD@Au NCs (incubation of $HAuCl_4$ + β-CD mixture at 90°C for 5 min, addition of NaOH, and heating at 90°C for 6 hours). b) The DA is transformed to dopaquinone and complexed by β-CDs in a buffer (pH 7), resulting in quenching of the fluorescence. Adapted from [HAL 18a]*

2.2.5. *Conclusion*

In contrast to metallic NPs of sizes greater than 3 nm, which are characterized by plasmonic-type absorption, metallic NPs of under 2 nm exhibit a molecular behavior, characterized by energy band gaps which widen as the size of the NP decreases. These energy characteristics mean that the particles are potentially luminescent. In reality, luminescence only becomes apparent in the presence of certain ligands, resulting from the stabilizing effects of the metallic core. Some of these NCs are extremely stable, such as those with 25 Au atoms. Simple one-pot production methods are often used. The ease with which these NCs can be functionalized makes them particularly suitable for applications in biomedical imaging or analysis.

2.3. Semiconductor quantum dots

Semiconductor QDs, which may also be considered as semiconducting NCs, are the result of binary, ternary and multinary combinations of different elements. These QDs constitute a new family of inorganic nanomaterials, with sizes of less than 10 nm. Much work has been published on this subject over the last 20 years, focusing both on preparation techniques and on the physicochemical properties of QDs. Their fluorescent properties have valuable applications in biomedical imaging, while their electroluminescence may be used in light-emitting diodes (LEDs), screens and photovoltaic devices [KOV 15].

These QDs are characterized by a band gap that can be modulated according to their size and composition, allowing color emissions ranging from UV-visible to IR and far IR. Most QDs are binary combinations of elements in columns II and VI of the Periodic table, known as chalcogenides (CdS, CdSe, ZnS, ZnSe, PbS, TeS, etc.). However, more recent developments have involved ternary and even quaternary combinations of metals with much lower toxicity than the original elements. These include combinations of type I-III-VI_2 ($CuInS_2$, also known as CIS, or $CuIn_{1-x}Ga_xSe_2$ (CIGS), chalcopyrites) and I_2-II-IV-VI_4 (Cu_2ZnSnS_4 or CZTS: kersterites). The bandgap widths of these compounds are well adapted to the solar spectrum, making them particularly suitable for photovoltaic applications.

Generally speaking, QDs are characterized by broad excitation bands, narrow fluorescence emission peaks and high intensities. Their QYs are comparable to those of organic fluorophores, but their extinction coefficients are 10–50 times higher; this strong propensity to absorb light makes them highly suitable for use in energy conversion systems. Furthermore, they are more chemically stable than organic fluorophores and are more resistant to photobleaching (loss of fluorescence during irradiation), performing 100 times better than organic molecules in this respect [YAO 14].

2.3.1. *Development of colloidal QDs*

The first syntheses of colloidal solutions of QDs such as CdS and CdTe, in which QDs are stabilized by the adsorption of organic molecules, were carried out in the 1980s. However, the fluorescence yields of these earliest solutions were very low. Significant progress was made in the 1990s by the Bawendi group, who were able to control the dimensions of CdS, CdSe and CdTe via a QD synthesis process involving the pyrolysis of organometallic reagents, injected into a high-temperature solvent with high coordination power [MUR 93][8]. Other synthesis methods have

8 This preparation technique, one of the "injection" methods of CdSe production, makes it possible to obtain CdSe of controlled size (between 1.5 and 11.5 nm) but is not suitable for large-scale production. It relies on the use of a reaction medium consisting of a mixture of solvents (trioctylphosphine (TOP) and trioctylphosphine oxide (TOPO)) that can be heated to very high temperatures (300°C). The Cd and Se reagents take the form of dimethylcadmium (Me_2Cd) and TOPSe, respectively, in solution in TOP. These are injected at room temperature into the TOP/TOPO solvent mixture. A rapid yellow-orange color change is observed due to incipient formation of CdSe NPs. The temperature is then progressively increased to

since been proposed to produce even larger quantities of QDs. The Cao group [YAN 05] notably developed a method for CdSe synthesis based on high-temperature reaction of a mixture of two precursors in solution (Cd Myristate and Se)[9].

A further major advance was made by the Brus group, who pioneered the synthesis of inorganic core-shell structures such as ZnS@CdSe in which the fluorescent QD (CdSe) is covered by a layer of a semiconductor such as ZnS, characterized by a band gap which is much larger than that of CdSe [KOR 90]. This has the effect of passivating the CdSe core and preventing any interaction with the external medium, considerably increasing fluorescence QY (of the order of 50%); concurrently, it also attenuates the fluorescence blinking observed with CdSe [KOR 90, NIR 96]. Mahler *et al.* [MAH 08] were able to practically eliminate fluorescence blinking by coating the emissive CdSe core with a thick layer of CdS (4–5 nm).

Another way of improving the fluorescence emissions of CdSe@CdS core-shell QDs involves progressively modifying the CdSe–CdS interface by introducing a layer of $CdSe_{1-x}S_x$ alloy. This has the effect of decreasing the crystallographic shift between CdSe and CdS, attenuating the stress at the CdSe–CdS interface and thus reducing the number of structural defects in the CdSe [BAE 13, OWE 17]. These CdS ($CdS@CdSe_xS_{1-x}@CdSe$), with an intermediate layer of $CdSe_xS_{1-x}$ alloy of controlled thickness, are obtained in a colloidal medium using a suspension of CdSe NPs (Figure 2.14)[10].

One beneficial effect of this type of structure is that it reduces fluorescence blinking, particularly prevalent at high excitation intensities, which results from an

230–260°C in order to “grow” the QDs. NC size is monitored over time by sampling and spectroscopic analysis.

9 Cd myristate (Cd salt of tetradecanoic acid-$CdC_{28}H_{54}O_4$) decomposes at 226°C; selenium powder has a melting point at 221°C, and becomes slightly soluble in a solvent such as 1-octadecene (ODC-$C_{18}H_{36}$) from 190°C. A reaction at a temperature of 240°C, in ODC medium in the presence of oleic acid, ensures optimal conditions for reacting Se with Cd and producing QDs with high monodispersity (size deviations of less than 5%) and very good fluorescence properties (QY of around 30%–40%) [YAN 05].

10 The $CdSe_xS_{1-x}$ layer is obtained by slowly reacting a mixture of Se trioctylphosphine (TOPSe) and 1-dodecanethiol (1-DDT) at high temperature (300°C) on CdSe NCs dispersed in a solution of Cd oleate ($Cd(OA)_2$) in a mixture of 1-octadecene (ODE) and trioctylamine (TOA). After around 30 minutes, the reaction results in the formation of an alloy layer of about 1.5 nm thickness on the CdSe NCs. A new CdS layer is then deposited directly onto the $CdSe_xS_{1-x}$ by injecting a mixture of $Cd(OA)_2$ and 1-DDT in solution in TOA directly into the $CdSe_xS_{1-x}$ suspension, while maintaining a constant temperature of 300°C.

Auger recombination process when the QD is ionized, annihilating the emission process[11].

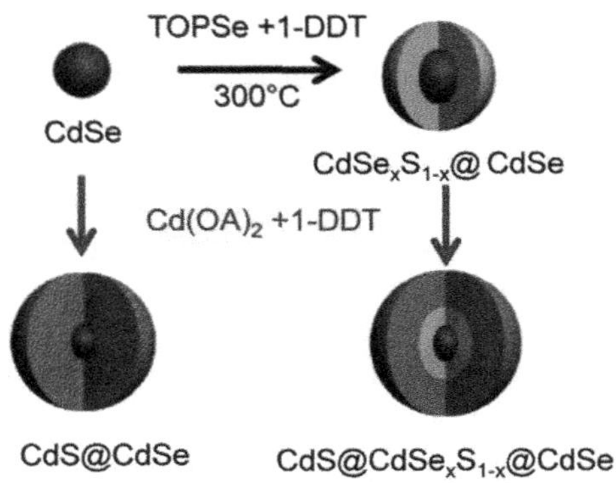

Figure 2.14. *Reactions from CdSe NPs showing the addition of an alloy layer (CdS) on the CdSe NPs, then the final three-layer structure CdS@CdSe$_x$S$_{1-x}$@CdSe. Adapted from [BAE 13]*

At high excitation intensities, another effect of the presence of the CdSe$_x$S$_{1-x}$ intermediate layer is observed. In the case of CdS@CdSe, at high excitation intensities, the fluorescence emission band centered at 615 nm (which corresponds to the CdSe core of the QD) is accompanied by a second fluorescence band, centered around 518 nm and attributable to the CdS (Figure 2.15a). In QDs with the intermediate layer of CdSe$_x$S$_{1-x}$, however, no emissions occur at 518 nm, even under high-intensity excitation. A single fluorescence band is present at 650 nm, slightly shifted toward the red compared to the CdS@CdSe QD; this band is characteristic of CdSe (Figure 2.15b).

---

11 All known fluorophores are affected by a fluorescence blinking phenomenon, assimilated to On-Off emission states corresponding to alternating periods of brightness and darkness. Many physical models have been proposed to account for the phenomenon [FRA 08, EFR 16]. One of the most widely accepted, proposed by Efros and Rosen, attributes this phenomenon to an Auger-type recombination phenomenon, which takes place from an ionized QD. The ionization of the QD is caused by the ejection of the electron from an exciton toward the surface of the QD, creating a hole (h^+) in the valence band of the QD. An exciton formed after QD ionization (QD^+) can deactivate in a non-radiative manner by recombining with the positive charge and transferring the recombination energy to the hole, thus maintaining the ionized state of the QD. This exciton deactivation process is repeated for as long as the QD remains ionized. Its return to the neutral state is necessary in order for the photoemission process to restart.

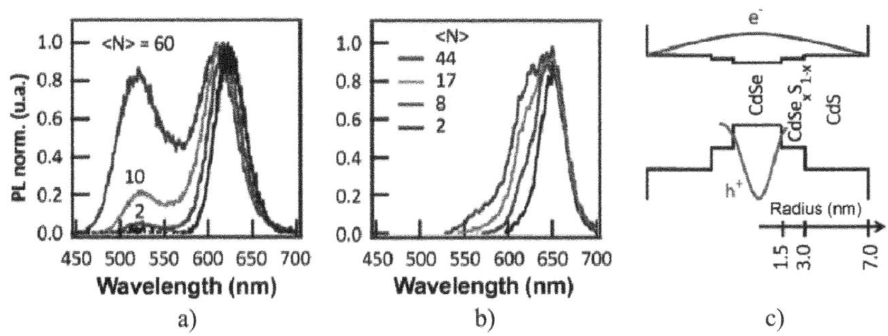

Figure 2.15. *Normalized photoluminescence (PL norm.) of CdSe@CdS and CdSe@CdSe$_x$S$_{1-x}$@CdS QDs as a function of excitation intensity at 400 nm (3.1 eV), measured as the average number <N> of excitons formed per QD and per time interval (0.1 ns)*

COMMENT ON FIGURE 2.15.– *a) Fluorescence of CdSe@CdS for increasing excitation intensity, with <N> varying from 2 to 60. Note the appearance of a second emission band at 518 nm once <N> exceeds 2, attributable to CdS, which grows as a function of excitation intensity; however, it never reaches the same intensity level as the first emission band at 615 nm, characteristic of CdSe. b) Fluorescence of CdSe@CdSe$_x$S$_{1-x}$@CdS under the same excitation conditions as for a). Note the absence of the second emission band at 518 nm. c) Simplified energy diagram of the CdSe@CdSe$_x$S$_1$-x@CdS heterojunction with a total radius of 7 nm. Adapted from [BAE 13].*

The energy diagram of the heterostructure (Figure 2.15c), which highlights the shifts between the different energy levels of the materials, shows a preponderant localization of holes (h^+) in the CdSe and a delocalization of the electrons (e^-) in the whole volume of the QD. This is why, at low excitation intensities, the preponderant emission comes from the CdSe core of the QD, and it is only at high excitation intensities (very high values of <N>) that some holes will occur in the CdS and CdSe$_x$S1$_{-x}$, contributing to the emission of the higher energy radiation corresponding to the 518 nm band. The presence of the intermediate layer reduces the amplitude of the initial shift between CdS and CdSe. It results in a "cascade" of progressive energy shifts of the successive valence bands of CdS, CdSe$_x$S$_{1-x}$ and CdSe, facilitating the migration of the holes formed in the CdS toward the CdSe. This results in the elimination of the emission band at 518 nm.

The QYs of CdS@CdSe QDs are roughly the same with and without transition layer, and decrease as the overall size of the QD increases. As the overall diameter of the QD increases from 8 to 14 nm, the QY decreases from 60% to 40%. The wavelengths corresponding to the fluorescence peaks shift slightly toward the red as QD size increases. For the size variations mentioned above, the fluorescence peak of

CdS@CdSe shifts from 620 to 630 nm, and that of CdS@CdSe$_x$S$_{1-x}$@CdSe shifts from 640 to 650 nm.

Colloidal QD production techniques have advanced considerably in recent years, and it is now possible to obtain QDs in large quantities and at an accessible cost. The emissions of these QDs cover a wide range of wavelengths, from visible to near IR [OWE 17, BAN 17, PU 18] (Figure 2.16).

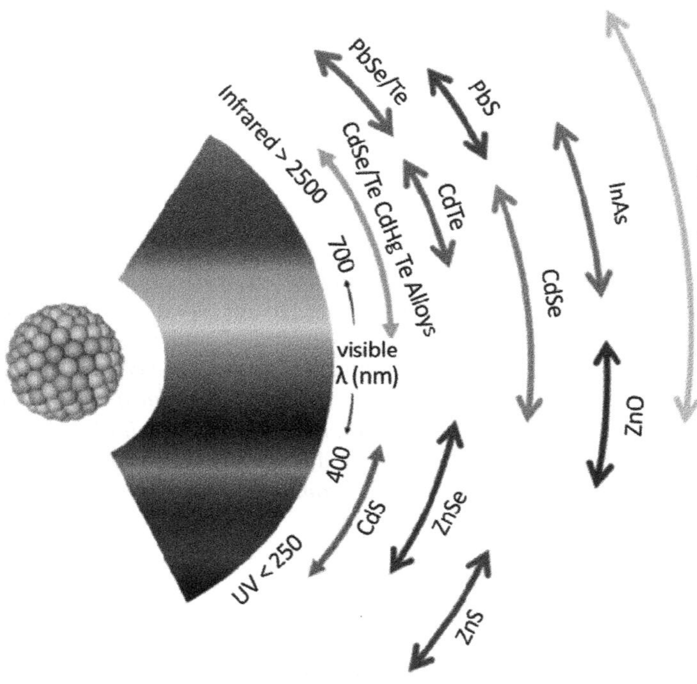

Figure 2.16. *Photoluminescence of QDs, specifying their color emission ranges. From [PU 18]*

QDs with fluorescence in the near IR (NIR) are particularly important for biomedical imaging applications. Some of these QDs are based on heavy metals such as lead, which makes it possible to reach the NIR-II window beyond 1000 nm; however, their toxicity means that they are unsuitable for biological uses. Alternatives include QDs made of Ag_2S NPs in an aqueous medium, which are biocompatible and non-toxic. These emit a fluorescence that can be controlled, from IR to NIR (700–900 nm) and up to the NIR-II window (between 1000 and 1400 nm), while maintaining good QYs of around 15%; this value is very high for IR emission domains beyond 1000 nm [HON 12, WAN 12a, YAN 13a]. Other formulations have been tested as replacements for Cd-based QDs, notably core/shell InP/ZnS [BAN 17] and CuInS/ZnS [KIM 16]; unfortunately, these do not attain the same yields and brightness values as CdSe/ZnS QDs [CHO 18a].

New semiconductor materials, such as perovskites, have also displayed good fluorescence properties that can be modulated over the entire visible spectrum, from 400 nm ($CsPbCl_3$) to 700 nm ($CsPbI_3$) [PRO 15]. However, their interest may be limited, due to their low stability in humid environments. This is a major obstacle to the development of perovskite photovoltaic cells. The presence of lead in their formulation poses additional problems.

QD production methods have also evolved; manufacturing in aqueous medium is now possible, a development which reduces the ecological impact of the process. Biosynthetic means of obtaining QDs have also been put forward [DUN 16], as described in [PU 18].

In conclusion, it is important to note the exceptional fluorescence properties of colloidal QDs. This is due to the fact that their emission wavelengths are scalable according to their size, that their QYs are close to one, and that their emission colors are very pure, especially in red, green and blue (the three primary colors of light). Furthermore, their high photostability and resistance to photobleaching set them apart from organic fluorophores. Finally, it is now possible to produce these QDs on a large scale and at low costs. These properties make QDs particularly suitable for use in display and lighting applications where electroluminescence, corresponding to the emission of light under the effect of an electric current, is required. This aspect of QDs will be examined in greater detail in Chapter 4 in the context of energy conversion.

2.4. Two-dimensional inorganic lamellar nanosheets

Since the discovery of graphene and the extraordinary conductivity properties arising from its two-dimensional (2D) structure, other 2D inorganic materials, such as transition metal dichalcogenides (TMDs) and a number of other oxides and hydroxides with lamellar structures comparable to graphite, have attracted significant attention; over the last 15 years, means of obtaining 2D nanosheets made up of one or more atomic layers of these materials have been developed. A detailed description of the optoelectronic properties and applications of the main TMDs, including MoS_2, $MoSe_2$, WS_2 and WSe_2 nanosheets, is available [WAN 12b].

Alongside this vast family of bidimensional dichalcogenides of transition materials, we must consider other lamellar materials, such as tin sulfide (SnS_2), which has proven to be a good anode material for sodium and lithium-ion batteries (LIBs) when used in the form of a composite with graphene (SnS_2-rGO) [QU 14]. Phosphorene, a recent discovery, is a semiconductor which consists of an assembly of black phosphor (BP) atoms that can also be exfoliated into nanosheets of sub-nm thickness. It is characterized by high charge carrier mobilities (~1000 cm^2/Vs) and an energy band gap that can be modulated between 0.3 and 2 eV; these properties make it suitable for use in field effect transistors (FETs) and other electronic devices [KOU 15]. Hexagonal boron nitride (hBN), a 2D material which is isomorphic to graphene but insulating, is another 2D material consisting of a hexagonal network of boron and nitrogen atoms. It is characterized by high thermal conductivity, very good mechanical properties and high chemical stability. hBN has been cited as a possible component for electronic devices, in association with other 2D materials such as graphene or TMDs [ZHA 17a]. Finally, a new emerging family of MDs, named MXenes, was discovered by Naguib *et al.* [NAG 11] in 2011. These take the general form $M_{n+1}X_nT_x$, where M is a transition metal, X is a carbon or nitrogen atom and T is an OH, O or F group. They are obtained in the form of 2D nanocrystals, exfoliated from MAX ternary compounds after heating in solution and extraction of metal A, which is usually aluminum [NAG 11]. MXenes are highly conductive, and as such constitute excellent additives for the fabrication of anodic electrode materials (see Chapter 3)[12].

Contrary to graphene, composed uniquely of carbon atoms, the great chemical diversity of atoms in TMDs leads to a much wider range of properties. These materials may be semiconducting (MoS_2, WS_2), semi-metallic (WTe_2, $TiSe_2$), metallic (NbS_2, VSe_2) or superconducting ($NbSe_2$, TaS_2) [LV 15]. Moreover, depending on the number of layers and their composition, semiconducting TMDs may feature a direct or indirect energy bandgap, and have n- or p-type charge carriers [CHH 13]. These properties pave the way for a wide range of applications in catalysis, energy storage and electronics. New FETs manufactured from TMDs have notably been used for analytical purposes, for the detection of chemical compounds, and in various optoelectronic devices and new logic circuits, described in the literature [TAN 15, ANI 18, XIO 19].

2.4.1. *Transition metal dichalcogenides*

Metal dichalcogenides such as molybdenum (MoS_2), tungsten (WS_2) and titanium (TiS_2) sulfides have attracted the most attention, with MoS_2 being

12 Gogotski and Barsoum proposed the name "MXene" in reference to the fact these materials have a structure similar to that of graphene [NAG 11].

considered the most accessible [HIN 15, ANI 18]. Unlike graphite, which consists of a stack of identical layers of graphene, TMDs comprise an assembly of hybrid layers in which a layer of a transition metal M is "sandwiched" between, and covalently bonded to, two outer layers of sulfur atoms. The outer planes of these hybrid layers are linked together by Van der Waals forces, which explains the ease with which they can be separated using exfoliation techniques (as in the case of graphite).

These TMDs take two main crystal structures: 1T (octahedral, the most stable) and 2H (trigonal prismatic). In the case of a MoS_2 crystal, the distance between two layers is 0.65 nm (Figure 2.17).

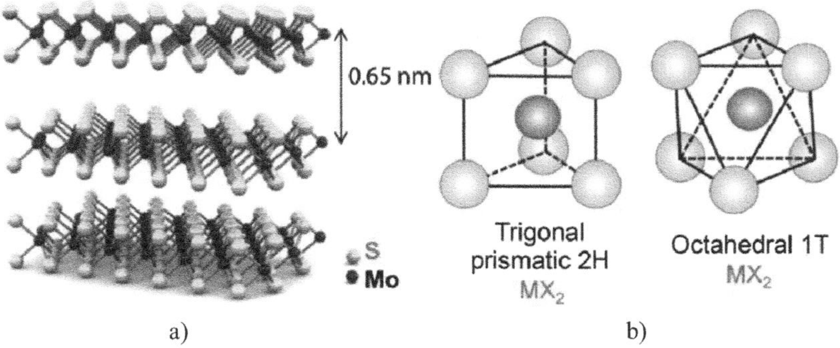

Figure 2.17. *Representation of the crystal structures of MX_2 type TMDs. a) Three-dimensional representation of the different planes of a MoS_2 crystal. Adapted from [RAD 11]. b) Representation of the two crystal forms: 2H and 1T. Adapted from [LV 15]*

There are two main methods used to produce nanosheets of this type: exfoliation techniques, applied to materials in the bulk state, or CVD (chemical vapor deposition). The first approach is more accessible in terms of material requirements, and consists of using an electrochemical reaction to intercalate Li^+ ions in a bulk TMD using a similar mechanism to that found in LIBs. The resulting product is then subjected to sonication in water or ethanol, leading to a release of hydrogen that causes the TMD to disintegrate, resulting in the formation of nanosheets. These are collected as a suspension in water with a yield of over 90% [HUA 13a]. The same direct sonication process can be used to induce TMD disaggregation using a solvent such as N-methyl pyrrolidone, which has good wettability with TMD. This method can be used to produce single and multi-atomic layer nanosheets of MoS_2, WS_2, $MoSe_2$, $NbSe_2$, $TaSe_2$, $NiTe_2$, $MoTe_2$, h-BN and Bi_2Te_3. Less widespread, a mechanical exfoliation technique has also been proposed [LI 14].

The CVD method for TMD preparation can be used to produce crystals of large lateral size and with controlled thicknesses of one or more atomic layers. MoS_2 nanosheets can be obtained in this way by heating sulfur and MoO_{3-x} precursors in powder form. The sulfur vapor reacts with the MoO_{3-x} suboxide, producing MoS_2, which condenses as a film on a Si/SiO_2 substrate. Many variations of the method have been proposed with a varied choice of precursors also making it possible to obtain large nanosheets.

One advantage of these TMDs, and of MoS_2 in particular, lies in the ease with which their surface can be functionalized with the addition of metals, the fact that their composition can be modified by doping or alloy formation, and the possibility of creating hybrid nanostructures by associating the TMD nanosheets with carbon-based or inorganic nanomaterials [TAN 15]. Notable examples of these varying forms include TMD nanocomposites in which the surface is doped using precious metals (Pt, Pd, Au and Ag) or alkalis, TMD alloys of the $MoS_{2(1-x)}Se_{2x}$ and $Mo_{1-x}W_xS_2$ type, and assemblies of TMD nanosheets with graphene, CNT or carbon fibers.

2.4.1.1. *Surface coating of TMDs with precious metals and metal oxides*

Very small NPs of precious metals (Pt, Pd) can be deposited on monolayer MoS_2 nanosheets, creating efficient catalysts for the electrochemical reduction of water. Composite nanostructures of this type are relatively easy to produce. For example, given a suspension of MoS_2 nanosheets in water, Pd NPs can be deposited on the surface of these nanosheets by simply reducing K_2PdCl_4 salts with ascorbic acid in the presence of a surfactant (cetyltrimethylammonium bromide). This results in Pd NPs, with sizes of around 5 nm, being deposited on the surface of the MoS_2, bonded to the sulfur atoms. A similar wet process can be used with Pt: in this case, K_2PtCl_4 is reduced by photoirradiation with a halogen lamp, leading to very small Pt NPs (around 3 nm) being deposited on the MoS_2 layers. The very high aspect ratio of MoS_2 composite nanostructures with precious metal coatings is a considerable advantage for catalytic purposes. MoS_2/Pt nanostructures are excellent catalysts for electrochemical water reduction and hydrogen production; while their Pt content is much lower than that of traditional Pt/carbon catalysts, their catalytic efficiency is just as good [HUA 13b], representing significant cost savings (Figure 2.18).

In addition to precious metals, oxides such as MoO_3 or Fe_2O_3 can also be deposited on the surface of TMD nanosheets in the form of nanometric NPs (2–10 nm); the same is true of various organic and organometallic compounds [TAN 15].

The formation of alloys, which consists of partially substituting another metal for the TMD transition metal, is another technique which may be used to modulate the optoelectronic properties of TMD nanosheets. In the case of MoS_2, the formation of

$Mo_{1-x}W_xS_2$ monolayer crystals makes it possible to modulate the width of the energy band gap, E_g, along with the conductivity of the TMD nanosheets. This may be achieved using either CVD or wet preparation methods.

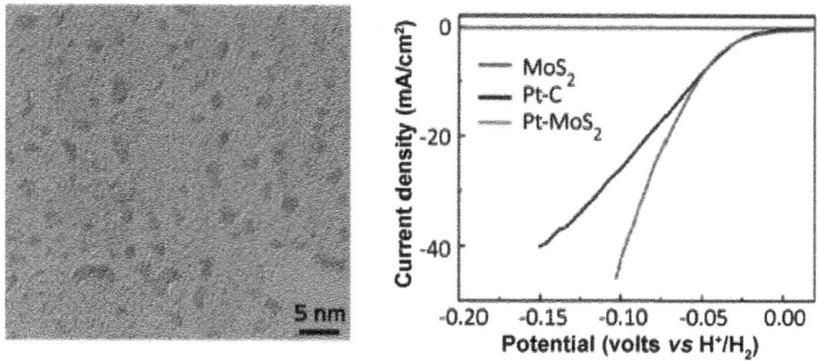

a) b)

Figure 2.18. *MoS_2/Pt catalyst for water reduction in an acidic medium*

COMMENT ON FIGURE 2.18.– *a) TEM image of a MoS_2 nanosheet modified by Pt NP deposits (1–3 nm). b) Comparison of water reduction curves using MoS_2, Pt-C and Pt-MoS_2 in an 0.5 M H_2SO_4 medium. Note that the overpotential in the case of Pt-MoS_2 is much lower than that observed with a commercial Pt-C catalyst. Without the addition of Pt, MoS_2 is practically inactive. Adapted from [HUA 13b].*

Doping using a metal (different from an alloy), where the metal is chemisorbed on the surface of the TMD (Figure 2.22c), is yet another way of modulating the electronic properties of TMDs. Potassium-doped MoS_2 and WSe_2 nanosheets in the vapor state, for example, can be transformed into degenerate n^+ semiconductors, characterized by very high surface electron densities ($10^{13}/cm^2$ and $2.5\times10^{12}/cm^2$ for MoS_2 and WSe_2, respectively). This alkali doping also modifies the width of the energy band gaps, which decreases by several hundred meV as a function of the doping rate. In the case of WSe_2 nanosheets, which are initially intrinsic semiconductors, potassium doping produces degenerate n-type SCs (n^+) with mobilities greater than 100 cm^2/Vs. This value is comparable to the hole mobility of undoped WSe_2 [FAN 13]. The degenerate n^+ semiconductors can be used in ohmic contacts with other metals. It is possible to produce FETs with very good electrical characteristics using a channel of MoS_2 or WSe_2 nanosheets, in which the extremities in contact with the electrodes (source and drain) are n^+ doped (Figure 2.19). Moreover, the fact that n and p zones can be created on a single nanosheet of a material such as WSe_2 makes it possible to couple n and p FETs in a simple manner in order to create voltage inverters [TOS 14]. One advantage of using MoS_2 or $WeSe_2$ nanosheets instead of silicon is that their surface is extremely homogeneous, and the thickness of the nanosheets is perfectly uniform along their entire length.

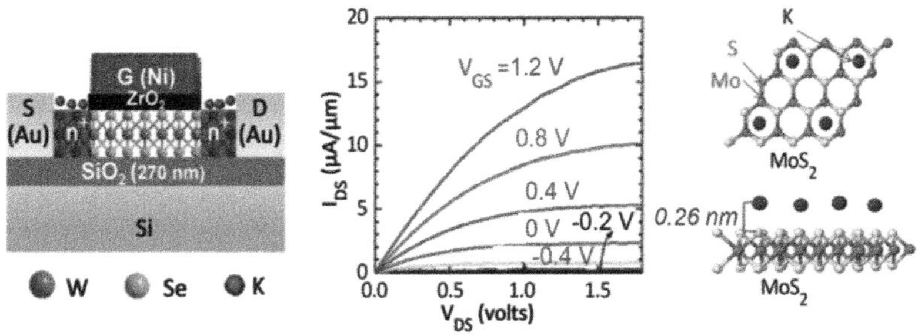

a) b) c)

Figure 2.19. *Field effect transistor (FET) produced using a WSe_2 nanosheet*

COMMENT ON FIGURE 2.19.– *a) Diagram of an FET with a WSe_2 channel (three monolayers, 6.2 μm length), n^+ doped with potassium at the extremities. b) Output characteristics of the FET as a function of different gate potentials, confirming the n character of the WSe_2 semiconductor and the ohmic nature of the contacts between the channel and the S and D electrodes. c) Typical structure of a MoS_2 nanosheet doped by an alkaline metal (K). Adapted from [FAN 13].*

2.4.1.2. *TMD-carbon and TMD-mineral nanocomposites*

Many research groups have combined MoS_2 nanosheets with carbon or mineral derivatives to produce composite anode materials for lithium insertion in batteries. MoS_2 nanosheets have been grafted onto carbon fibers, graphene and CNT with the aim of improving the electrical connection between particles and the diffusion of lithium ions inside the material. Carbon fibers obtained by electrospinning – a process that can be adapted for large-scale production – have also been used to create MoS_2/FC composite nanostructures. Zhu *et al.* [ZHU 14a] showed that a composite nanostructure of this type is characterized by a specific capacitance of around 1100 mAh/g, obtained for a discharge rate of 0.5 A/g; this value decreases, but only to 600 mAh/g, for a high discharge rate of 10 A/g. These results are very promising, particularly given the stability of the operation over 1000 charge-discharge cycles, with respect to their potential use for making anodes with far better performance than common graphite electrodes[13].

13 The theoretical specific capacity of a graphite anode is 372 mAh/g, much lower than that of the MoS_2/FC composite.

Composite nanostructures of MoS_2 nanosheets and graphene have also been proposed for use in LIBs. MoS_2 nanosheets grafted onto nitrogen-doped graphene perform particularly well as anode materials for lithium storage, and have been discussed extensively in the literature.

Xia *et al.* [XIA 18a] show that MoS_2 nanosheets, vertically grafted onto N-doped r-GO (reduced graphene oxide) fragments, constitute ideal nanostructures for Li^+ ion insertion. The process, which can be easily scaled up, is carried out thermally using a one-pot reaction. Polyethylene imine (PEI), sodium molybdate (MoO_4Na_2) and thiourea are added to a suspension of r-GO particles in water. Prolonged heating of the mixture at 200°C results in the direct formation of N-rGO/MoS_2 nanostructures in the form of a black precipitate; this is then annealed at 600°C to improve the crystallinity of the MoS_2. Charge/discharge (C/D) cycles performed with C/D rates increasing from 100 mA/g to 1000 mA/g (i.e., approximately C/10 to C) are characterized by a specific capacity ranging from 1050 mAh/g to 650 mAh/g; these values are significantly higher than those obtained using anodes made of rGO/MoS_2 nanostructures in which the r-GO is not doped by nitrogen. The authors attribute the improvement in performance to the vertical alignment of the MoS_2 nanosheets which are grafted to the r-GO sheets via the imine (NH) groups of the PEI chains. These react with a Mo atom in the MoS_2, resulting in a "honeycomb" structure. This architecture considerably facilitates the diffusion of Li^+ ions inside the material; it is also responsible for the high reversibility of the lithiation and de-lithiation reactions of MoS_2.

Compounds other than carbon nanomaterials have also been used to realize hybrid MoS_2 nanostructures. Taberna *et al.* [TAB 06] first tested ferric oxide, Fe_3O_4[14], as an anode material for Li^+ ion insertion in 2006; this compound has since been used for lithium anodes in the form of composite nanostructures, where ultrafine Fe_3O_4 NPs (size ~3 nm) are adsorbed on the faces of MoS_2 nanosheets[15] [CHE 14b]. In the 2006 study, Fe_3O_4 was deposited on copper nanosheets, while in the case where MoS_2 was used, Fe_3O_4 NPs adhered to the walls of MoS_2 nanosheets. The results obtained in this latter case were better in terms of C/D cycle longevity

14 Specific capacities of around 1000 mAh/g were obtained with a nanostructured anode consisting of copper nanotips of about 2–3 μm in length and 200 nm in diameter, on which Fe_3O_4 is deposited as a more or less homogeneous film. The copper nanotips are produced by electrolysis through the pores of an alumina membrane, resulting in the rods being oriented perpendicular to the collector electrode.

15 The method used to obtain MoS_2 nanosheets is relatively simple, consisting of thermally reacting thioacetamide (C_2H_5NS) with sodium molybdate (Na_2MoO_4) at 220°C. Fe_3O_4/MoS_2 composites are obtained by adding ferric chloride ($FeCl_3$), sodium bicarbonate ($NaHCO_3$) and ascorbic acid ($OHCH_2CH(OH)C_4H_3O_4$) to a suspension of MoS_2 nanosheets in water, all subjected to prolonged heating at 150°C.

(low variation of the capacity after 1000 C/D cycles), and the capacities were slightly higher than those obtained using copper NRs (1000 mAh/g for variable C/D speeds between C/10 and 2C) (Figure 2.20a).

Surprisingly, the capacities obtained with MoS_2 alone are very low; according to the authors, this difference in behavior between MoS_2 and Fe_3O_4/MoS_2 is due to the fact that the Fe_3O_4 NPs adsorbed onto the MoS_2 prevent the MoS_2 nanosheets from stacking together during the C/D cycles (Figures 2.20b and 2.20c). This method of electrode preparation is relatively simple, and would be suitable for larger scale production. The materials used are widely available, and inexpensive; however, more reliability tests are still needed in order to confirm the interest of the material before commercial production can begin.

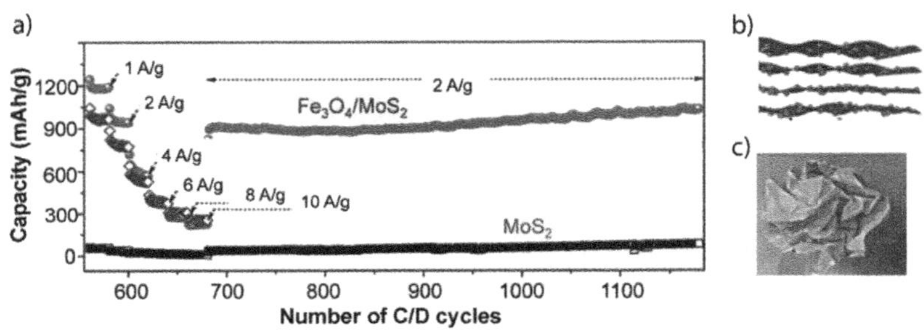

Figure 2.20. *Electrochemical lithiation and de-lithiation of MoS_2 and Fe_3O_4/MoS_2 nanostructures*

COMMENT ON FIGURE 2.20.– *a) Evolution of the specific capacitances of MoS_2 and Fe_3O_4/MoS_2 nanostructures as a function of the number of C/D cycles. b) Conservation of the lamellar structure of the Fe_3O_4/MoS_2 nanocomposite after cycling. c) Destruction of the MoS_2 lamellar structure after cycling (responsible for the low capacitance value). Adapted from [CHE 14b].*

Alongside these developments of composite nanostructures for energy storage, other structural configurations involving TMD nanosheets have also been considered for use in photocatalysis reactions. Water photolysis (water splitting), which uses light irradiation of semiconductors, is a major theme in current research; one key aim is to develop catalysts with the capacity to convert solar energy into chemical energy with high yields.

The development of photocatalysts which does not require the use of precious metals is an important challenge. The essential aim is to replace Pt, currently considered as the best catalyst for electrochemical water reduction and hydrogen production, with a low-cost alternative. As early as 1972, Fujishima and Honda

[FUJ 72] showed that electrochemical photolysis of water could be achieved using a UV-irradiated TiO_2 semiconductor electrode. Since then, many solutions have been proposed using various semiconductors; one of the crucial points of this research is to minimize the electron-hole recombination resulting from the excitons which form after light absorption by the semiconductors. To this end, the semiconductors must be associated with a co-catalyst, which (i) optimizes the separation of the electron-hole charges of the exciton by capturing the electrons and (ii) is likely to provide proton reduction sites favorable to hydrogen production. Platinum meets these criteria perfectly, but, as we have seen, it is extremely costly. For this reason, it is used sparingly in a highly divided form, dispersed onto conductive but catalytically inactive carbon materials.

Early work showed that MoS_2 nanosheets could be used as catalysts for the reduction of protons to hydrogen, thanks to the sulfur atoms located at the periphery of the nanocrystal. Unlike the core atoms, which are inactive because of the saturation state resulting from tri-coordination, the peripheral sulfur atoms are unsaturated; they thus constitute active sites on which protons are adsorbed. These protons can then be reduced to H_2 by electrons of electrochemical or photochemical origin.

Chang *et al.* [CHA 14] proposed a hybrid photocatalyst made of MoS_2 sheets (sized between 100 and 300 nm), adsorbed onto graphene (G) and onto which CdS semiconductor NPs (sized around 10 nm) are grafted (Figure 2.21).

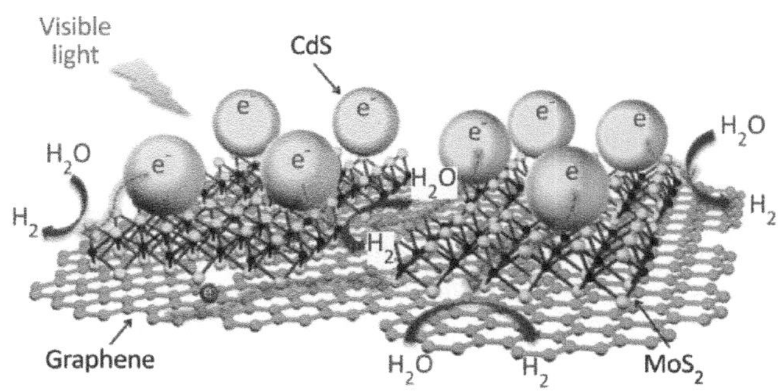

Figure 2.21. *Schematic representation of the graphene/MoS_2/CdS nanostructure*

COMMENT ON FIGURE 2.21.– *The photoelectrons leaving the CdS NPs are transmitted to the MoS_2 sheets and to the underlying graphene, which ensures good electronic conduction between the different sheets. The reduction of protons to H_2 (H_2O in acidic medium) occurs at the level of the sulfur atoms located on the periphery of the MoS_2 sheets. The positive charges from the CdS are neutralized using lactic acid. Adapted from [CHA 14].*

When immersed in a lactic acid solution[16] and irradiated with visible light (wavelength 420 nm), a significant release of hydrogen (QY of 28%) occurs; the amount collected increases linearly with light exposure time (Figure 2.22). Compared to Pt/CdS, MoS_2/CdS, G/CdS and CdS photocatalysts, MoS_2/G-CdS appears to be the most efficient by far. Its catalytic power is much higher than that of the Pt/CdS composite (about 10 times higher in terms of H_2 production), and is also higher than that of MoS_2/CdS (1.4 times higher H_2 production). The photocatalytic efficiency of G/CdS and CdS is negligible compared to that of the two previous materials.

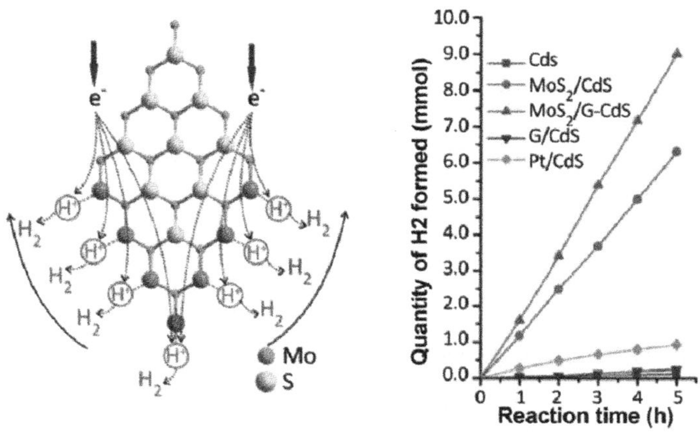

a) b)

Figure 2.22. *MoS_2/graphene-CdS composite nanostructure*

COMMENT ON FIGURE 2.22.– *a) MoS_2 nanosheet showing active sites corresponding to the unsaturated (active) sulfur atoms on which protons, reduced by photoelectrons, are adsorbed. b) Comparison of hydrogen formation rates through photolysis on different types of nanostructured surfaces in a lactic acid medium. Note the very low reactivity of CdS, G/CdS and Pt/CdS compared to MoS_2/CdS and MoS_2/G-CdS. Adapted from [CHA 14].*

These results clearly show the presence of MoS_2 nanosheets to be responsible for the catalytic effect, not the graphene; in this case, the role of graphene is simply to improve the electronic connectivity between the MoS_2 sheets. The G/CdS nanostructure, like CdS and Pt/CdS, is characterized by low photocatalytic efficiency. It is also important to note that the water splitting yields obtained using

---

16 Lactic acid ($C_3H_6O_3$) is used to neutralize the holes from excitons formed during CdS irradiation.

this ternary nanostructure (graphene-CdS-MoS_2) outperform those of Pt-CdS; this further confirms the high catalytic power of MoS_2 resulting from the external sulfur atoms. This phenomenon had been demonstrated previously by Chorkendorff's group (see section 5.1).

2.4.2. *Conclusion*

TMD nanosheets are a new family of 2D nanostructures with superior electrical and photonic properties which are easy to produce and to functionalize. This makes them a viable alternative to other 2D nanomaterials developed under much more difficult conditions. Due to their semiconducting properties, they are more suitable than graphene for use in transistors. The applications described above are just a few examples of an immense variety of potential uses, likely to lead to important advances in the fields of plastic electronics, energy and analytical chemistry, via the design and creation of innovative sensors.

2.5. Hybrid metal-organic frameworks

Metal-organic frameworks (MOFs) are directly formed by self-assembly reactions between metal ions or metal clusters (nodes) and organic compounds with reactive functions in specific directions (multitopic ligands). Similar to porous crystal, these materials have an advantage over other porous inorganic materials (zeolites, silicas, etc.), in that it is possible to modulate porosity and "host-guest" interactions by the choice of reagent used in the synthesis process. Interest in MOFs has continued to grow since their discovery several decades ago, giving rise to a growing field of applications, ranging from enhanced hydrocarbon recovery and gas capture or separation to sensor design. They also have potential applications in the biomedical field, notably in imaging, drug delivery, and the encapsulation and storage of biomolecules for safer handling.

2.5.1. *MOF production*

Two main methods are used to produce MOFs. The first involves a cross-linking reaction (reticulation) between predefined inorganic blocks and organic ligands, which leads to hybrid crystalline structures with controlled porosity. The second method, of equal importance, consists of using an existing MOF and carrying out transformations by ligand substitution in order to obtain new functionalities.

2.5.1.1. *Cross-linking method*

The use of cross-linking techniques for direct production of MOFs has been developed, on a large scale, by the Yaghi group in the United States and Ferey in France. The synthesis of hybrid, crystalline and robust materials with controlled and predefined porosity has been described in several journal articles [YAG 03, CÔT 07, FÉR 08a]. Using this method, MOFs are the result of self-assembly between pre-constructed inorganic blocks known as secondary building units (SBU) and rigid, directional organic ligands (Figure 2.23).

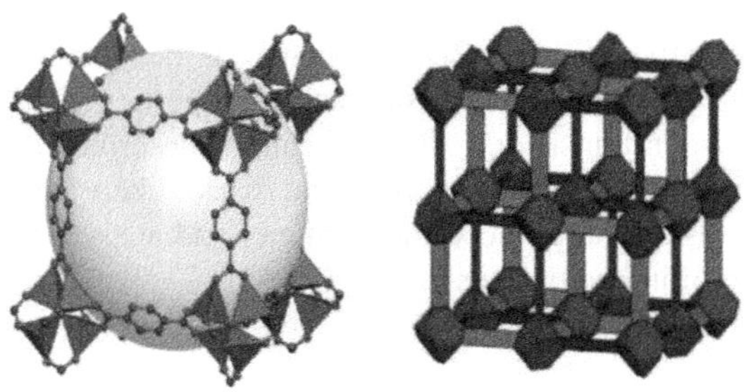

a) b)

Figure 2.23. *Structure and topology of MOF-5*

COMMENT ON FIGURE 2.23.– *a) Representation of the elementary lattice obtained by reacting the zincates (blue tetrahedra) and carboxylate groups in 1,4-benzene dicarboxylate (BDC; oxygen atoms are shown in red and carbons in black). The yellow sphere represents the maximum occupancy volume of the internal cavity (diameter ~1.2 nm) defined by eight groups of four zincates and 12 BDCs. b) Representation of the cubic network of MOF-5. Assemblies formed by four zincates (SBU) are shown as red truncated tetrahedra; links between zincates correspond to the BDCs, and are shown in blue. Adapted from [YAG 03].*

Remarkably, the SBU blocks retain their original geometric structure during the cross-linking reaction; given the range of rigid organic compounds and inorganic SBUs that are available, it may be possible to obtain an infinite number of different materials of this type, each with predefined porosity.

These synthesis principles may be observed in one of the very first examples of MOF production (MOF-5) described by Li *et al.* [LI 99], in which zincates arranged in the form of a super-tetrahedron-like cluster (SBU) react with rigid organic ligands (1,4-benzene dicarboxylate: BDC). Cross-linking takes place under heat in an organic medium (between 60°C and 120°C) in a closed reactor. The result is a cubic lattice MOF, in which the four zincates are linked together by a common oxygen (O) and six carboxylate ligands (CO_2) from the six BDCs; overall, this corresponds to the $Zn_4(O)(BDC)_3$ stoichiometry.

In practical terms, MOF-5 is obtained by reacting zinc nitrate with the diacid H_2BDC in a solvent mixture (N,N dimethylformamide/chlorobenzene) and in the presence of triethylamine, which deprotonates the diacid. A small amount of H_2O_2 is also added to favor the formation of O^{2-} at the center of the four ZnO_4 units. After reaction, analysis reveals the presence of DMF and chlorobenzene, corresponding to the composition $Zn_4O(BDC)_3(DMF)_8(C_6H_5Cl)$; these solvent molecules, embedded in the cavities, can be easily removed by prolonged heating at 300°C [LI 99].

These porosity properties can be enhanced still further by choosing larger organic ligands, while maintaining the topology of the crystal (the "isoreticulation" principle). A few years after the work cited above, Yaghi *et al.* synthesized a series of MOFs with the aim of producing porous systems with cavity dimensions that can be predetermined by the choice of ligands linking the inorganic SBUs [YAG 03].

Using the same inorganic cluster formed by the four ZnO_4 derivatives (identical to that of MOF-5), and choosing linear, rigid organic ligands of increasing lengths, ranging from fumarate to TPDC (triphenyl-4,4″-dicarboxylate), the authors were able to considerably increase cavity size to a volume 7.8 times larger than that found in an MOF obtained with fumarate (Figure 2.24)[17]. Intermediate sizes were obtained with ligands such as BDC (phenyl-1,4-dicarboxylate, MOF-5) and BPDC (biphenyl-4,4'-dicarboxylate, IRMOF-10) for which the inner volume enhancement factors are 1.7 and 4.0, respectively, compared to the fumarate MOF.

Other teams have created a wide range of different structures by altering the nature of both the ligands and the inorganic SBUs. Over 20000 MOFs have been synthesized over the last 10 years, and these materials have a wide range of applications. The record for BET adsorption measurements is currently over 7000 m^2/g, and theoretical values of over 10000 m^2/g have been predicted [FAR 12]. It is also important to note that these MOFs can be synthesized in the form of nanoMOFs (n-MOFs), allowing them, for example, to operate in a constrained micellar

17 Compared to "conventional" microporous materials such as zeolites (alumino silicates AlO_4^{5-} and SiO_4^{4-}), silicas (SiO_2) or active carbons, for which gas adsorptions measured by the BET method are between 900 and 1,200 m^2/g, and which have respective porosity volumes of 0.3 g/cm^3, 1.15 g/cm^3 and 0.60 g/cm^3, MOFs have significantly higher gas adsorption capacities, from over 2000 m^2/g with MOF-5 to values in excess of 7000 m^2/g for NU-110, with very low mass densities [FUR 13]. The BET (Brunauer-Emmett-Teller) method consists of determining the equivalent surface area of a porous material from the amount of nitrogen adsorbed in the pores of the material at a temperature of 77 K.

environment; this results in luminescent n-MOFs, which, once again, may have significant applications in the field of biomedical imaging [CAR 11].

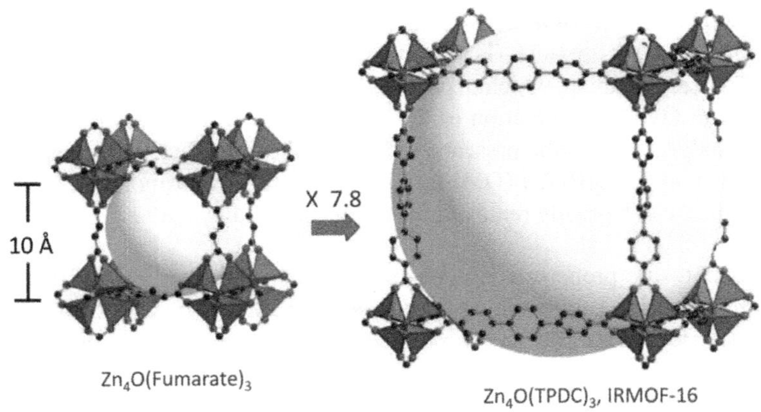

Figure 2.24. *Isoreticulation principle*

COMMENT ON FIGURE 2.24.– *The use of fumarate followed by Terphenyl-4,4"-dicarboxylate (TPDC) increases the length of the bonds between the SBUs (ZnO_4) while maintaining the initial topology. This has the effect of increasing the size of the cavity at the center of the cubic lattice of the MOF. As in the previous figure, the yellow sphere represents the maximum size of a chemical species that can be inserted into the cavity. Adapted from [FUR 13].*

2.5.1.2. *Epitaxial method of MOF growth*

In attempting to obtain MOFs with even larger porosities, researchers have encountered a problem due to the interpenetration of a secondary network which inevitably reduces pore size and the space available inside the MOF. This challenge can be countered by means of a new synthesis protocol in which organic ligands are added to a suitably functionalized solid support, enabling the organic ligands involved in the formation of the MOF to be grown epitaxially [SHE 09] (Figure 2.25). This technique, known as liquid phase epitaxy synthesis, involves a layer-by-layer approach; it is mostly used to produce MOF films adsorbed on a solid support [OHN 15].

Functionalizing the substrate with long-chain hydrocarbon molecules, capable of self-assembly on the surface and with a reactive end (such as –COOH) that can bind with the ligand, promotes the growth of an MOF film, resulting in a "SURMOF" anchored to the substrate.

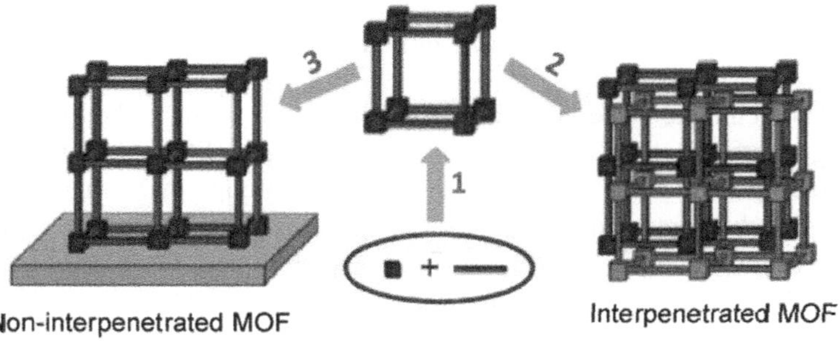

Figure 2.25. *Principle involved in the production of "interpenetrated" MOFs (1→2 volume growth of two networks) and "non-interpenetrated" MOFs (1→3 epitaxial growth of a single network on a functionalized surface). Adapted from [SHE 09]*

For instance, the SURMOF film with formula $Cu_3(TMA)_2$, denoted HKUST-1[18], which is widely used in the creation of thin-film optoelectronic devices, is obtained via a series of immersion, rinsing and drying cycles applied to a gold surface functionalized by a self-organized layer of 16-mercaptohexadecanoic acid. The first cycle consists of immersion in a copper acetate solution, followed, after rinsing and drying, by a second immersion in a 1,3,5-benzene tricarboxylic acid solution (trimesic acid-TMA), resulting in the production of a first layer of the SURMOF. This layer has a thickness of around 2 nm when the operations are conducted at 25°C. The cycle of operations described above can be repeated several times, and the thickness of the film grows in an approximately linear manner as a function of the number of cycles [OHN 15]. This technique is particularly interesting as it enables the application of an MOF film of controlled thickness and dimensions to a suitably functionalized surface. It is thus possible to design various detection devices based on optical or electronic responses, with modulable porosity, which may be used for selective absorption of gases and molecules [FAL 14].

2.5.1.3. *Post-synthetic methods*

Starting with a sufficiently robust MOF, several different transformation reactions can be applied to produce further MOFs with the same topology but new functionalities [ISL 17]. These transformations may involve ligand incorporation or

18 Most MOFs are identified by acronyms relating to the laboratory or university where they were first discovered and produced. HKUST is the acronym of Hong Kong University of Science and Technology.

exchange, taking place in solution, or ALD (Atomic Layer Deposition)[19] in the vapor phase of a metal that can subsequently receive other complexing groups.

For this technique to be applied, the starting MOF must be thermally resistant, have large pores so that the precursor molecules can diffuse rapidly inside the crystal (mesoporous MOF) and possess functional groups that can react with the precursors, such as free and reactive hydroxyl groups. MOFs with zirconium cluster nodes, linked together by carboxylic ligands, notably meet these conditions. MOF NU-1000 (NU for Northwestern University), which has the stoichiometric formula $Zr_6(\mu3\text{-}OH)_8(OH)_{8\text{-}}(TBAPy)_2$ (TBAPy: tetrabenzoic pyrene acid), has a mesoporous structure, hydroxyl groups pointing into the pores and is resistant to temperatures of 500°C; this makes it particularly suitable for ALD metallization reactions. NU-1000 MOFs have been metallized with aluminum and zinc using this method [MON 13].

2.5.1.4. *Organometallic channel formation by self-assembly of low-dimensional organometallic entities*

This method is a variant of the general MOF formation method, consisting of controlled creation of channels in an organometallic structure by the supramolecular self-association of low-dimensional entities. The method used to create these MOC (Metal Organic Channel) compounds, described by Noh and Yang [NOH 16], essentially involves inducing the self-assembly of entities consisting of a primary assembly of metal ions and organic ligands (Figure 2.26).

The forces responsible for this assembly may result from electrostatic-type interactions, Van der Waals forces or interactions between aromatic nuclei (π...π stacking), as well as possible hydrogen bonds. The structures obtained by this method have properties comparable to those of MOFs. Figure 2.26 shows a typical example obtained by self-assembly of primary entities consisting of planar macrorings, which create a channel inside the structure by stacking in an eclipsed arrangement.

---

19 ALD is a vapor deposition technique in which reactive chemical species (precursors in the vapor state) react with functional groups on a surface. In general, the technique involves two precursors, which act sequentially by reacting with one another on the surface. Other forms involve a single precursor which reacts with surface functional groups, forming a mono-molecular layer. Note that this technique differs from CVD (chemical vapor deposition), in which the precursors are introduced simultaneously, and where the thickness of deposits is more difficult to control.

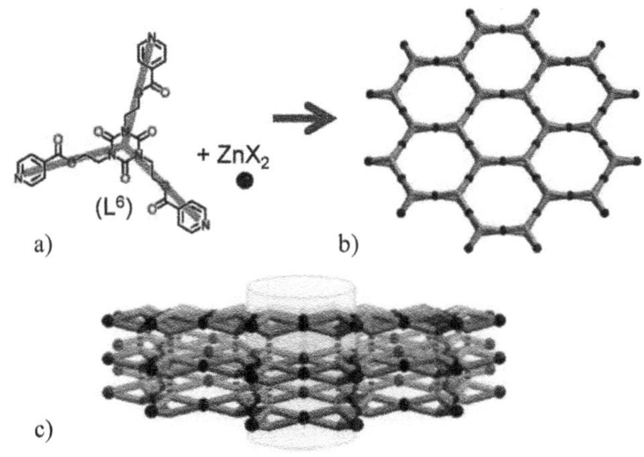

Figure 2.26. *Principle of MOC production by means of a self-assembly reaction. a) Multipole ligand. b) Coordination polymerization between the ligand and the Zn^{2+} cation, leading to the eclipsed stacking of macrorings. c) MOC structure highlighting the central channel (yellow cylinder). Adapted from [NOH 16]*

2.5.2. *Potential applications of MOFs*

A vast range of different structures may be obtained by adjusting the nature of the ligands and/or of SBUs. These materials have various applications in medical imaging and drug delivery [HOR 10], catalysis, optoelectronic device design and the improvement of proton conduction in H_2/O_2 fuel cells. MOFs have also been used for gas storage and separation and molecule identification, and may be functionalized to create multiple luminescence centers; some examples of these applications are described below.

2.5.2.1. *Gas storage and separation*

Gas separation is usually carried out using porous and adsorbent materials such as activated carbons, zeolites, silica gels or activated aluminas. The mechanism used to separate a gas mixture relies on a combination of different effects. First, there is a "sieve" effect, directly linked to pore size, which determines whether or not a molecule can penetrate a material. Second, there is a thermodynamic effect, corresponding to an adsorption equilibrium of the molecules in the pores, and, third, a kinetic effect corresponds to the variable adsorption speeds of different compounds. Finally, there is a "quantum" effect, resulting from isotopic differences of elements; this is observed in the case of H_2/D_2 mixtures, for example [LI 09a].

As we have seen, MOF porosity can easily be adjusted across a wide range. This is not the case of zeolites, for which the pore sizes are very small and hard to alter.

The nature of the functional groups lining the interior of these pores is another factor which affects gas adsorption and desorption properties, and these parameters are key to the "host/guest" interactions between molecules entering an MOF or porous solid.

The wave of activity in MOF synthesis in the 2000s resulted in a large body of work, much of which focused on potential applications for gas separation and storage in industrial contexts, along with processes to capture CO_2 emissions resulting from the combustion of fossil fuels. A number of promising systems have emerged; these have been evaluated and described in detail in the literature [LI 09a, SUM 12, LI 18a].

In an industrial context, efficient separation and storage mechanisms are notably required for CO_2, H_2, CH_4 and the toxic gases CO and NH_3 [LI 18a]. The production of CO_2 from combustion and steam methane reforming (SMR) constitutes a significant environmental problem, and research into efficient means of CO_2 capture and separation is still ongoing[20]. Some MOFs have demonstrated significant potential in this respect; important progress has recently been made in terms of capturing carbon dioxide (CO_2) using an MOF, with performance rivaling that of conventional industrial techniques.

The MOF used is structurally comparable to an MOF-74, and includes Mg^{2+} nodes linked by ligands bearing two carboxylic groups. It has the formula Mg_2 (dobpdc), where dobpdc is 4,4'-dihydroxy-3,3'-biphenyl dicarboxylic acid [MCD 15]. It is post-functionalized with N,N′-dimethyl ethylenediamine (mmen), of which one -NH(CH_3) group binds to an unsaturated site on the Mg while the other is pendant in the MOF channel. A CO_2 molecule will adsorb on this amino pendant group, accomplishing the capture function. The mechanism proposed by the authors involves a chain process, in which a proton transfer occurs simultaneously between two NH groups, followed by electrophilic addition of CO_2 onto the NH group that lost a proton. The proton transfer takes place between the NHs of two neighboring amines (noted 1 and 2 in Figure 2.27a): a hydrogen from the NH group of amine 1 (bound to metal M) is transferred to the free NH group of amine 2, which is transformed into NH^{2+}. A CO_2 molecule, which is highly electrophilic, reacts with the deprotonated nitrogen of amine 1, leading to the formation of a carbamate; at the same time, the bond between N and metal M is broken, and a new bond is formed between the oxygen in the carbamate and the metal (Figures 2.27b and 2.27c).

20 One of the main techniques currently used to capture CO_2 in industrial settings involves reacting the gas with amines in aqueous solution. Gases from coal combustion contain 70% nitrogen, 15%–16% CO_2, 5%–7% water vapor and 3%–4% oxygen by volume, under an overall pressure of one bar and at temperatures between 30°C and 40°C [MCD 12].

This cooperative mechanism propagates along the channel, resulting in rapid CO_2 adsorption for specific pressure and temperature thresholds. This mechanism, which involves a change of phase in the adsorbent, differs from classic Langmuir adsorption. It also implies a perfect match between the length of the diamine chain and the position of the metal centers (M) inside the MOF channel.

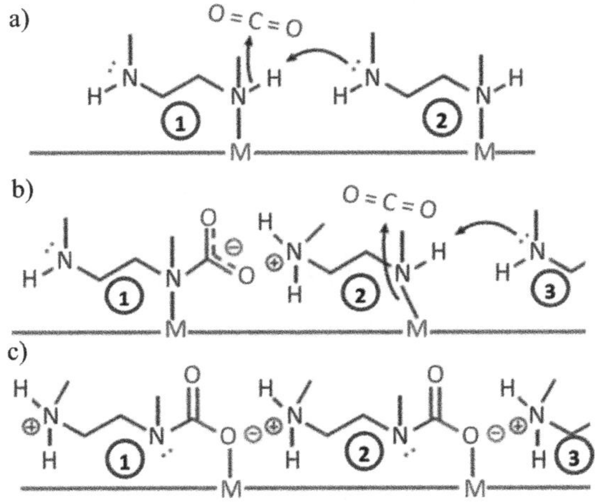

Figure 2.27. *Interaction of CO_2 molecules with $NH(CH_3)CH_2CH_2NH(CH_3)$ diamines adsorbed on a metal site M in the MOF channel. a) Simultaneous reaction between a proton transfer and nucleophilic addition of CO_2. b) The resulting carbamate function. c) Replacement of N-M bond by an O-M bond. Adapted from [MCD 15]*

This process is also likely to be responsible for the unusual shape of the isothermal and isobaric adsorption-desorption curves observed with MOF-Mg. In both cases, an abrupt change of adsorption occurs as a result of very small variations in pressure or temperature; this is very different from the classical forms observed with other materials, which display more progressive adsorption-desorption variations (Figure 2.28).

These properties, which are specific to magnesium and less striking in the case of manganese, disappear entirely when other metals such as nickel, zinc or iron are used. These characteristics are particularly favorable for dynamic regime CO_2 capture-desorption cycles; furthermore, they use less energy for CO_2 capture than traditional amine solution carbonation methods.

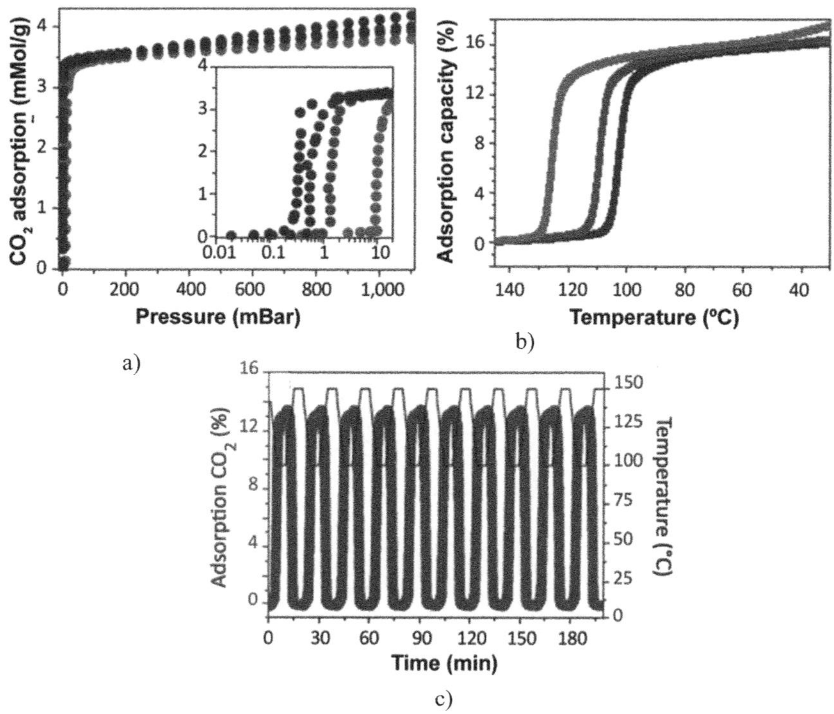

Figure 2.28. *Adsorption of CO_2 on MOF-Mg-Diamine*

COMMENT ON FIGURE 2.28.– *a) Adsorption isotherms as a function of pressure for different temperatures: 25°C (blue), 40°C (mauve), 50°C (burgundy), 75°C (red). Note the quasi-saturation of CO_2 adsorption once pressure exceeds 0.2 and 10 mbar at temperatures of 25°C and 75°C, respectively. b) Adsorption isobars for different CO_2-N_2 mixtures as a function of temperature and for a total pressure of 1 bar: 100% CO_2 (red), 15% CO_2 (green), 5% CO_2 (blue). c) Adsorption (at 100°C)-desorption (at 150°C) cycles of CO_2 in a CO_2 -N_2 mixture with 15% CO_2. Note the high stability of the MOF following around 10 cycles of 15 min each. Adapted from [MCD 15].*

In conclusion, this unusual adsorbent structure shows that a novel adsorption-desorption mechanism, involving an adsorption reaction similar to that used for CO_2 capture with amines in solution, can occur in the absence of solvent, and at much lower desorption temperatures. The energy cost of the desorption process is much lower than that associated with the amine solution technique, and the CO_2 adsorption capacities of these materials are high. Significant research efforts are currently being devoted to these materials, including by Long's laboratory at Berkeley, where the project was initiated. The compound described above possesses exceptional characteristics for CO_2 capture; furthermore, it is relatively simple and cost-effective to produce. However, the stability and longevity of the cited properties require more extensive testing before the product can be commercialized.

2.5.2.2. *MOF luminescence and devices for chemical species detection*

The combination of scalable porosity with luminescence, which may be obtained by selecting suitable ligands and metal clusters, makes MOFs suitable for use in a wide range of applications. Among these potential applications, we have chosen to focus on the development of photoluminescent devices for chemical species detection, notably for detecting explosive substances (nitrate derivatives). Electroluminescent lighting devices using MOFs have also been developed, but at present, these cannot compete with commercial LEDs based on electro emissive semiconductor compounds.

Luminescence in MOFs can have various origins, due to the ligand, the metal, the insertion of fluorescent species in the MOF or antenna effects between the metal and the ligand as a result of energy transfers.

These luminescence properties can be used in various ways. In the case where the ligand and the metal are both fluorescent, the porosity of the MOF can be exploited by inserting a particular chemical species; if this species is known to quench the fluorescence of the ligand or metal, then the altered MOF can be used to detect the species in question. This property has the potential to be used in detecting explosives on the basis of their vapor tension, which is very low.

Another way of using MOFs, for biomedical imaging purposes, is to insert elements with very narrow emission spectra into their nodes. MOFs with lanthanides (Ln) in their nodes produce intense light emissions. High-intensity emission spectra in red and green have been obtained using europium and terbium ions (EuIII, TbIII), respectively (Figure 2.29).

Surprisingly, these high emission intensities are not due to direct absorption of light by the lanthanide (Ln), but to an antenna effect between the MOF ligand and the lanthanide. Direct absorption toward the excited state of a lanthanide corresponds to a forbidden transition which can only produce low emission intensities. In the case of an energy transfer between a ligand and the lanthanide ion, the Ln produces intense emissions, as shown in Figure 2.29.

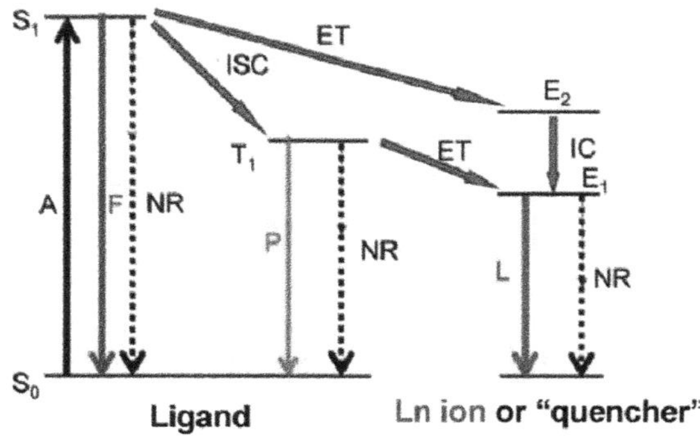

Figure 2.29. *Absorption, photoemission and energy transfer between a ligand and a chemical species inserted into an MOF (lanthanide or "quencher" ion)*

COMMENT ON FIGURE 2.29.– *After excitation of the ligand from the ground state S_0 to the singlet state S_1, there are three possible means of S_1 deactivation: (i) direct fluorescence emission (F) or non-radiative return (NR) to S_0; (ii) intersystem crossing (ISC) energy transfer to the triplet state of the ligand (T_1) followed by phosphorescence emission (P) or non-radiative return to S_0; iii) energy transfer (ET) from S_1 or T_1 to the respective E_2 or E_1 excited states of the lanthanide ion or quencher, followed by an internal conversion (IC) to a lower energy electronic (or vibrational) level E_1, giving rise to light emission L (in the case of a Ln ion) or NR deactivation in the case of a quencher. Adapted from [CUI 12].*

Furthermore, since the different Ln ions are characterized by the progressive filling of the 4f orbital from $4f^0$ (La^{3+}) to $4f^{14}$ (Lu^{3+}) and have a high number of electronic states, several emissions at different wavelengths may exist. In the case of an MOF with an Ln ion, the principal emission is a multi-step process; this begins with absorption of light by the ligand, followed by intersystem crossing (ISC) to the triplet state of the ligand, T_1. This is followed by an energy transfer from the triplet state to an energy level of the Ln ion ($T_1 \rightarrow E_1$), then by emission of light from the Ln^{3+} ion. It seems likely that these electronic transitions will only take place if the electronic levels of the ligand and the lanthanide ion are arranged in a particular way, notably including the presence of a triplet state of the ligand with higher energy than that of the lanthanide ion. This has implications for the choice of ligand.

Férey's group synthesized fluorescent MOFs by partially substituting the main metal of node cluster of an MOF with rare earth ions (Ln), using benzene-1,3,5-tricarboxylate as the organic ligand [SER 04]. The corresponding MOF, MIL-78[21]

21 The MOFs synthesized by the Férey group at Versailles all have the prefix MIL (*Matériaux de l'Institut Lavoisier*) followed by a number.

(Y, Ln), with general formula Y_{1-x} Ln_x $((C_6H_3)–(CO_2)_3)$ with x ~ 0.023 and Ln = Eu (europium), Tb (terbium) or Dy (dysprosium), is a fluorescent solid with red (Eu), green (Tb) or blue (Dy) emissions [SER 04] (Figure 2.30).

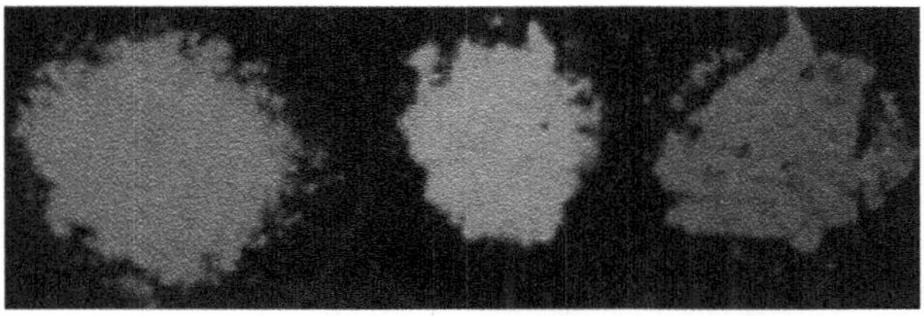

Figure 2.30. *Fluorescence of MIL-78 MOFs (Y, Ln) by irradiation at 252 nm. Red, green and blue fluorescence for lanthanides Ln = europium (Eu), terbium (Tb) and dysprosium (Dy), respectively. Adapted from [SER 04]*

The diagram in Figure 2.29, describing the emission of an Ln ion, is also applicable in the case where the ligand is strongly emissive and interacts with a non-emissive chemical species (quencher) with the potential to deactivate the excited state of the ligand. As before, this deactivation takes place from S_1 to an excited state of the molecule (E_2 or E_1), which returns to its ground state by non-radiative energy dissipation. This results in a decrease of the fluorescent intensity of the ligand; this decrease is more marked for higher concentrations of the quencher species.

Allendorf *et al.* [ALL 09] described a large number of emissive MOFs where this type of mechanism is at play. Ligand fluorescence quenching effects, caused by the insertion of non-emissive molecules in the MOF pores, have been used to detect a variety of different chemical species, notably explosives, which are characterized by very low vapor pressure values [HU 14b]. One of the advantages of using MOFs for this type of detection, compared to many other techniques, lies in their adjustable porosity. MOFs can be selected and synthesized according to the molecules to be detected, and fluorescence measurements can be used to attain high levels of detection sensitivity.

Lan *et al.* demonstrated the potential of MOFs for detecting explosives using a luminescent MOF with the overall formula $[Zn_2(bpdc)_2(bpee)]$ (bpdc = 4,4'-biphenyldicarboxylate; bpee = 1,2-bipyridylethene) (Figure 2.31a). This MOF permits rapid and reversible detection of DNT (2,4-dinitro toluene) and DMNB (2,3-dimethyl 2,3-dinitro butane) in the vapor state, with extreme sensitivity [LAN 09][22].

22 DNT is a by-product of TNT and is always present alongside it. As the vapor pressure of DNT is 20 times higher than that of TNT, DNY is easier to detect. DMNB is a marker added to each type of explosive (at a concentration of 0.1%–0.5%) to facilitate detection.

It is microporous (pore volume 0.17 cm^3/g, pore diameter ~0.75 nm, BET surface area of 328 m^2/g), and luminescence, due to the bpdc and bpee ligands (emission band centered at 420 nm), is observed after UV excitation (between 250 and 300 nm).

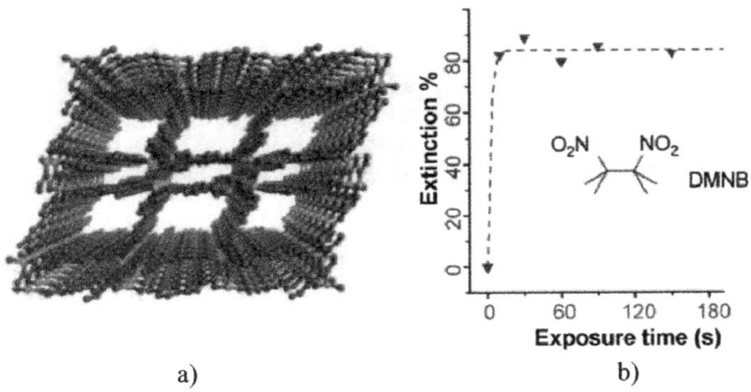

Figure 2.31. *Detection of DMNB by fluorescence quenching of MOF [$Zn_2(bpdc)_2(bpee)$]*

COMMENT ON FIGURE 2.31.– *a) View of MOF along the crystallographic axis b, showing the different channels (turquoise: Zn; red: O; gray: C; blue: N). b) Quenching of the intensity of the MOF fluorescence peak (between 420 and 460 nm) as a function of exposure time to DMNB vapors. MOF fluorescence is obtained by excitation at 320 nm. Adapted from [LAN 09].*

These MOFs, deposited in thin layers a few microns thick on a glass slide, react very rapidly in the presence of very low concentrations of DNT (0.18 ppm at 25°C) or DMNB (2.7 ppm at 25°C) vapors. This results in significant quenching of the intensity of luminescence, with a quenching rate of over 80% (Figure 2.31b). The MOF can be regenerated by simply heating it to 150°C for 1 minute.

This application of MOFs remains highly relevant, and improvements have been made in terms of detection selectivity; the technique is likely to become widespread in the near future, given the many advantages it presents (high sensitivity, ease of detection, cost-efficiency and portability).

2.5.2.3. *Protonic conduction*

Proton conduction is a key application of MOFs, and is currently the focus of much active research. The aim is to produce porous materials with proton conduction capacities greater than or equal to those of Nafion[23] membranes, which are currently the best commercially available option for producing hydrogen fuel cells.

Nafion has a certain number of limitations, notably in terms of reduced operation at temperatures above 80°C, the need for a membrane humidification system and the high manufacturing costs associated with the use of fluorine. The need for new protonic conduction membranes with all the advantages of Nafion but none of its drawbacks – notably in terms of humidity rates and cost – is thus evident.

Research into materials with protonic conductivities between 10^{-2} and 10^{-1} S/cm – known as superprotonic conductors – has accelerated over the last few years, notably following the 2009 discovery by Sadakiyo *et al.* [SAD 09] of an MOF bridged by oxalate ions, presenting protonic superconductor characteristics with a conductivity of the order of 10^{-2} S/cm. These properties result from the presence of protonated groups such as NH_4^+ and H_3O^+ in the MOF channels [SAD 09].

More recently still, Shimizu's group in Canada obtained conductivities of over 10^{-1} S/cm (at a temperature of 85°C and relative humidification of 90%) with MOFs containing different proton-carrying groups; the authors defined a number of physicochemical criteria for high proton conductivity [KIM 18]. According to these authors, conductivity can be improved by acting on three parameters derived from the Nernst-Einstein equation: increasing in the number of charge carriers ($-SO_3H$, $-PO_3H_2$ and $-COOH$), decreasing the activation energy corresponding to proton transfer between two neighboring site, and increasing the movement entropy of the charge carriers. These conditions may be obtained using MOFs that are particularly well suited for structural modifications, as they retain their original crystal architecture and porosity [CHA 19].

In 2019, Mukhopadhyay *et al.* [MUK 19] in India published a very fine analysis of the proton conductivity of an MOF. They demonstrated that the proton conductivity of an MOF with a pendant $-SO_3H$ group, linked to a node in the MOF via an alkyl chain, can be considerably increased by optimizing the length of the chain. To this end, they used an MOF (UiO-66)[24] in which a ligand bearing an NH_2

23 Nafion is the trade name of a fluorinated polymer discovered in the 1960s by DuPont; it is composed of tetrafluoroethylene fluoroethoxy groups with sulfonic groups at the chain ends which are responsible for proton conductivity.

24 UiO-66 (UiO = *University in Oslo*) is an MOF which is widely used for its robustness and proton conduction properties [KAN 10]. It is made up of $Zr_6O_4(OH)_4$ nodes linked together by benzene dicarboxylate (BDC) ligands. One of the characteristics of this MOF is that it retains its structural properties when the BDC ligands are functionalized with NH_2, NO_2 or Br groups.

group is chemically modified with an n-carbon cyclic sultone. This amounts to grafting an -NH-$(CH_2)_{n}$-SO_3H chain, located in the MOF channel, onto the ligand (Figure 2.32).

It is important to note that the initial crystal structure of the MOF is not altered by this transformation, which, in fact, is only effective for half of the NH_2 groups. One unexpected result is that the proton conductivity σ of the MOF is a function of the length of the chain separating the SO_3H group from the NH. This is maximized for the MOF (1) ($\sigma = 1.64\times10^{-1}$ S/cm), for which the alkyl chain corresponds to n = 3; for MOF (2) with n = 4, $\sigma = 4.66\times10^{-3}$ S/cm.

The reason for this increase in conductivity between MOF (1) and MOF (2) is not obvious; the authors cite several potential contributing factors. The higher acidity of MOF (1) compared to MOF (2) (pKa (1) = 3.47 and pKa (2) = 4.91), accentuated by the higher hygroscopicity of MOF (1) compared to MOF (2), could be the main reason for the higher conductivity observed with MOF (1). The fact that the cavities of MOF (2) are occupied by a larger hydrophobic volume (four hydrophobic CH_2 groups in MOF (2), compared to three in the case of MOF (1)) may be responsible for depleting the amount of water present in the cavities, resulting in lower dissociation of acids.

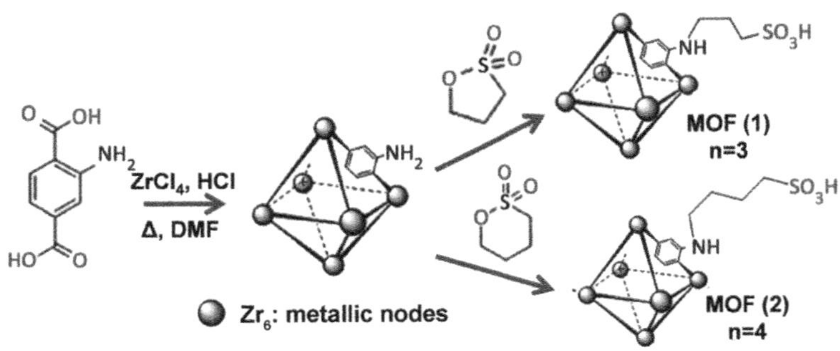

Figure 2.32. *Synthesis of MOF UiO-66-NH_2 and transformation by 3- and 4-methylene group cyclic sultones into MOF (1) and MOF (2). Adapted from [MUK 19]*

2.5.3. *Conclusions*

MOFs, which are synthesized by assembling metal ions with organic ligands, constitute a new class of 3D crystalline and hybrid porous materials. These are distinguished from traditional porous materials (zeolites, activated carbons, porous silicas, etc.) by an infinitely modulable porosity and by the option to introduce various functionalities into their pores, making them suitable for a multitude of applications. As production methods are relatively simple, carried out at relatively low temperatures (condensation generally takes place in an organic environment at temperatures between 100°C and 200°C), MOFs are more accessible than many comparable materials. A wide range of potential applications have been envisaged, notably in the fields of gas separation and storage, proton conduction in hydrogen fuel cells and the detection of chemical species through fluorescence.

2.6. Conclusions on inorganic nanomaterials

The number of potential applications of inorganic nanomaterials is far greater than that of carbon nanomaterials due to the almost unlimited range of possible compositions resulting from a wider choice of constituent elements. Three main families of NPs have been discussed here: metal nanoparticles (MNPs), semiconductor NPs or QDs, and the most recent 2D nanomaterials formed from transition metal dichalcogenides (TMDs). Gold metallic nanoparticles (Au NPs) have been the focus of a large body of work, and are used in many different applications. The optical properties of these particles vary considerably depending on their size. Au NPs smaller than 3 nm absorb light in a similar way to molecular compounds; for larger sizes, the metallic character becomes preponderant, and absorption is of the plasmonic type. Luminescence is observed in very small NCs, with intensities depending on the mode of functionalization of the NPs. Like carbon-based nanomaterials, and because of their non-toxic nature, gold NCs are used essentially for their fluorescence properties; these are valuable in biomedical imaging applications and for analyzing compounds at very low concentrations. Considerable progress has been made in recent years concerning the preparation and growth mechanisms of metallic NPs, and it is now possible to precisely control the morphology of these NPs (spheres, rods, cubes, etc.). Bimetallic core/shell structures have also been developed, making it possible to obtain useful catalytic properties at lower costs than before.

Semiconductor QDs, which are also synthesized using a wet process, are another family of nanomaterials. These are particularly valued for their exceptional fluorescence properties, far better than those of organic fluorophores. In the core/shell form, the fluorescence QYs of CdSe/ZnS type QDs are almost 100%; these materials are highly photostable and not subject to blinking emission, and, furthermore, highly pure fluorescent colors can be obtained depending on the composition and size of the QDs. These luminescence properties remain present in electroluminescent devices, making QDs an excellent material for producing color displays and, more generally, for the design and production of optoelectronic devices.

Two-dimensional TMD nanosheets, including MoS_2 and WS_2, are easy to produce in relatively large quantities. Interest in these materials is growing, and is currently at a level comparable to that of graphene in the period following its initial discovery. Their semiconducting and intercalation properties make them ideal for use in applications ranging from plastic electronics to energy storage. They can also be used to create various optoelectronic sensors for detecting gases and trace pollutants.

In terms of hybrid and composite materials, 3D organometallic networks (MOFs) constitute an additional family of important nanoporous compounds. These new materials present a significant advantage over zeolites in terms of the energy required to synthesize them. The structure of these compounds can be modified in an almost infinite number of ways, and pore size and crystallographic characteristics can be selected in advance; this means that MOF structures may be optimized for specific catalytic or separation functions.

PART 2

Nanotechnology and Nanomaterials for Energy

3

Energy Storage

Energy is a major social and societal issue, and the move to more sustainable forms of energy production is at the heart of environmental protection programs. The latest recommendations issued by the COP 21 (Paris 2015) issue a stark warning: we must act now to control global warming and ensure that global temperatures do not increase by more than 2°C before the end of the 22nd century. Specialists at the IPCC (International Panel on Climate Change) consider that this goal will be very difficult to achieve. All forecasts, using a range of different models and the data currently available, point to a global temperature increase of 1.5°C by 2030–2050. Given the constant increase in energy requirements, it will be very difficult to keep warming below 0.5°C for the next 50 years without immediate and drastic measures to reduce emissions of greenhouse gases, notably CO_2. The development of solar and wind power as means of producing electricity is crucial, but the effectiveness of these solutions depends on the parallel development of reversible energy storage systems. Developments in the field of rechargeable battery technology are badly needed, as the present state of operations remains largely unsatisfactory.

3.1. Worldwide energy use

The world's energy needs have been growing steadily since the advent of the industrial era at the beginning of the last century, with a marked acceleration in recent years. In a report published in 2017, the International Energy Agency (IEA) indicated that, between 1973 and 2015, the world's total primary energy supply

For a color version of all the figures in this chapter, see: www.iste.co.uk/lacaze/nanotechnology.zip.

(TPES) increased from 6101 Mtoe (1973) to 13647 Mtoe[1] (2015), an average annual increase of around 3% [IEA 17].

The majority of this energy comes from fossil fuels, which accounted for approximately 87% of TPES in 1973, and remained at around 81% in 2015. The contribution made by renewable energy sources, such as wind and solar power, was practically negligible in 1973 (0.1%) and did not exceed 1.5% in 2015. Hydroelectric generation is only marginally increasing, remaining below 2.5% in 2015. These statistics show how difficult it will be to replace fossil fuels (coal, oil and natural gas) with renewable energies, given constantly increasing energy demands; forecasters predict that around 18000 Mtoe will be required by 2050 (Figure 3.1).

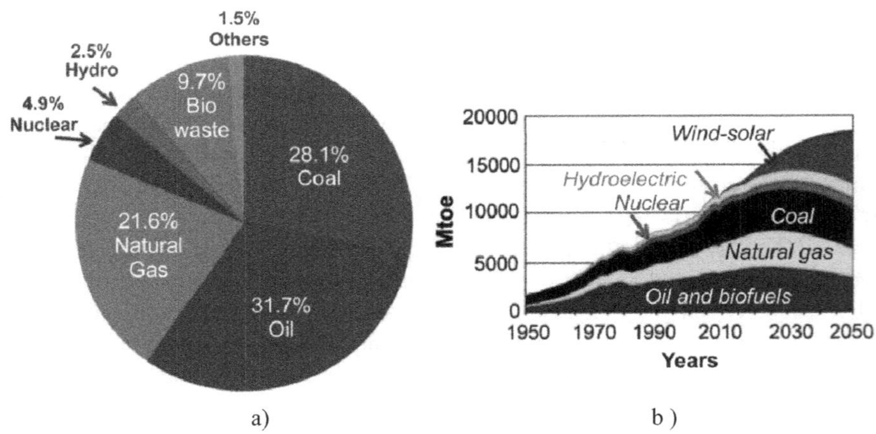

Figure 3.1. *Worldwide energy sources. a) Distribution and contribution of different energy sources to global TPES in 2015 (COP 21, Paris). b) Evolution of global consumption by energy source, including forecasts up to 2050. a) Adapted from an IEA report [IEA 17]. b) Adapted from a Seeking Alpha 2017 report [LI 17b]*

These energy sources, with the exception of nuclear and renewable forms (hydro, solar and wind), are the main cause of increasing CO_2 emissions. Energy demand is unevenly distributed around the world: at present, the United States and China are

1 The toe (ton oil equivalent) is a unit of energy corresponding, as its name suggests, to the energy contained in one ton of oil. 1 toe is equivalent to 11,630 kWh.

the biggest consumers[2], and are thus responsible for the highest shares of CO_2 emissions.

Electrical energy, which is, by definition, non-polluting, represents only 15%–16% of the total energy consumed in 2015: approximately 24300 TWh (24300 billion kWh). Unfortunately, most of this electricity (68%) is produced by thermal conversion processes that involve burning fossil fuels (coal, oil, natural gas and biofuels) with a consequent release of CO_2. Non-carbon sources of electricity generation, such as nuclear and renewable energy, account for only 11% and 20%, respectively. In the latter category, the contribution of hydroelectricity (dams) is by far the most important (16%), while solar and wind power contribute less than 4%, although a significant increase is expected in the coming years [UNI 18].

All of the efforts currently directed toward reducing CO_2 emissions require action to be taken on multiple levels. These include reducing fossil fuel consumption and saving energy across all sectors, notably in housing by improving insulation. Furthermore, the development of renewable energies, including solar, wind, geothermal and biological, is essential. In the case of wind and solar power, where generators are intended to be connected to the distribution network, the natural intermittence of their operation is a major issue[3], meaning that parallel improvements in energy storage technology are crucial.

3.2. Energy storage systems

A variety of systems are currently used for energy storage. These can be grouped into two distinct categories: electrochemical and chemical systems (the most widespread) and "other" systems, including mechanical, electrical, magnetic, thermal and biological approaches.

3.2.1. *Non-chemical/electrochemical storage*

Non-chemical, non-electrochemical storage systems draw on various phenomena related to mechanical, thermal and electromagnetic energy. As nanotechnologies play no role in these systems, we shall limit our discussion to a brief overview. The most widely used mechanical energy systems (characterized by their highest

---

2 According to statistics from the IEA and Berkeley National Laboratory, the TPES of the United States and China in 2015 are around 2,200 and 3,000 Mtoe, respectively; together, they account for over one-third of global TPES.

3 A variation of around 20% in the electricity supply to the distribution network is sufficient to disrupt the power balance, resulting in general dysfunction.

maximum energy capacity, MEC) include hydraulic pumping associated with dams (PHES, Pumping Hydro Energy Storage), which provides the highest MEC of (10^4 MWh). Next come compressed air energy storage systems (CAES) with MECs between 100 and 1000 MWh), flywheel energy storage (FES, MEC: 5 MWh) and thermal energy storage devices (TES, MEC of approx. 10 MWh), obtained by solar heating of fluids or solid materials in towers. The storage of magnetic energy in superconducting coils (SMES, Superconducting Magnetic Energy Storage) is another option which is currently being developed; these devices permit the storage of electrical energy with MECs between 10 and 1000 MWh. For more details concerning the operation of these systems and their energy characteristics, see the articles by Sabihuddin *et al.* [SAB 15], Luo *et al.* [LUO 15] and Guney and Tepe [GUN 17].

3.2.2. *Chemical and electrochemical storage systems*

The formation of chemical bonds in molecules with high potential energy, such as H_2 and CH_4, means that energy can be stored chemically. Water splitting, a process whereby produced electricity is transformed into hydrogen by electrolyzing water, is likely to play a major role over the coming years. Hydrogen can be stored indefinitely and is easily transportable; it can be converted back into electrical energy in a fuel cell, and its combustion is non-polluting. Hydrogen can also be stored in the form of metal hydrides; this is a tried and true storage method with many practical applications. The best storage rates are obtained with magnesium hydrides (MgH_2), capable of storing 7.6% hydrogen by weight; unfortunately, these systems release H_2 at high temperatures (600 K). Other materials, in the form of alloys ($LaNi_5(H_6)$, $TiV_2(H_4)$, $ZrV_2(H_{5,5})$ and $FeTi(H_2)$) have lower storage rates (between 1 and 3 wt%), but release H_2 at almost room temperature (300 K); their equilibrium pressures vary between 1 and 10 atmospheres [ABE 19]. Although the development of new hydrides will not be discussed in this chapter, it is important to note that research is still ongoing in this area, with a main focus on identifying suitable nanostructured materials [LEY 14, YAR 19].

Water splitting – the transformation of water into hydrogen and oxygen – can be carried out in electrolyzers with good energy yields (around 65%) and can be implemented directly at the output points of photovoltaic or wind power plants. However, cost constraints prevent large-scale application; current technologies rely on the use of precious metal catalysts (platinum, iridium or ruthenium), to produce high energy yields. The same cost problems are encountered in the operation of fuel cells, for which the reverse reaction of converting dihydrogen and dioxygen into water and producing electricity is also dependent on precious metal catalysts. The push to identify transition metal catalysts which provide similar catalytic properties at much lower costs is thus understandable, and many laboratories are involved in this work. This cost factor is also the reason why the transformation of methane (CH_4) into hydrogen by industrial steam reforming is still the most widespread technique (representing 95% of hydrogen production); unfortunately, the low financial cost is offset by high levels of CO_2 emissions[4]. Furthermore, these elements provide motivation for work on valorizing CO_2, based on catalytic transformation of CO_2 into products with high added value or high energy potential, such as CH_4. It should be noted that at the time of writing, the conversion rates obtained using these forms of CO_2 transformation have yet to reach suitable levels for large-scale operations. Nevertheless, many projects are still ongoing around the world, grouped under the umbrella term "carbon capture, utilization and storage" (CCUS) [INS 19].

Electrochemical cells offer a second method of storing electricity. The market is currently dominated by two main types of system: rechargeable batteries and supercapacitors. Both technologies are the subject of intense research in the energy sector, and both stand to benefit from the use of new nanomaterials.

A rechargeable battery consists of two electrode materials (cathode and anode) which can be reversibly oxidized and reduced after the passage of a determined quantity of electricity Q, corresponding to the charge of the battery. This charge, measured in relation to the mass of the battery or that of an electrode material (expressed in specific capacity, mAh/g or Ah/kg, or in capacity density, mAh/cm^3 or Ah/L), is obtained at an average voltage V which is characteristic of the electrode materials. It corresponds to a quantity of stored energy QV[5]. In its charged state, the battery can release this energy by connecting the electrodes together; this amounts to applying a discharge current, corresponding to the reverse chemical charge reaction, resulting in a restoration of the battery to its initial state. The environmental sustainability of batteries is a further important point to consider and has been analyzed in detail by Grey and Tarascon [GRE 17].

---

4 Steam methane reforming (SMR) involves two main reactions. A first endothermic reaction, carried out at high temperature (between 840°C and 950°C) and under pressure (20–30 bar), in the presence of a nickel-based catalyst, results in a mixture of carbon monoxide and hydrogen via the reaction $CH_4 + 2\ H_2O \rightarrow 3H_2 + CO$. A second, weakly exothermic reaction between CO and H_2O, carried out at lower temperatures, results in a mixture of $CO_2 + H_2$. The final result is a mixture of carbon dioxide and hydrogen, which also contains small quantities of CO.

5 This QV energy, stored by an electrode material (anode and cathode), is characterized in relation to the mass or volume of the material and corresponds to the mass and volume densities of energy, respectively; it is expressed in Wh/kg (or mWh/g) and in Wh/L (or mWh/cm^3), respectively. The specific power which determines the intensity of the current supplied under the characteristic discharge voltage is expressed in W/kg (or mW/g) and in W/L (or mW/cm^3).

Supercapacitors use a simpler system, relying on the discharge of a capacitor due to the adsorption of ions on the surface of each electrode; for this reason, they may also be referred to as electrical double-layer capacitors (EDLCs). Unlike batteries, where the charge and discharge (C/D) processes take place slowly and are controlled by the diffusion rates of ions through the thickness of the electrode materials, the C/D processes for EDLCs are extremely rapid due to the adsorption and desorption of ions on the electrode surfaces. This is the reason why supercapacitors are suitable for use as power devices. On the other hand, the charge and the energy are confined to the surface of the electrodes, meaning that supercapacitors have low specific capacities, not exceeding a few Wh/kg; the specific capacities of rechargeable batteries, such as lithium-ion batteries (LIBs), are in the range of 200–250 Wh/kg. The energy storage levels of supercapacitors can be improved by combining the capacitive process with a redox process in systems known as pseudocapacitors. Pseudocapacitors combine very long service life with extremely short recharging times, important advantages that determine different areas of application. The recharge time of a battery depends on the charging rate and the nature of the materials, and may take several hours, whereas a supercapacitor can be recharged in a few minutes. Moreover, the average life expectancy of a battery is a few thousand C/D cycles, whereas supercapacitors can perform hundreds of thousands of C/D cycles without loss of efficiency.

These considerations mean that power, reversibility and longevity are the main characteristics that will determine the choice of an electrochemical system for a given application. Understandably, energy requirements differ depending on the system, from cell phones to electric vehicles to electrical substations connected to wind and photovoltaic fields. In the same way, some cases may require rapid delivery of an energy load, and in this case, supercapacitors are more suitable. The battery/supercapacitor dilemma is explored in detail in a widely cited article that summarizes the situation perfectly: “Where do batteries end and supercapacitors begin?” [SIM 14].

In this chapter, we will show how nanomaterials have been used to improve the characteristics of batteries and supercapacitors. In the case of rechargeable batteries, we will examine progress made in the charging and discharging processes through the use of nanomaterials. Similarly, we will show how the introduction of new carbon nanomaterials, such as graphene and carbon nanotubes (CNTs), have significantly improved the storage performance of supercapacitors and pseudocapacitors.

3.2.3. *Rechargeable batteries*

At present, the capacity of rechargeable battery systems is not sufficient, and significant losses in efficiency result in limited lifetime in terms of the number of C/D cycles. This is currently one of the major problems facing the scientific community and one which, indirectly, constitutes a barrier to the development of renewable energies such as solar and wind power. Unlike nuclear power, these technologies depend on intermittent energy sources (sun and wind); the resulting fluctuations make it impossible to connect them directly to the electrical grid. Storage systems are needed to regulate this intermittent energy by storing electricity produced during the active period and supplying it to the grid at times when the system is no longer producing. These storage solutions must combine high capacities with response times adapted to the fluctuations of the power grid.

Lead-acid batteries (LABs), which account for a large part of the market, are mostly used for the electrical components of vehicles with internal combustion engines; however, they are also used as a source of energy in hybrid vehicles (HEV) or those equipped with stop-start devices. Given their high reliability, recent advances in manufacturing techniques and their low cost, they are also used, in some countries, in electricity substations where volume and weight are not an issue[6]. Redox flow batteries (RFBs)[7] are also used in this way.

Lithium-ion batteries (LIBs), which consist of different cathode materials for lithium insertion and a graphite anode, are currently the most widely used solution for equipping electric vehicles and the majority of connected portable devices. Their high reliability justifies their use in many fields of application.

3.2.3.1. *Operation of a lithium-ion battery*

The development of LIBs was driven by three key players Whittingham, Goodenough and Yoshino, joint laureates of the 2019 Nobel Prize for Chemistry. Whittingham, in the 1970s, was the first to recognize the value of lamellar materials such as titanium sulfide ($LiTiS_2$) with the ability to electrochemically and reversibly

6 The new generation of lead-acid batteries (UltraBattery™) ensures longer lifetimes and high power [MOS 15]. These batteries are frequently used for on-site storage of electricity generated by solar power plants. An important advantage of lead-acid batteries is the extremely high material recycling rate, up to 99% [MAY 18].

7 Redox flow batteries are an old technology, but one which has recently enjoyed an upsurge in popularity for large-scale static energy storage. The technique is based on redox reactions of compounds in solution. The dominant technology uses vanadium compounds [BRY 18], and does not seem likely to profit from the use of nanomaterials. The evolution of this technology, alongside descriptions of the different redox systems used, can be seen in the literature [LI 17a, PHA 20].

insert lithium ions at room temperature while maintaining their structure [WHI 76]. Subsequently, Goodenough proposed a different lamellar cathode compound, $LiCoO_2$, as a replacement for $LiTiS_2$; this resulted in a cell with twice the previous power, combining a graphite anode and a $LiCoO_2$ cathode [MIZ 80]. Some years later, in a patent [YOS 87], Yoshino *et al.* described the use of an anode based on carbon compounds, notably graphite[8], alongside a carbonate mixture as the solvent. The main effect of the solvent is to produce a thin protective layer on the surface of the anode, preventing the formation of Li dendrites, which can cause short-circuits between electrodes. The final assembly of two electrodes made up of lamellar materials, $LiCoO_2$, and graphite, along with a carbonate mixture used as a solvent, resulted in the first truly successful LIBs. One further innovation involved the use of aluminum as a current collector and as a support for CoO_2 (positive electrode) in the place of precious metals. Goodenough [GOO 13] and Tarascon [TAR 18], key figures in the field, have provided a detailed description of the different stages and the successive advances leading up to the commercialization of LIBs, which still constitute the main solution for powering electric vehicles.

During discharge[9], the reaction at the anode [3.1] consists of a release of Li^+ ions inserted into the graphite into solution (soln) (reaction 1); at the $Li_{1-x}CoO_2$ cathode [3.2], these same Li^+ ions which are released into the solution are inserted (reaction 2):

$$\text{anode}: Li_xC_6 \rightarrow xLi^+(soln) + 6C + xe^- \qquad [3.1]$$

$$\text{cathode}: Li_{1-x}CoO_2 + xLi^+(soln) + xe^- \rightarrow LiCoO_2(s) \qquad [3.2]$$

The battery is recharged by imposing a voltage of opposite sign between the two electrodes, thus reversing the direction of reactions [3.1] and [3.2]; maximum charge is attained once the anode is C_6Li and the cathode is CoO_2, a state corresponding to $x = 1$. Due to the back-and-forth movement of Li^+ ions in the electrolyte medium (a mixture of aprotic organic solvents containing a salt such as $LiPF_6$), these batteries were initially referred to as lithium rocking-chair batteries (Figure 3.2).

---

8 Yazami *et al.* [YAZ 83] were the first to show that graphite could constitute an anode, functioning as an insertion compound for Li^+ ions, in a solid electrolyte electrochemical cell (polyethylene saturated in $LiClO_4$).

9 Note that in a battery, in discharge regime, the anode – where an oxidation reaction takes place – has negative polarity, while the cathode – where a reduction reaction takes place – is positive. The situation is reversed with electrolyzers, in which a current is used to transform products: the anode and cathode have respective positive and negative polarities, corresponding to oxidation and reduction reactions in the same way as before.

In terms of storage capacities[10], we note that the insertion of Li^+ ions is limited to one Li for six C; from this, we derive a theoretical specific gravimetric capacity of 372 Ah/kg for the graphite electrode and of 135 Ah/kg for the $LiCoO_2$ electrode (insertion-de-insertion of 0.5 Li); this gives us a practical specific capacity of around 147 Ah/kg for the whole battery. For an average output voltage of 3.6 V, this should lead to a specific energy of 530 Wh/kg. The actual capacity of the battery is highly dependent on the discharge rate and, in practice, the specific energy of this type of battery does not exceed 300–400 Wh/kg [GUO 17b].

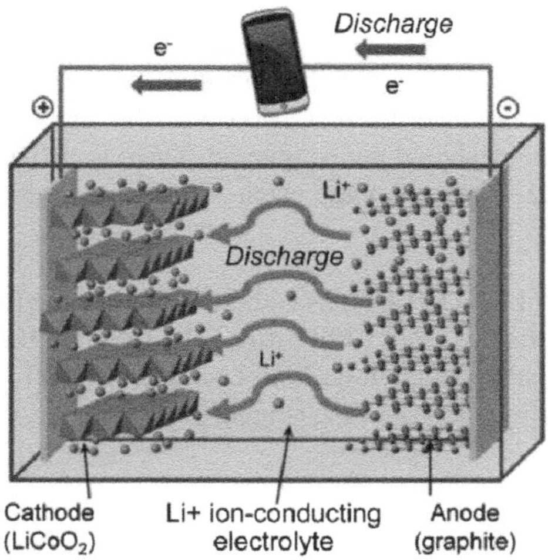

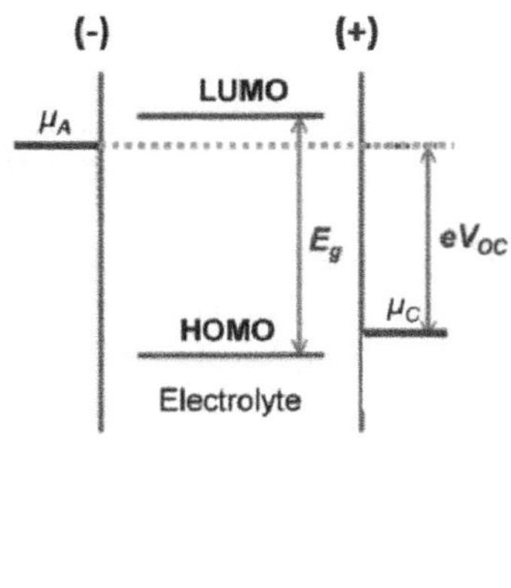

Figure 3.2. *Operating principle of a lithium-ion battery in discharge regime*

COMMENT ON FIGURE 3.2.– *a) Diagram of an electrochemical cell showing the $LiCoO_2$ lamellar cathode material (positive electrode), alternating layers of Li^+ ions and $[CoO_2]$-, and the graphite anode material (negative electrode) charged with Li^+ ions. The usual electrolyte is PF_6Li, dissolved in a mixture of aprotic solvents (ethylene or propylene carbonates). A membrane, permeable to Li^+ ions (not shown), separates the two electrodes. The pink arrows indicate the direction of Li^+ ion migration in the electrolyte, while the blue arrows indicate the movement of electrons in the external circuit. Adapted from [ISL 14]. b) Energy diagram of a lithium-ion battery. μ_A and μ_C are the electrochemical potentials of the anode*

10 The specific capacity C (Ah/kg) of a battery consisting of an anode and a cathode characterized by specific gravimetric capacities C_A and C_C, respectively, is given by the relation $C = 1/(1/C_A + 1/C_C + 1/C_M)$, where C_M (Ah/kg) is a corrective parameter corresponding to materials other than the anode and cathode [KAS 07].

(graphite) and cathode ($LiCoO_2$), respectively. E_g is the energy window of the solvent and indicates its chemical inertia with respect to the two electrode materials. V_{OC} is the voltage across the open circuit electrodes when the battery has just been charged. Adapted from [GOO 13].

3.2.3.2. *Importance of LIBs versus other rechargeable batteries*

Li-Co batteries enjoyed immediate commercial success. In Japan, Sony began large-scale manufacturing in 1991, and the batteries are now used in smartphones, laptops and most mobile devices. Compared to LABs, which still dominate some markets, these batteries have much higher specific energies due to the lightness of the electrode materials and the high oxidation-reduction potentials of the cathode materials; furthermore, they have a much longer lifespan. However, the high cost of cobalt and the likelihood of future scarcity should not be ignored. There is also a risk of dangerous malfunctions, and the batteries may catch fire as a result of overcharging or overdischarging [GOL 14].

There are four main alternatives to $LiCoO_2$ (LCO) as cathode materials for LIBs. These include $LiMn_2O_4$ (LMO), specifically $LiNi_{0,33}Mn_{0,33}Co_{0,33}O_2$; $LiFePO_4$ (LFP); and $LiNi_{0,8}Co_{0,15}Al_{0,05}O_2$ (NCA). Global production in 2016 exceeded 180000 tons, and is expected to exceed 400000 tons by 2025 [PIL 17] (Figure 3.3).

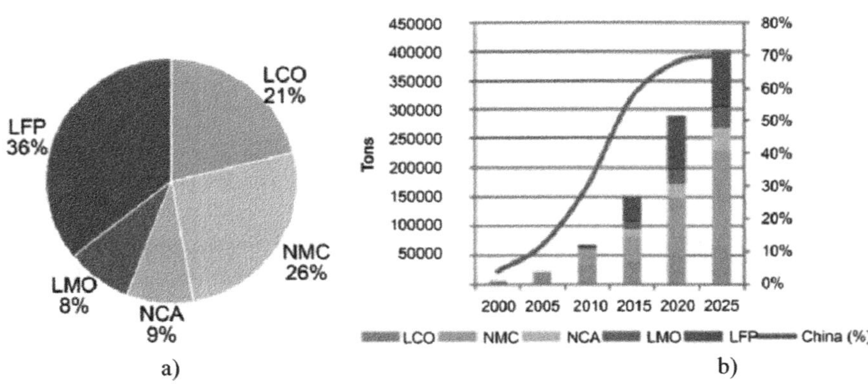

Figure 3.3. *Comparison of global tonnages of the main active cathode materials produced for different types of LIBs. a) Mass distribution of the five main cathode materials in 2016. b) Evolution of tonnages between 2000 and 2025 (forecast values). Note the rapid growth of NMC compared to LCO. The brown curve represents the production share, in %, of China. Adapted from [PIL 17]*

Each of these materials has certain advantages and disadvantages compared to $LiCoO_2$, and is best suited to different applications depending on factors such as cost, longevity, power, energy capacity and, notably, safety of use. A detailed evaluation of these characteristics, in both theoretical and practical terms, can be found in an article by Liu *et al.* [LIU 16a].

In combination with a graphite anode, Li-Co (LCO) batteries have a specific energy of between 150 and 200 Wh/kg and an average operating voltage of 3.6 V. However, they cannot deliver high currents (maximum 1 C)[11]; this limits their use as power batteries, and means that they can only be used in devices that do not require high discharge currents, such as cell phones, tablets, laptops and cameras. The high cost of cobalt is a further disadvantage, meaning that other types of cathode materials containing little to no Co may be preferred.

This is particularly true of Li-manganese oxide (LMO) batteries: unlike LCO (2D lamellar), these have a three-dimensional (3D) spinel structure and are characterized by greater robustness and higher conductivity than LCO. LMO batteries can deliver high current intensities (from 1 to 10 C), making them suitable for use as power batteries. However, their energy capacity is lower than that of LCOs (between 100 and 150 Wh/kg). These batteries are mainly used in power tools, medical devices and electric vehicles. The characteristics of LMO batteries can be improved by mixing manganese with nickel and cobalt. Thus, nickel manganese cobalt oxide (NMC) batteries, in which the cathode is made up of equal parts of Ni, Mn and Co, perform significantly better, in terms of power (with the possibility of discharge at 2 C), specific energy (comparable to LCOs, between 150 and 220 kWh/kg) and lifetime. This material is becoming increasingly dominant in the LIB market (26% of total tonnage in 2016, expected to increase to 41% by 2025) [PIL 17]. NMC batteries have been selected for use in certain electric vehicles, such as the Nissan Leaf, the Chevy Volt and the BMW-i3. Another, cheaper cathode material is iron-Li phosphate ($LiFePO_4$ or LFP). Batteries with this type of cathode are capable of delivering high currents (up to 25 C) without causing dangerous overheating, and have good lifetime. Unfortunately, their specific capacities are low (between 90 and 120 Wh/kg), as is the operating voltage, between 3.2 and 3.3 V. LFP batteries are used in devices requiring high charging currents. Finally, Li nickel cobalt aluminum (NCA) batteries, first developed in 1999, may be considered as a development of NMC batteries, substituting aluminum for manganese. These batteries are characterized by high specific energy (between 200 and 260 Wh/kg), high lifetime and acceptable power characteristics allowing the delivery of 1 C

11 Charge or discharge rates are measured in multiples of C, corresponding to current intensities. The value of C (in current intensity) is equal to the specific capacity of the battery expressed in Ah/kg. Thus, a discharge rate of 2 C for a battery with a specific capacity of 2 Ah/kg corresponds to a current of 4 A; this also indicates that the battery will be completely discharged after half an hour of use.

currents at an average voltage of 3.6 V. However, they are very expensive and present a non-negligible safety risk.

Overall production of LIBs, including all of the different cathode materials, has increased steadily in terms of energy volume since they were first introduced in 1991. Between 1991 and 2016, there was an average annual increase of 22%, showing the market to be extremely dynamic. However, production levels are still well below those of lead-acid batteries (LABs)[12], a much older technology used to fulfill the electrical requirements of conventional combustion-powered vehicles (Figure 3.4).

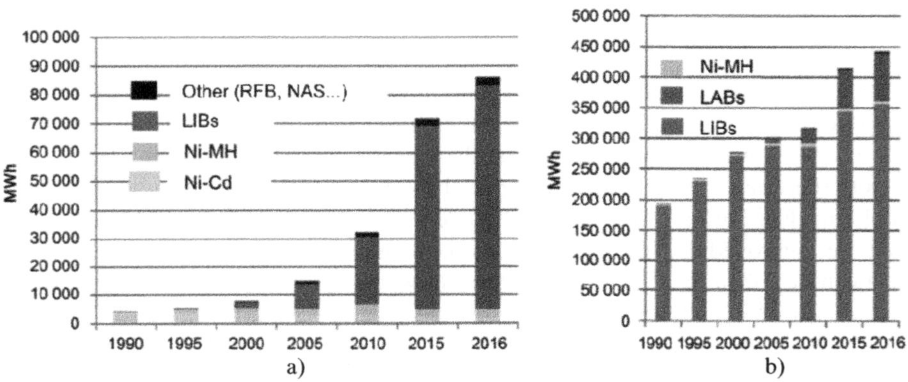

Figure 3.4. *Evolution of energy quantities produced by different types of rechargeable batteries. a) Comparative growth in energy volume of Ni-Cd, Ni-MH, LIBs and other batteries. b) Comparison between energy volumes of lead-acid batteries (LABs) and lithium-ion batteries (LIBs). Adapted from [PIL 17]*

Other rechargeable batteries, such as nickel/cadmium (Ni-Cd), nickel/metal hydride (Ni-MH) and, to a lesser degree, RFBs and sodium-sulfur batteries (SBS), used for specific applications, occupy a much smaller share of the market. Their production has varied very little between 1990 and 2016, remaining at around 5000 MWh per year. Total energy production across all commercial rechargeable

12 The LAB market remains significant, despite their low specific energy (around 40 Wh/kg) and their negative environmental impact. The volume of energy produced in 2016 was considerable (350,000 MWh compared to 80,000 MWh for LIBs), but the average annual growth rate is only 3%. The new generation of LABs (UltraBattery™) promises a longer lifetime and high performance [MOS 15]. These batteries are suitable for on-site storage of energy generated by solar power plants, where weight and volume considerations are of secondary importance. The lead-acid battery market in 2015 totaled $32 billion, while the LIB market was still only $27 billion [PIL 17].

batteries amounted to approximately 450000 MWh in 2016, and is expected to continue to grow in the coming years [PIL 17][13].

All of the batteries described above are characterized by specific energies of between 150 and 200 Wh/kg. These specific energies evidently limit their use for electric-powered transport. "All-electric" vehicles currently have a maximum autonomous operating distance of around 350–400 km; for a medium-powered car (such as the Renault ZOE), this requires on-board energy of at least 40 kWh, that is, a battery of around 260 kg in weight. There is an imperative need to increase the specific energy of these batteries in order to increase operating autonomy well beyond the current 300 km limit, without exceeding masses of over 300 kg.

3.2.3.3. *Improving the efficiency of lithium-ion batteries*

The low lithium insertion capacity of the anode and cathode materials used in LIBs is one of the main reasons for their limited specific capacity. Both the graphite anode and the CoO_2 cathode can accept only one Li atom for masses equivalent to C_6 (72 g) and CoO_2 (91 g), respectively; this gives them theoretical maximum capacitances of 372 Ah/kg and 272 Ah/kg, respectively. In reality, the specific capacitances shown by experimental discharge curves are lower. In the case of $LiCoO_2$, the actual specific capacitance is around 135–140 Ah/kg, due to the insertion of only 0.5 Li (Figure 3.5).

For other cathode materials, such as LMO ($LiMn_2O_4$), the estimated actual specific capacitance of 130 Ah/kg is relatively close to the theoretical value (154 Ah/kg); the same is true of NCA cathodes ($LiNi_{0,8}Co_{0,15}Al_{0,05}$), for which the measured specific capacity of 190 Ah/kg is not too much lower than the theoretical value of 221 Ah/kg, probably indicating lithium insertion rates close to stoichiometry.

The prevailing idea for improving battery capacity involves the use of different cathode and anode materials, with higher specific capacities than the current graphite anodes and transition metal oxide cathodes.

For anode materials, metal lithium, characterized by its lightness (atomic weight of 7 g) and by good reversibility of the Li^+/Li^0 redox couple in an aprotic organic medium, appears to be ideal for use in high-capacity anodes. Its theoretical value is 3830 Ah/kg, more than 10 times the capacity of a graphite anode. The use of lithium anodes was first proposed long before the first LIBs were developed. However,

---

13 For comparison purposes, the 450,000 MWh/year corresponding to the total energy available from all the batteries produced on the market is equivalent to a power of around 50 MW; a nuclear power plant reactor of the EPR (European Pressurized Reactor) type provides a power of 1,300 MW.

metal lithium is difficult to use due to its very high reactivity. It is also likely to react with the electrolyte medium, resulting in the formation of dendrites on its surface, which can cause short circuits between the electrodes. Nevertheless, new solvents and new ways of using lithium have gone some way toward solving these problems, resulting in a resurgence of Li metal anodes [ABR 15, GUO 17b, LIN 17b].

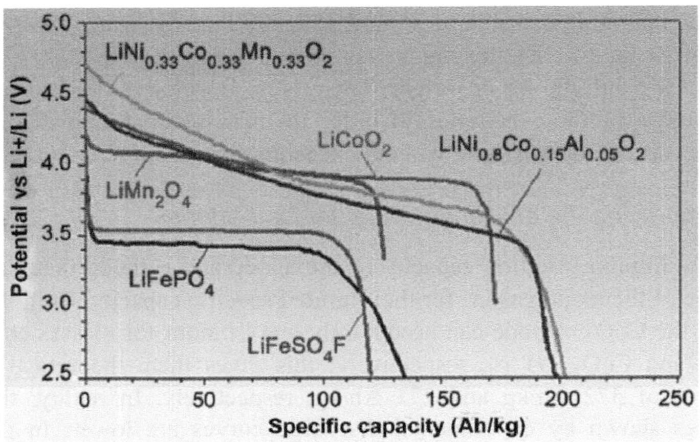

Figure 3.5. *Discharge curves of some typical cathode materials used in LIBs*

COMMENT ON FIGURE 3.5.– *These discharge curves are obtained for a particular current intensity. The measured charges correspond to the integration of the current as a function of discharge time. The specific capacity is determined at the point where the cathode voltage drops off, corresponding to the total deinsertion of lithium from the cathode material. Note that these curves, taken from several different publications, are obtained for unspecified discharge rates; hence, the measured capacities can only be indicative. Adapted from [NIT 15].*

The diagram in Figure 3.6 provides an overview of the potential energy gains which could be achieved by using lithium and sodium as anodes and sulfur or oxygen as cathodes in new batteries.

Sulfur (in the form of S_8) and oxygen are two key elements that have been used as cathodes, resulting in Li/O_2 or Li/S_8 batteries. Both elements are characterized by redox couples involving a two-electron exchange, giving them theoretical specific capacities of 1672 Ah/kg; however, their redox potentials are lower than those of cobalt-based cathodes (2.9 V for O_2 and 2.15 V for S_8 vs. Li^+/Li^0).

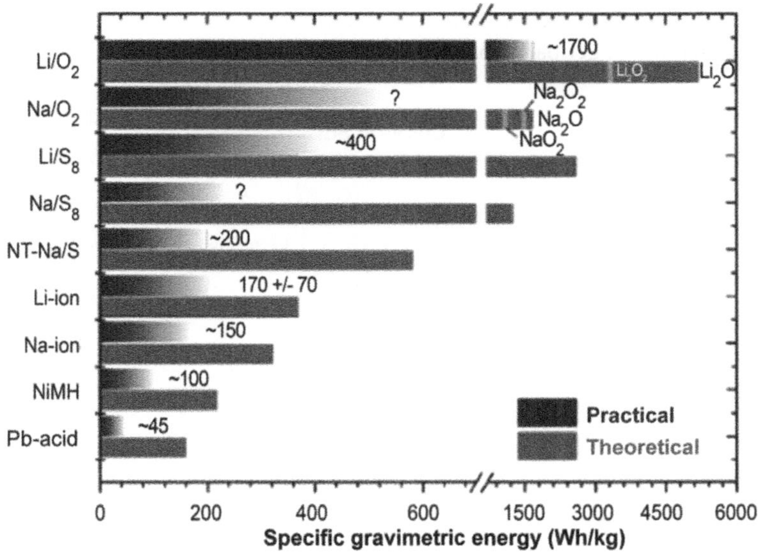

Figure 3.6. *Theoretical and practical specific energies of different types of rechargeable batteries*

COMMENT ON FIGURE 3.6 – *Sodium batteries (Na-ion and Na-metal), which have a lower specific energy than lithium batteries, are of economic interest due to their lower cost and to the fact that sodium supplies (in the form of NaCl) are almost inexhaustible. Adapted from [ADE 15].*

3.2.3.4. *Lithium metal anodes: malfunctions and solutions*

Batteries designed with a Li metal anode are prone to malfunction, due to the extreme reactivity of Li with the organic electrolyte medium. Several different causes of malfunction have been identified and analyzed [TIK 16]. One major cause is the formation of Li dendrites, resulting in short circuits between the two electrodes. Dendrite formation originates from cracks in a very thin solid layer, the SEI (solid electrolyte interphase), which is an electronic insulator but is permeable to Li^+ ions. This SEI layer is initially a few nm thick, and forms instantly when the Li is brought into contact with the electrolytic liquid medium. The successive deposition and dissolution of Li causes strong mechanical stresses in the SEI layer, which eventually cracks (step 1, Figure 3.7). These cracks constitute privileged sites for Li growth, resulting in the formation of Li dendrites (step 2). The dendrites eventually detach from the electrode and contribute to the formation of Li "islands" (step 3) which no longer participate in the energy balance[14].

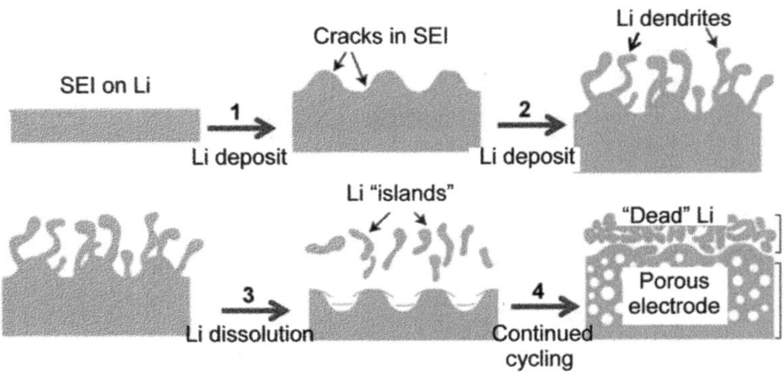

Figure 3.7. *Surface evolution of a lithium electrode, initially covered with a thin SEI layer (yellow), during deposition and dissolution of lithium metal (light blue)*

COMMENT ON FIGURE 3.7.– *(1) formation of cracks in the SEI layer; (2) appearance of dendrites; (3) detachment of dendrites from the electrode during a Li dissolution reaction; (4) thickening of the SEI layer, increase in Li porosity and increase in volume of dead Li. Adapted from [LIN 17b].*

Charge/discharge cycles, which cause strong variations in volume, accentuate the formation of these islands which accumulate to form a layer of "dead" Li; at the same time, the electrode becomes increasingly porous (step 4). To avoid the formation of dendrites, it is important to obtain a SEI layer that allows a homogeneous diffusion of Li^+ ions and avoids the development of "hot spots" on the SEI surface where dendrites might form. Various solutions have been put forward to counteract this breakdown of the lithium electrode.

One main focus of research in this area concerns increasing the robustness of the SEI layer, while maintaining its permeability to Li^+ ions. Promising results have been achieved using fluorinated solvents, and through the synthesis of artificial SEI layers meeting the required criteria.

14 Brissot *et al.* [BRI 99] studied the formation of dendrites on a Li electrode in the 1990s. They showed that filament development was related to local increases in the electric field, resulting from a space charge; this charge itself resulted from high depletion of Li^+ ions on the electrode surface, occurring at high current intensities, corresponding to very high Li^+ ion diffusion/migration rates.

Fan *et al.* [FAN 18a] have shown that by using high (10 M) concentrations of a fluorinated electrolyte (lithium bis(trifluorosulfonyl)imide: LiTFSI) in solution in a carbonate solvent such as dimethylcarbonate (DMC), propylene carbonate or an ethylene carbonate/DMC mixture, it is possible to create a battery with a Li anode and a NMC622 cathode ($Ni_{0.6}Mn_{0.2}Co_{0.2}$) which is capable of enduring a significant number of C/D cycles without failure (Figure 3.8).

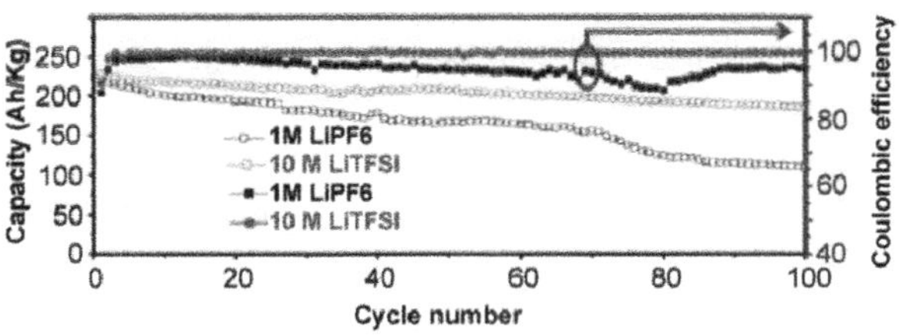

Figure 3.8. *Specific capacities and coulombic efficiency of Li//NMC622 batteries*

COMMENT ON FIGURE 3.8. – *The charge/discharge (C/D) rates are performed at C/2 in a solvent mixture of ethylene carbonate and dimethyl carbonate. After 100 C/D cycles, the capacity of the battery containing 10 M LiTFSI is still 175 Ah/kg, equivalent to 80% of its initial value, whereas using 1 M $LiPF_6$ electrolyte, this value drops to 110 Ah/kg, or 50% of the initial value. Adapted from [FAN 18].*

According to the authors, this improvement is essentially due to passivation of the cathode, which is protected by a thin fluorinated organic layer; this has the effect of stopping the decomposition reaction between the solvent and the salt. The SEI layer adjacent to the Li metal, which consists of products of decomposition of the electrolyte, also contains fluorinated compounds that contribute to its stabilization. The high concentration of LiTFSI is essential: the authors also state that this property decreases the reactivity of the solvent. Lower concentrations (1 M) of electrolyte salt result in passivation layers with less fluorine derivatives, reducing the performance of the battery.

These batteries are characterized by an initial capacity of 220 Ah/kg; after 100 C/D cycles, this capacity is still 83% and the coulombic efficiency[15] between charge and discharge remains at 100%.

15 The coulombic efficiency of an electrode is given by the ratio Q_D/Q_C. Q_D and Q_C are the quantities of electricity corresponding to the total discharge and charge of the electrode, respectively.

Another way of preventing dendrite formation consists of artificially creating an SEI layer which is much more stable than the one formed spontaneously by decomposition of the electrolyte. Solutions of this type have been tested by several research groups. Liu *et al.* [LIU 17b] deposited a film of "Cu_3N" nanoparticles (NPs) coated with a styrene-butadiene polymer binder (SBR) on a lithium electrode[16]. On contact with lithium, the Cu_3N layer is transformed into Li_3N and Cu; the high ionic conduction of the layer permits a homogeneous diffusion of Li^+ ions, preventing dendrite formation. Furthermore, the mechanical properties of the binder – notably its elasticity – prevent the film from dislocating under the effect of the volume variations caused by C/D cycles. A total of 300 C/D cycles were applied using this anode with a solvent mixture of 1,3-dioxolane and 1,2-dimethoxyethane in the presence of LiTFSI; the final coulombic efficiency was still more than 98%.

Similar results were obtained by depositing a Li_3PO_4 film of about 50 nm thickness onto lithium. This film, obtained by directly reacting polyphosphoric acid (PPA) on lithium [LI 16b], is mechanically resistant and a good ionic conductor. A battery made with this PPA-Li anode and a $LiFePO_4$ (LFP) cathode and with $LiPF_6$ electrolyte in solution in a dimethyl carbonate/ethylene carbonate (DMC:EC) mixture did not display any malfunctions after 200 C/D cycles. The specific capacity of around 140 Ah/kg was maintained at a C/D rate of 0.5 C. A Li/S battery made with the PPA-Li anode and a sulfur cathode also performed significantly better than an equivalent battery using an untreated lithium anode. However, the specific capacity of the battery drops considerably between the first and the hundredth cycle, decreasing from 1600 to 600 Wh/kg, despite a low C/D rate of C/10.

A relatively simple and seemingly effective process for stabilizing the Li/electrolyte interface has recently been developed by a Chinese group [LI 18b]. This process consists of treating the lithium surface with polyacrylic acid, leading to the formation of a thin (~20 nm), compact and homogeneous layer of lithium polyacrylate (LiPAA) with very high elasticity. It ensures a very good contact with the lithium, and withstands volume variations in the lithium electrode during C/D reactions over a period of 700 hours with no degradation. Surface protrusions remain below 100 nm after 10 minutes of C/D cycles at a rate of 0.2 mA/cm^2; for lithium without LiPAA film, these protrusions can measure more than 1 μm under the same conditions [LI 18b].

16 "Cu3N" nanoparticles of less than 100 nm in size are obtained by a condensation reaction between copper dimethylate ($Cu(OMe)_2$) and benzylamine. They are then dispersed in tetrahydrofuran to form a colloidal solution, which is applied to a lithium film, itself deposited on a copper collecting electrode.

Endurance tests carried out on a battery with a $LiFePO_4$ (LFP) cathode and a LiPAA-modified lithium anode[17] show that a capacity of around 120 mAh/g is maintained after over 500 cycles at a rate of 1 mA/cm^2. This performance is likely explained by the high elasticity of the LiPAA layer, which can be stretched by nearly 600% without breaking, and has a Young's modulus of around 0.11 GPa.

These developments are just a few examples of the solutions which may be used to improve the stability of the SEI layer, preventing the formation of cracks which lead to malfunctions in the Li electrode. However, although endurance tests have been carried out, the numbers of cycles involved remain small and further tests are needed before these solutions are adopted on a larger scale. Furthermore, the resistance of these artificial layers has only been evaluated for low C/D rates, and few indications have been given concerning their behavior under high-rate C/D regimes.

Other solutions have been developed to improve the reliability of Li electrodes, notably by eliminating the risk of dendrite formation while maintaining high storage capacities. The most widespread idea involves using materials with the ability to form alloys (or insertion compounds) with lithium and with insertion rates well above those of graphite. Silicon, which is very light (atomic weight of 28 g), widely available and "eco-compatible", is one element which can form an alloy with lithium; this alloy can contain significant amounts of Li, up to $Li_{4.4}Si$. This alloy itself releases Li^+ ions, with a theoretical specific capacity of 4,200 Ah/kg (more than 10 times that of a graphite anode) at a potential close to that of the Li^+/Li couple (0.4 V to Li^+/Li). These characteristics make silicon a prime candidate for use as anodes in LIBs with high specific capacity. Extensive research is currently being carried out in this area.

3.2.3.5. *Lithium-ion batteries with silicon anodes*

As we have seen, the use of silicon anodes for lithium insertion appears promising at first glance. However, the silicon begins to break down quickly, after only a few C/D cycles, due to the significant variations in the volume of the material caused by lithium insertion/de-insertion: these variations can exceed 400% at full load [RYU 04]. Early work using Si in the 2000s was abandoned for this reason. Recent developments, showing the potential improvements to be made by using

17 The battery consists of a cathode made up of LFP cathode (80 wt%) mixed with C-Super P® black (10%) and 10% PVDF solubilized in N-methylpyrrolidone (NMP). The surface amount of LFP is 4 mg/cm^2. The electrolyte is a mixture of ethylene carbonate/dimethyl carbonate/diethyl carbonate (EC/DMC/DEC, 1:1:1 by volume). The quantity of lithium greatly exceeds that of the cathode material. A Celgard 2325 membrane (25 μm thick, 28 nm pore diameter) is used as a separator between the anode and cathode.

divided silicon as an anode material, have resulted in a resurgence of interest in the material.

Since 2008, a considerable number of studies have been published proposing the use of silicon in the form of nanotubes, nanostems, nanospheres, core/shell (yolk-shell[18]) nanoparticles or thin films. These have led to effective improvements in the performance of lithium electrodes [ZUO 17b, LI 18c].

The idea of replacing solid Si by a nanostructured material, featuring an arrangement of vacant spaces, was proposed by Cui's group at Stanford [WU 12a]. The authors showed that silicon nanotubes with a thin surface coating of silica (SiOx) constitute a suitable nanostructure (Si@SiOx) which can withstand increases in the volume of the silicon with no damage to the material (Figure 3.9).

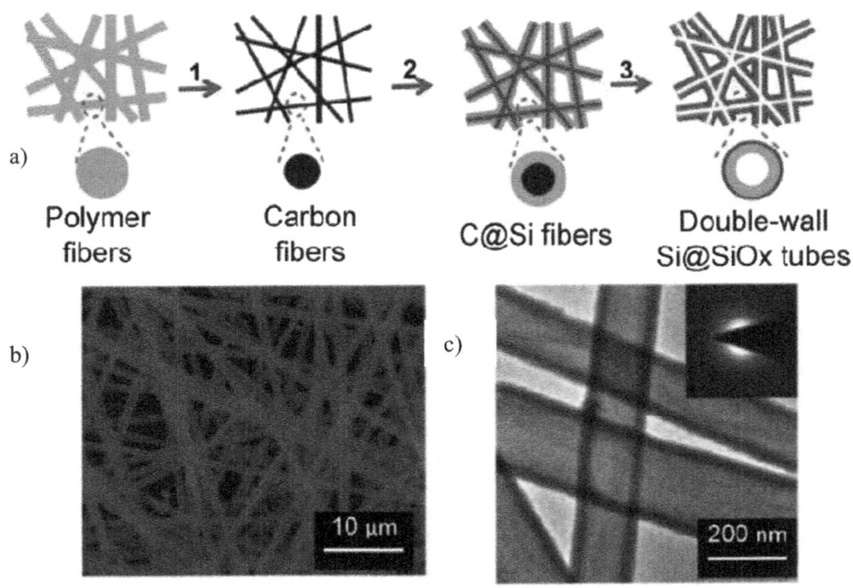

Figure 3.9. *Manufacture of double-walled Si@SiOx tubes*

COMMENT ON FIGURE 3.9.– *(a) Different steps in the formation of Si@SiOx tubes. (1) Pyrolysis: conversion of polymer fibers to carbon fibers. (2) CVD deposition of Si, resulting in C@Si fibers. (3) Further pyrolysis in air (500°C) of the C@Si fibers, resulting in the double-walled Si@SiOx tube structure. b) SEM image of*

18 A term coined by Cui *et al.* by analogy with the structure of an egg structure. In this case, the "yolk" is Si, contained inside a carbon shell [LIU 12].

double-walled nanotubes. c) TEM image of Si@SiOx showing the thinness of the SiOx layer (in black). Adapted from [WU 12a].

This architecture establishes a barrier, in the form of a mechanically resistant SiOx layer in this specific case, between the Si and the electrolyte; the barrier is, however, permeable to Li^+ ions. The tubular shape of the material results in free space on the inside: this allows it to withstand the volumetric expansion of Si during Li insertion and prevents dendrite growth. These nanotubes are obtained by a series of transformations applied to polyacrylonitrile fibers, which are produced using an electrospinning technique. The first transformation, by pyrolysis, results in carbon fibers. The second stage is a chemical vapor deposition (CVD) process in which Si is deposited. The final process consists of a heat treatment in air at 500°C, resulting in surface oxidization of the Si and destruction of the carbon. The result is tubular Si microfibers, covered with a thin 30 nm film of SiOx.

This type of nanostructuring means that, unlike simple Si or C fibers, the double-walled tubular material is able to withstand a very large number of C/D cycles (more than 5000 cycles) while maintaining a specific capacity of more than 80% (Figure 3.10).

However, note that although the specific gravimetric capacities are relatively high (between 1300 and 1500 Ah/kg at C/5), they remain well below the theoretical values. They are also dependent on C/D rates, falling to around 600 Ah/kg for a C/D rate of 12 C, corresponding to a complete discharge time of 5 min.

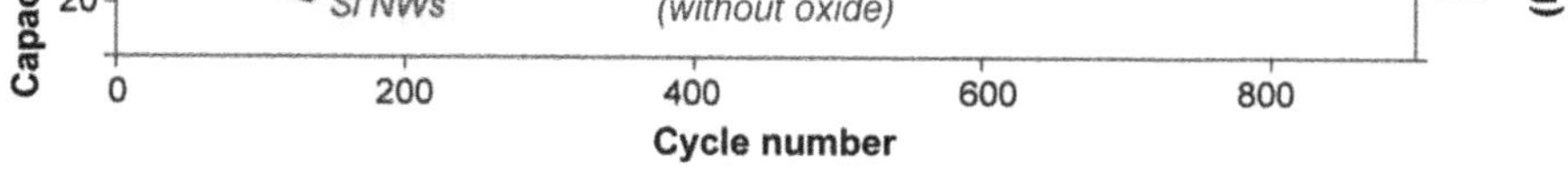

Figure 3.10. *Evolution of the capacities of different Si nanostructures during charge-discharge cycles performed over a period of more than 12 months, at a charge/discharge rate of C/5*

COMMENT ON FIGURE 3.10.– *The specific capacities of Si nanowires (Si NW, red) and oxide-free Si NT (blue) are seen to drop off rapidly to 20% and 50%,* *respectively, after 200 and 400 cycles. With SiOx-coated SiNTs, the electrode exhibits more robust behavior, retaining 80% of its capacitance up to and beyond 1000 cycles. The C/D reactions took place in an ethylene carbonate/diethyl carbonate medium with 1 M $LiPF_6$. Adapted from [WU 12a].*

These results are particularly interesting from a fundamental standpoint, but the process involved is too complex to be easily applied in an industrial setting. Other, more accessible solutions based on the same concept have been proposed, using nanoporous Si anodes [WAD 14] or nanoporous Si-Ni alloys [WAN 15c].

New approaches have been developed based on the use of Si particles coated with various carbon-based binders. These coatings are able to withstand variations in the volume of the silicon over the course of C/D reactions, and the material is relatively simple to produce, leading to a recent upsurge in interest [LI 18c]. However, as Kovalenko *et al.* [KOV 11] have noted, these binders must meet several criteria in order to be effective. They must protect the silicon particles from the solvent while allowing the diffusion of Li^+ ions; withstand the volume variations of the Si, which requires a certain elasticity of the binder surrounding the Si particles; and, finally, have sufficient electrical conductivity to allow the Si particles to remain electrically connected to one another.

Alginates appear to meet these conditions when used as binders for silicon nanoparticles. Alginates have a high elasticity modulus of between 3.5 and 5 GPa; the presence of carboxyl groups allows Li^+ ions to diffuse toward the Si via a "hopping" phenomenon which takes place between the carboxyl and hydroxyl groups; and, according to the authors, the alginate layer is thin enough not to cause high electrical resistance between particles [KOV 11]. Compared to traditional binders such as Polyvinylidene Difluoride (PVDF), polyacrylic acid (PAA) or carboxymethylcellulose (CMC), alginate results in improved performance in the Si anode. This appears to be due, at least partly, to the superior mechanical properties of these binders[19] (Figure 3.11).

Discharge curves for different rates show that the silicon anode behaves very well: its capacity remains at around 2 Ah/g for a discharge rate of 4.2 A/g and an initial charge of 3 Ah/g, that is, a discharge rate of 4/3 C. Note, however, that these values were obtained for very small masses of material arranged in thin layers, and, in practical terms, may not necessarily apply to electrodes with thick layers of anode

19 Alginates are natural polysaccharides extracted from brown algae, corresponding to copolymers of β-D mannuronic and α-L guluronic acids. Their elasticity modulus, between 3.5 and 5 GPa, is significantly higher than that of PVDF, which is of the order of 0.01–0.02 GPa in the wet state and 0.4–1 GPa in the dry state.

material; in this case, conductivity may fall off sharply and degrade electrode function.

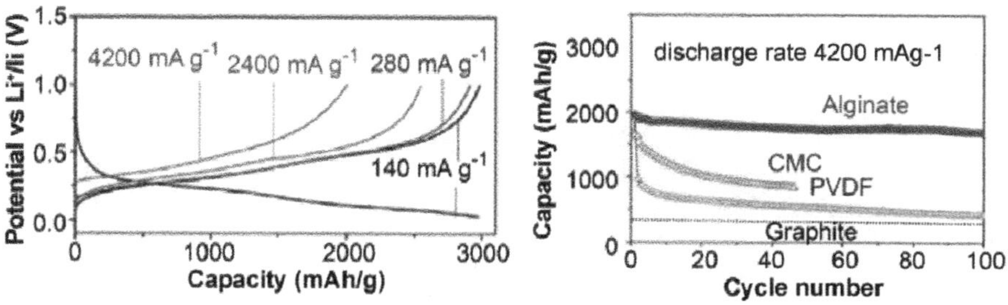

a) b)

Figure 3.11. *Evaluation of the storage properties of a silicon nanopowder mixed with alginates*

COMMENT ON FIGURE 3.11.– *a) Discharge curves for a silicon anode at different rates. The Si anode was charged at a rate of 140 mA/g and its charge was fixed at 3030 mAh/g. Discharge was stopped when the potential reached 1 V. b) capacity variations over a period of 100 C/D cycles for silicon anodes modified with PVDF, CMC and alginate binders. For comparison purposes, the theoretical capacitance of a graphite anode is 372 mAh/g. Adapted from [KOV 11].*

Many teams have also suggested using silicon composites coated with conductive carbon derivatives to minimize loss of conductivity in the core of the cathode material [LI 18c]. The processes involved are easily scalable for industrial use, involving techniques such as micro-bead milling, CVD, spray-drying or thermal reduction of silica by Mg (magnesiothermic reduction) [WAN 15b].

More recently still, Si composites doped with nickel, boron or phosphorus and combined with carbon-based materials have been put forward as possible anode materials; however, even these materials fail to resolve the issue of the initial drop which occurs in the first C/D cycle. Huang *et al.* [HUA 17] showed that silicon microparticles doped with equal parts of red phosphorus and graphite constitute excellent lithium anodes (Figure 3.12).

The production process is simple, consisting of pulverizing a mixture of silicon and phosphorus (at a ratio of 99:1) in a ball mill. The resulting product is then mixed with graphite and subjected to a further, longer milling process, giving a phosphorus/silicon/graphite (PSG) composite with a particle size of between 200 and 300 nm. The best results in terms of specific capacitances are obtained when the phosphorus-doped silicon is mixed with graphite in equal parts (PSG55). However, a significant drop in coulombic efficiency (CE = 64%) is noted in the first C/D cycle; the authors attribute this decrease to the formation of an SEI layer on the PSG particles. The CE values recover over the following cycles, up to a value of 93%, before decreasing steadily; after 200 and 500 cycles, specific capacities of 883 and 600 Ah/kg, respectively, are observed.

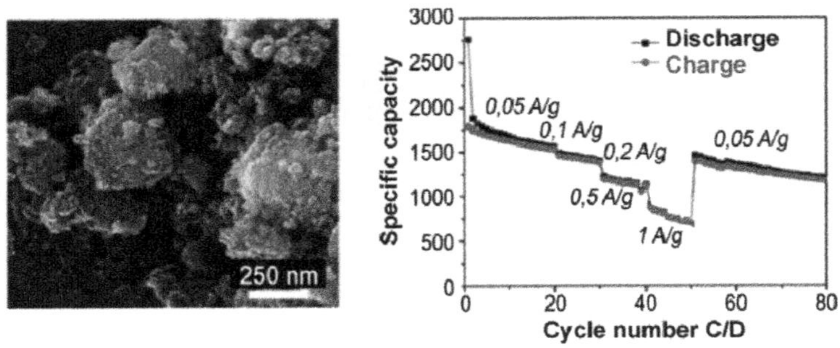

a) b)

Figure 3.12. *P-Si-graphite (PSG) composites and storage properties. a) SEM image of PSG particles: phosphorus-doped silicon-graphite composite. b) Variation of the specific capacities of the PSG55 composite measured for different C/D rates and as a function of the number of C/D cycles. Adapted from [HUA 17]*

Good electrical connectivity between silicon particles and the collector electrode is crucial in order for electrochemical exchanges to take place throughout the whole volume of the anode material. The difficulty of obtaining storage capacities with near-theoretical values is probably a result of low interfacial conductivity between Si particles during repeated C/D cycles.

Polypyrrole [DU 14], poly(3,4-dioxythiophene):poly(styrene sulfonic acid) (PEDOT:PSS) [HIG 16] and graphene have been used as binders between Si particles with the aim of improving connectivity and forming a mechanically resistant envelope around the particles. Each of these binders results in clear improvements when compared with conventional composites, in which Si nanoparticles are mixed with carbon-based binders such as carbon black. Thus, in the case of PEDOT:PSS binders, the first 100 electrode lithiation-delithiation cycles are characterized by a decrease in capacity from 2865 to 2000 mAh/g; in the same circumstances, the values for Si NPs coated with carbon black decrease from 2800 to 1600 mAh/g. Another interesting fact, which confirms the correlation between high storage capacities and electrical connectivity between particles, is the increase in anode capacity when the PEDOT:PSS is made more conductive[20].

---

20 The addition of formic acid (10% by weight of PEDOT:PSS) to Si NP and PEDOT:PSS (20% by weight of Si) mixture results in an increase in the conductivity of the PEDOT; this results in a significant increase in silicon de-lithiation .

Similarly, the use of conductive 2D MXenes as binders for silicon NPs shows that improved connectivity can significantly increase the capacity and lifespan of a silicon anode. For example, Zhang *et al.* [ZHA 19a] recently showed that 2D Ti_3AlC_2 titanium carbides can be easily exfoliated in an aqueous HF medium to produce $Ti_3C_2OH(F)$ nanocrystals that, after transformation into a conductive ink (MX-C), can be used as binders for silicon particles.

A silicon-based lithium anode was produced by mixing Si powder with the MX-C conductive ink containing the $Ti_3C_2OH(F)$ nanocrystals. The Si powder is made of either nSi NPs or Si NPs with a graphene coating (Gr-Si). Deposition on a copper foil, followed by drying, results in the formation of an nSi/MX-C or GrSi/MX-C anodic material with a thickness of around 350 μm. The resulting film is flexible, mechanically strong and extremely conductive (around 3500 S/m for Si/MX-C, increasing to more than 5000 S/m for GrSi/MX-C structures).

Remarkably, near-theoretical specific capacities have been observed using these relatively thick (350 μm) anodic films. This is due to the use of MXenes, highly conductive binders which provide excellent connectivity between silicon particles and the current-collecting electrode. In combination with graphene NPs, a protective envelope is also formed around the particles. These factors result in very high areal capacities (23.3 mAh/cm^2 for around 12 mg of active material), compared to values not exceeding 4 mAh/cm^2 for conventional systems using C/Si composites.

The areal capacity, measured in mAh/cm^2, varies linearly with the active surface mass of the silicon introduced into the nSi/MX-C and Gr-SiMX-C mixtures, although increased capacity is noted in cases where graphene is present in the mixture (Figure 3.13a). When the total mass of anode material is taken into account ($M_{Total} = M_{Si} + M_C + M_{Gr}$), the variation of the surface capacity as a function of the active mass ceases to be linear, but attains near-theoretical values, demonstrating the total penetration of the Li^+ ions in the anodic layer. This is a result of the high electrical connectivity between particles (Figure 3.13b)

However, microscopic observation of the electrode after around 20 C/D cycles with a high concentration of active material shows surface degradation, even at low discharge rates. Thus, despite the high levels of connectivity, Li-Si are still subject to reliability issues which, at the time of writing, are not close to being solved.

The examples described above show that silicon anodes perform better in terms of both lifespan and storage capacity if better connectivity between Si composite particles is established and if the structural integrity of these particles is maintained over the lithium C/D cycles. This may be achieved by coating Si particles with a material that is both conductive and sufficiently elastic to reversibly absorb

Si volume variations and prevent the growth of an SEI layer, allowing lithiation/de-lithiation reactions to take place throughout almost all of the volume of the anode.

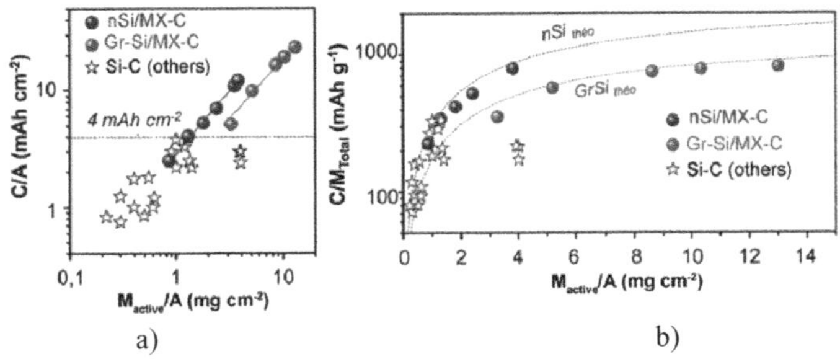

Figure 3.13. *Electrochemical characterization of nSi/MX-C and Gr-Si/MX-C anodes and comparison with other Si-C carbon binders*

COMMENT ON FIGURE 3.13.– *a) Variation of the areal capacity (C/A) as a function of the amount of active material per cm^2. b) Specific capacities calculated for the total weight (C/M_{Total}) of the composite anode material as a function of the density of the active weight (Si). The dashed curves represent the theoretical values of the capacities. Adapted from [ZHA 19a].*

The phenomena described above, which may cause malfunctions in silicon anodes, all relate to the presence of organic solvents which are not totally inert in contact with lithium. As we have just seen, the partial decomposition of these solvents leads to the formation of the SEI, which may grow and crack, leading to failure and the loss of storage capacity. This issue can be eliminated, along with the problem of flammability, by replacing organic solvents with a solid electrolyte. This is a recent development, and research is still ongoing; however, it could become widespread in lithium batteries. The chances of success are linked to the conductivity properties of solid electrolytes, their chemical inertia with respect to lithium, and to the quality of the interface contacts between the electrodes and the electrolyte [XIN 17].

3.2.3.6. *Lithium metal anodes with solid electrolytes*

The first attempts to identify suitable Li^+ or Na^+ ion-conducting membranes for use in all-solid-state lithium batteries (ASSLBs) were made in the 1980s. These attempts were quickly abandoned due to the low ionic conductivity ($<10^{-3}$ S/cm at 25°C) of most of the available electrolytes, and to the high interfacial impedances between electrodes and electrolytes. The only material which came close to the conductivities of organic solvent-based electrolyte solutions (of the order of 10^{-2} S/cm) was lithium nitride (Li_3N), discovered in the 1970s, which is characterized by a conductivity of 6×10^{-3} S/cm; however, it is unstable at high positive potentials. Research into ASSLBs only resumed following the discovery of a whole group of solid-state electrolytes (SSEs), with ionic conductivities approaching 10^{-2} S/cm at room temperature; furthermore, these materials are stable and can be used at high potential, making them compatible with Li metal and Li^+ insertion cathodes which are characterized by high positive redox potentials. The majority of these SSEs are oxide-based ceramics (typically lithium garnet $Li_7La_3Zr_2O_{12}$-LLZO), sulfur-containing glass-ceramics, LISICON (lithium super ionic conductor) and NASICON (sodium super ionic conductor) and polymers (Figure 3.14).

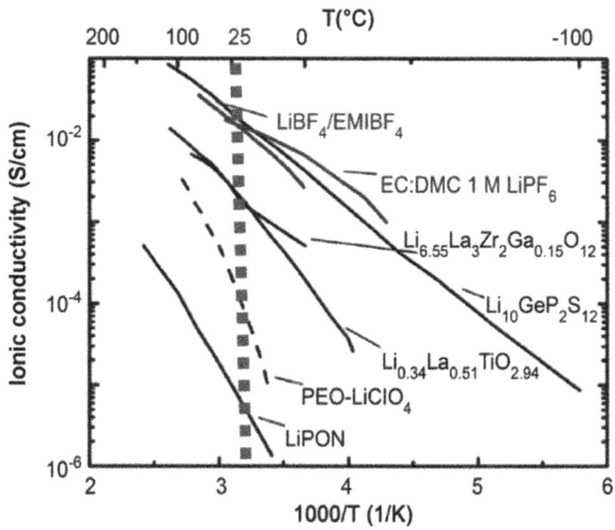

Figure 3.14. *Variation curves of Li+ ionic conduction in a selection of solid and liquid electrolytes*

COMMENT ON FIGURE 3.14.– *At 25°C, the liquid electrolyte EC:DMC (ethylene carbonate:Dimethylcarbonate) and the ionic liquid $EMIBF_4$ (1-ethyl-3-methylimidazolium tetrafluoroborate) have conductivities greater than 10^{-2} S/cm, while the solid electrolyte LiPON (Li_2PO_2N), the perovskite $Li_{0,34}La_{0,51}TiO_{2,94}$, the garnet $Li_{6,55}La_3Zr_2Ga_{0,15}O_{12}$ and the polymer electrolyte PEO-$LiClO_4$ (polyethylene oxide-$LiClO_4$) have conductivities much lower than 10^{-2} S/cm. The only solid electrolyte with a conductivity above 10^{-2} S/cm is thio-LISICON, $Li_{10}GeP_2S_{12}$. Adapted from [BAC 16].*

These solid electrolytes are described in detail in the literature [REN 15b, BAC 16, JUD 18, NIE 18, ZHA 18e, ZHE 18, ZHA 19b, ABA 20]. A brief chronology of discoveries relating to different SSEs and their use in producing ASSLBs is also described by Xia *et al.* [XIA 19a]; it is clear from this analysis that very few SSEs exceed conductivities of 10^{-2} S/cm at 25°C.

The Kanno group in Japan was among the first to synthesize an ionic superconductor based on sulfur, phosphorus and germanium: thio-LISICON, formula $Li_{10}GeP_2S_{12}$ (LGPS), which has a conductivity 1.5×10^{-2} S/cm [KAN 01, KAM 11]. The same group went on to discover another sulfur electrolyte (glass-ceramic Li_2S-P_2S_5, conductivity 1.7×10^{-2} S/cm) [SEI 14]. One of the major obstacles currently limiting the development of ASSLBs is the difficulty of producing low-resistivity solid/solid interfaces. The discontinuity of contact observed at Li/SSE or Cathode/SSE interfaces is one of the problems which can lead the batteries to malfunction; it appears to be caused by the lack of plasticity of the electrolytes. Unfortunately, sulfide electrolytes, which are highly conductive and are considered to be sufficiently flexible, are highly sensitive to humidity and are often incompatible with the very high redox potentials of cathode materials. Oxide-based electrolytes have a better stability, but are very rigid, and thus exhibit very high contact resistances [LI 19b].

One solution to these drawbacks is to improve the wettability of the SSE with respect to the anode and the cathode. The use of composite solid electrolytes (CSEs), which combine inorganic and polymer electrolytes, is a new approach developed by multiple teams in recent years [ZHA 17d]. CSEs generally consist of a solid polymer electrolyte, usually lithium polyethylene oxide (Li-PEO), and a ceramic or other inorganic solid electrolyte.

Among the most recent and successful implementations, Zhang *et al.* showed that solid electrolyte pellets, LAGP and LATP[21], surface-modified with composite layers of PEO and an inorganic mixture of lithium, sulfur, phosphorus and oxygen (LPOS)[22], were particularly suitable for reproducible lithium C/D reactions over long periods of time [ZHA 17e].

A symmetrically constructed battery featuring two lithium electrodes [Li//CSE-LAGP-CSE//Li], in which both sides of the solid electrolyte LAGP are coated with CSE, operates in a perfectly reversible manner when subjected to C/D cycles at ± 0.1 mA/cm^2 for more than 1000 hours at 60°C. The same battery without the CSE

21 LAGP ($Li_{1.5}Al_{0.5}Ge_{1.5}(PO_4)_3$ and LATP ($Li_{1.5}Al_{0.5}Ti_{1.5}(PO_4)_3$ are two widespread, commercially available NASICON-type electrolytes. They are stable in air and water, but like most oxide electrolytes, the electronic contacts between these electrolytes and cathode electrode materials such as $LiFePO_4$ (LFP) are poor, resulting in battery dysfunctions.

22 The LPOS mixture used to improve the wettability of the LAGP electrolyte with respect to lithium contains 75 wt% Li_2S, 24% P_2S_5 and 1% P_2O_5.

coating will rapidly malfunction in the same conditions, due to a rapid increase in the contact resistance between the LAGP and Li (Figure 3.15).

The interfacial resistance between lithium and PEO modified using the LPOS mixture presents an impedance of the order of 25–40 Ω, which remains broadly constant over a period of about 15 days. Without the addition of LPOS, the impedance increases progressively from 75 Ω to more than 150 Ω, reflecting a significant degradation of the interface.

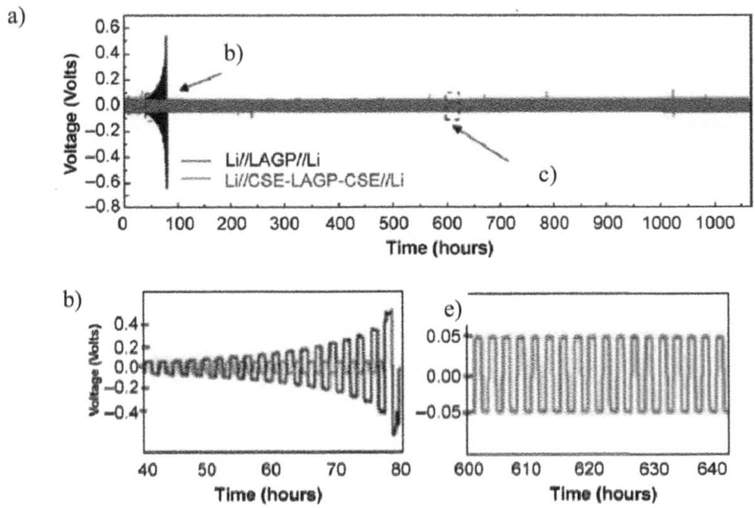

Figure 3.15. *Galvanostatic charge-discharge (C/D) curves at 0.1 mA/cm^2 for symmetrical all-solid state Li//LAGP//Li and Li//CSE-LAGP-CSE//Li batteries*

COMMENT ON FIGURE 3.15.– *a) C/D curves performed over a period of more than 1000 hours, with each C/D cycle lasting around 2 hours. b) and c) Respective galvanostatic C/D curves for Li//LAGP//Li and Li//CSE-LAGP-CSE//Li batteries, shown over a dilated time scale. Note the increase in resistance of the Li//LAGP//Li battery (b) after 80 hours of operation; conversely, in (c), extracted from figure (a), the galvanostatic C/D curves for the Li//CSE-LAGP-CSE//Li battery between 600 and 640 hours are perfectly reproducible. Adapted from [ZHA 17e].*

This behavior of the Li//CSE-LAGP interface can also be seen in batteries using $LiFePO_4$ (mixed with Li-PEO) as the cathode and LAGP or LATP as the solid electrolyte, modified, as before by the LPOS mixture. The initial discharge capacitance of the Li//CSE-LAGP//LFP battery (153.4 mAh/g) is practically unchanged (99.9%) after 200 C/D cycles at 0.1 C; for the same battery without the CSE composite added to the LAGP, the initial specific capacity of 158 mAh/g drops off rapidly, declining to 1.6% of its initial value after the same number of C/D cycles.

Other experimental results performed at higher discharge rates (1 C) confirm the high ionic conductivity of the CSE// electrolyte–electrode assembly. After 1000 C/D cycles, the battery capacity remains at 127.8 mAh/g, which represents a very high retention rate of 96.6% of the initial value (134.8 mAh/g). Its performance also remains high over a temperature range of 40–100°C; at 100°C, an even higher initial specific capacity value is observed, reaching 160.6 mAh/g, which is very close to the theoretical value of 170 mAh/g.

At this stage, it is important to note that the main problem at present concerns the creation of a perfect Li-electrolyte interface, that is, one that is perfectly stable over time, highly conductive, capable of withstanding high C/D rates and suitable for large-scale production. All of these parameters are difficult to combine in a single system; this is why various composite electrolytes have been proposed. Many new research pathways have been identified in recent years, notably involving a wide range of inorganic electrolytes and functionalization possibilities.

Efforts to improve conductivity generally rely on chemical interactions between the electrolyte and the electrodes. Zhang *et al.* [ZHA 17d] show that polyvinylidene fluoride (PVDF) and LLZTO[23] ceramic interact with each other to produce a composite electrolyte (CSE) with high ionic conductivity (5.10^{-4} S/cm at 25°C). Lithium-metal batteries with a $LiCoO_2$ cathode and this composite electrolyte [Li//PVDF-LLZTO//$LiCoO_2$] are characterized by a specific capacity of 150 mAh/g which, after 120 C/D cycles with a discharge rate of 0.4 C, retains 98% of its initial value [ZHA 17d].

The same LLZTO solid electrolyte was used by Goodenough and coworkers mixed with LiTFSI-PEO [CHE 18b]. The composite membrane obtained in this way has an ionic conductivity toward lithium which is a function of LLZTO content and temperature. At 60°C, a membrane made up of 10 wt% LLZTO and 90 wt% LiTFSI-PEO has an ionic conductivity of 7.10^{-4} S/cm, while a membrane with 80% LLZTO reaches only 10^{-4} S/cm. The mechanical properties of the membrane also depend on the proportions of LLZTO in the composite. With 10% LLZTO (ceramic-in-polymer), the membrane is very flexible, while in the case of 80% LLZTO (polymer-in-ceramic), it is much stiffer and somewhat brittle. However, the latter membrane can be made more flexible by adding a small amount of PEG (5%) to the

---

23 LLZTO is a ceramic, of composition $Li_{6.75}La_3Zr_{1.75}Ta_{0.25}O_{12}$, obtained in powder form. The LLZTO-PVDF composite membrane has good mechanical strength properties (ultimate tensile strength of 200% before breakage, corresponding to an elasticity coefficient of around 3 MPa) and good thermal stability.

mixture [CHE 18b]. The two composite electrolytes were tested in a battery assembly by inserting them between a $LiFePO_4$ cathode and a lithium anode[24]. The specific charges obtained correspond to predictions, and the C/D cycles applied to the case of the ceramic-in-polymer composite show a significant improvement compared to the LiTFSI-PEO single polymer electrolyte. The capacity retention rate after 100 C/D cycles at 0.2 C and 55°C was 94% in the case of the composite (10% LLTZO), but only 50% with the simple LiTFSI-PEO electrolyte, for an initial specific capacity of 150 mAh/g in both cases (Figure 3.16).

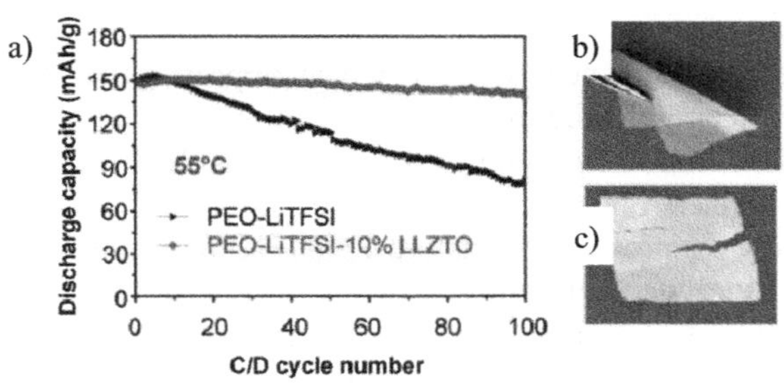

Figure 3.16. *Discharge characteristics of all-solid batteries [Li//electrolyte//LiFePO4] and images of solid composite membranes*

COMMENT ON FIGURE 3.16.– *a) Charge/discharge (C/D) curves obtained at 55°C for batteries with the simple PEO-LiTFSI electrolyte (black) and the composite PEO-LiTFSI-10% LLZTO electrolyte (red). b) and c) Images of the ceramic-in polymer (10% LLZTO, above) and polymer in ceramic (80% LLZTO, below) membranes. The first membrane is flexible, whereas the second is brittle. Adapted from [CHE 18b].*

These results confirm the importance of the composite electrolyte: unlike simple polymer electrolytes, these composites prevent dendrite formation and exhibit high levels of capacity retention over the course of multiple C/D cycles. However, tests have been limited to a relatively short period of 100 cycles.

24 The composite membrane inserted between the Li anode and $LiFePO_4$ cathode of the battery was around 100 μm thick, formed from micron-sized LLZTO particles (average diameter 6 μm), embedded in PEO mixed with LiTFSI lithium salt (one lithium per eight EO groups). The cathode material was a mixture of $LiFePO_4$, carbon black and polyvinylidene fluoride (80:15:5%) and contained 2–3 mg/cm^2 of active $LiFePO_4$ material. The anode was a 20 μm-thick lithium film.

Goodenough's group performed another similar study using an LSTZ perovskite ($Li_{3/8}Sr_{7/16}Ta_{3/4}Zr_{1/4}O_3$), which has an ionic conductivity of around 10^{-3} S/cm and is stable in humid air, as an electrolyte. Adding LiTFSI-PEO to create a composite electrolyte results in ionic conductivities of 5.4×10^{-5} S/cm at 25°C and 3.5×10^{-4} S/cm at 45°C, values similar to those obtained for LLZTO-PEO-LiTFSI. A battery made from $LiFePO_4$ and the LiTFSI-PEO-LSTZ composite performs at a level comparable to the previous battery, confirming the quality of the contacts between electrolyte and electrodes; in comparison to ceramics, it also offers improved stability with respect to air and humidity [XU 19b].

Generally speaking, any new solution which improves the interfacial contact between the solid electrolyte and the electrodes will provide a positive contribution to the development of ASSLBs. In this sense, Li *et al.*'s [LI 19b] proposed epitaxial deposition of a solid electrolyte onto a cathode constitutes a novel approach, and one which considerably improves the nature of the interfacial contacts between electrodes and electrolyte composites. High-temperature sintering of an LLTO perovskite ($Li_{0.33}La_{0.56}TiO_3$) on a lamellar cathode material made up of lithium-enriched titanium oxide particles ($0.54Li_2TiO_3$-$0.46LiTiO_2$) results in an interface which is homogeneous and continuous on an atomic scale; this permits very high-intensity discharges, comparable to those obtained with liquid electrolytes [LI 19b].

The examples cited above provide only a glimpse of the extensive research work carried out in this area; however, they are representative of efforts to obtaining stable, non-dendrite-generating interfaces between electrodes and electrolytes. In 2019, nine laboratories with an active interest in this area published a collaborative paper listing the main parameters for improvement (mostly technological) which would permit the creation of all-solid state batteries, using a lithium intercalation cathode with high redox potential (NMC type with high nickel concentration), capable of exceeding the current limit of around 350 Wh/kg. The goal is to reach a specific energy of 500 Wh/kg at a cost of less than 100 US$ per kWh; this is considered as a floor standard, necessary for wider diffusion and acceptance of lithium intercalation cathodes and their use in electric vehicles [LIU 19].

Another strategy which may contribute to reaching the 500 Wh/kg threshold involves replacing the intercalation cathodes with conversion cathodes, in which the (reactive) cathode material reacts with lithium. This results in specific capacities that are much higher than those of lithium intercalation materials, albeit with lower redox potentials, which ultimately attain in specific energies well above the 500 Wh/kg threshold. This approach may play a role in the development of a new generation of batteries, crucial for increasing the autonomous range of electric vehicles (Figure 3.17).

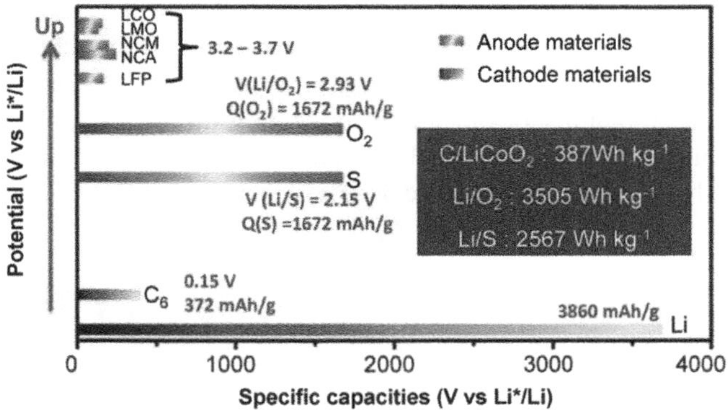

Figure 3.17. *Specific capacities and redox potentials of anode (graphite, lithium) and cathode (sulfur, oxygen and intercalation materials (LCO, LMO, NCM, NCA and LFP)). Inset: theoretical specific energies of C/LiCoO2, Li/O2 and Li/S batteries. Adapted from [GUO 17b]*

3.2.3.7. *Development of high-energy Li metal and conversion cathode batteries*

Associated with lithium metal, there are two main cathode systems which offer high specific energy capacity, using sulfur and oxygen. These systems offer significant theoretical gains compared to conventional insertion cathodes.

None of these systems has yet been commercialized, given the many reliability issues involved. However, recent developments, essentially relating to the use of nanomaterials, appear promising.

3.2.3.7.1. Li/S batteries and optimization of the sulfur cathode

The very high theoretical energy capacitance of Li/S batteries (close to 2600 Wh/kg), together with the cost-effectiveness of the electrode materials, provide strong motivation for research into these systems. However, a certain number of problems, described in the literature [BRU 12, WIL 15, FOT 17][25] and [YU 17], need to be solved before these batteries can be made commercially available. One major issue concerns the solubility of intermediate sulfur polysulfides, which dissolve in aprotic solvents; this results in a rapid drop in capacity and coulombic efficiency. Furthermore, the insulating character of sulfur means that large volumes of conductive additives are required. Finally, the wide volume variations (of almost

25 A consortium of university and industrial laboratories directly involved in a project to develop Li/S batteries: OXIS Energy, Culham Science Centre, Abingdon and the Advanced Vehicle Engineering Centre, Cranfield University.

180%) resulting from the successive stages of sulfur reduction from S_8 to Li_2S are also problematic. The reduction and oxidation reaction corresponding to the global transformation $S_8 + 16\ Li^+ \leftrightarrow 8\ Li_2S$ occurs at an average potential of 2.2 V versus Li^+/Li, passing through a series of intermediate polysulfides Li_2S_6, Li_2S_4 and Li_2S_2, several of which (mainly Li_2S_4 and Li_2S_6) are soluble in the aprotic solvents that constitute the liquid electrolyte (Figure 3.18a).

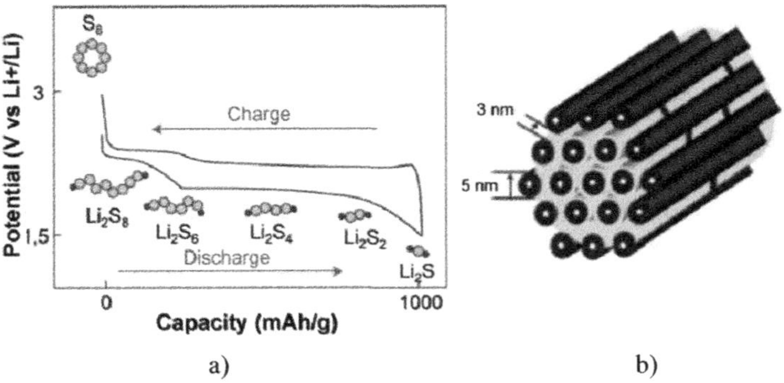

Figure 3.18. *Charge/discharge (C/D) curves of a sulfur-carbon composite cathode in the presence of a liquid electrolyte*

COMMENT ON FIGURE 3.18.– *a) The cathode is a composite mixture of mesoporous carbon and sulfur. Adapted from Bruce et al. [BRU 12]. b) The S-C composite is obtained by introducing molten sulfur (yellow) into mesoporous carbon (CMK-3) consisting of a network of carbon nanotubes (black) linked together by carbon fibers. Adapted from [JI 09].*

As we have seen, one of the main causes of the drop in battery capacitance over successive C/D cycles is the solubility of polysulfide anions that migrate from the positive to the negative lithium electrode through the separator, and are reduced in the form of Li_2S_2 and Li_2S precipitates. These precipitates are deposited on the lithium surface, where, first, they contribute to making the discharge irreversible, and, second, they lead to a loss of mass, resulting in a rapid decrease in capacity after around 100 C/D cycles.

One strategy used to avoid polysulfide migration in the electrolyte consists of encapsulating the sulfur in a sufficiently conductive and resistant structure, which prevents (or slows down) polysulfide migration in the electrolyte; this is similar to techniques used with lithium.

This approach produces significantly better results than those obtained with "non-protected" sulfur, and was first described by Nazar and coworkers [JI 09]. The authors propose the use of CMK-3 mesoporous carbon (commercially available, although very expensive), which consists of an ordered assembly of hollow carbon rods (6.5 nm in diameter), spaced 3–4 nm apart and linked together by carbon microfibers. Molten sulfur can be inserted into this network to form an S-C composite material (Figure 3.18b) [JI 09]. Highly reproducible cyclability, with discharge rates higher than C/10, was observed over around 20 C/D cycles, with an average discharge capacity value of 1100 mAh/g. Loss of sulfur mass in the cathode was also reduced (30% loss after 30 C/D cycles, compared with 80% loss in the case of a simple sulfur electrode), demonstrating the merits of the concept.

Cui's group applied the same concept, developing yolk-shell structures in which the sulfur is confined inside a TiO_2 capsule [SEH 13][26]. In this case, the cathode material was made up of TiO_2/S nanostructures mixed with carbon black and a fluorinated polymer binder. The electrode retained high capacity when immersed in a liquid electrolyte[27] and subjected to C/D reactions. Starting with a specific capacity of 1030 mAh/g, after 1000 C/D cycles at 0.5 C, the capacity was still around 700 mAh/g (67% of the initial value), representing an average decrease of 3.3% per 100 C/D cycles. The authors showed that the presence of a vacant space between the TiO_2 and sulfur helps the material to withstand the volume variations of the sulfur during lithiation/de-lithiation reactions. Similarly, the small pore size of TiO_2 (~3 nm) minimizes the diffusion of polysulfides into the electrolyte and improves the reproducibility of sulfur C/D reactions.

Compared to sulfur, or sulfur coated with a TiO_2 film (with no gap between the TiO_2 and the sulfur), yolk-shell cathodes appear to offer much better electrochemical performance (Figure 3.19). After 200 C/D cycles, the capacitance retention of the S-TiO_2 yolk-shell cathode was more than 80%, compared to 65% for the S-TiO_2 core-shell cathode and 50% for the sulfur-only cathode. This improvement in terms of capacitance retention is accompanied by a reduction in sulfur mass (19% loss in the case of the yolk-shell structure, compared to 44% for the core-shell structure and 66% in the case of sulfur alone).

In conclusion, the presence of a vacant space around the sulfur appears to be beneficial in terms of the performance of sulfur composite cathodes. However, these results only concern cathode behavior, and are not sufficient to evaluate a complete

---

26 The TiO_2/S nanostructure is obtained by reacting a titanium salt (diisopropoxide bis acetylacetonate $[(CH_3)_2CHO]_2Ti(C_5H_7O_2)_2$) on a dispersion of sulfur nanoparticles, resulting in a core-shell structure S@TiO_2. A yolk-shell structure is obtained by partial dissolution of sulfur in toluene-isopropanol medium. The sulfur NPs are obtained reacting HCl with sodium thiosulfate, $Na_2S_2O_3$.

27 A solution of lithium bis-(trifluoro methane sulfonylimide) (Li TFSI) and $LiNO_3$ in a mixture of 1,2-dimethoxyethane/1,3-dioxolane, with equal parts by volume.

Li/S cell. Moreover, from a practical standpoint, these materials are relatively complex to produce, and this may hinder future commercial production.

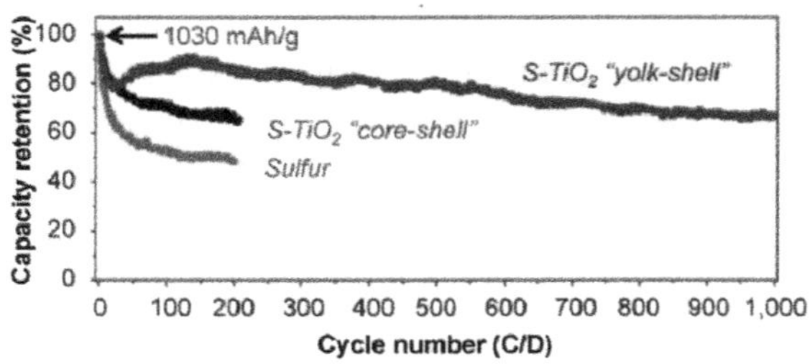

Figure 3.19. *Capacity retention of different sulfur composite cathodes as a function of the number of C/D cycles. The initial capacity is 1030 mAh/g; after 1000 C/D cycles, the value for an S-TiO2 yolk-shell cathode discharged at a rate of 0.5 C is 67% of the initial value. Adapted from [SEH 13]*

Zhang's group recently applied the same concept, using bacterial membranes for sulfur encapsulation. *Staphylococcus aureus* bacteria, which feature a natural carbonated membrane wall containing a variety of nitrogen- and carbon-based substances, can be transformed by a series of simple chemical operations into a spherical nanostructure, closed by a bi-layer TiO_2/carbon bi-layer cap, in which sulfur is confined [WU 18] (Figure 3.20).

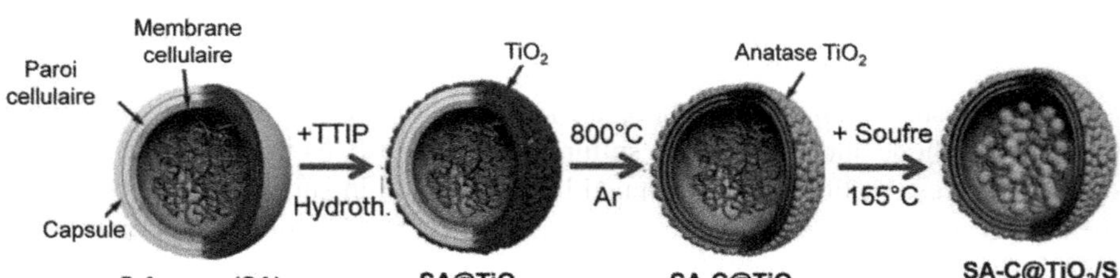

Figure 3.20. *Simplified diagram of the transformation of* S. aureus *bacteria. SA@TiO2 is obtained after a first hydrothermal treatment in the presence of TTIP (titanium tetraisopropoxide)*

COMMENT ON FIGURE 3.20.– *Heat treatment at 800°C results in the conversion of the TiO_2 layer into anatase crystal. Carbon and nitrogen compounds in the bacteria are converted to carbon and nitrogen-doped carbon sublayers (SA-C@TiO_2). Molten sulfur is then introduced into the microstructure (SA-C@TiO_2/S). Adapted from [WU 18].*

A first hydrothermal treatment of *S. aureus* (SA) with titanium tetraisopropoxide (TTIP) results to the deposition of a thin layer of TiO_2 (~20 nm thick) on the surface of the bacterium (SA@TiO_2). Heat treatment at 800°C results in the transformation of this TiO_2 layer into a uniform layer of anatase nanocrystals, adhering to a nitrogen-doped carbon-based sublayer: this consolidates the whole structure (SA-C@TiO_2). Molten sulfur (150°C) is then introduced into the microcapsule, resulting in a spherical microstructure with a diameter of around 800 nm (SA-C@TiO_2/S), used to create the sulfur cathode. Note that this microstructure is very similar to the yolk-shell structures seen earlier, leaving a vacancy around the sulfur in the capsule; this helps the material to withstand volume variations during C/D reactions. The specific capacitances of an SA-C@TiO_2/S cathode obtained at low discharge rate (0.1 C) in a conventional electrolytic medium (Li-TFSI + $LiNO_3$ in 1,3 dioxolane-dimethoxymethane medium) are comparable to those obtained with the previous yolk-shell structures. An initial capacitance of 1200 mAh/g is observed for a discharge rate of 0.2 A/g and, after about 100 cycles, this value remained at 1100 mAh/g. At higher speeds (1.5 A/g, i.e., above 1 C) and after a very large number of cycles (1500) the capacitance fell to 567 mAh/g, representing a relative loss per cycle of around 0.024%.

Other processes have been put forward for the development of sulfur cathodes, based on mixtures of sulfur and conductive carbon-based additives, or on the use of solid electrolytes [YU 17, UME 19]. The latter approach appears most promising as a means of preventing polysulfides from dissolving in the electrolyte.

Other authors have mixed sulfur with various carbon-based additives. Zhang *et al.* [ZHA 18c] recently obtained excellent performance in terms of cyclability and capacity, showing that sulfur mixed with a LiPS[28] solid electrolyte and carbon nanomaterials (an assembly of multi-walled carbon nanotubes (MWCNT), ultra-long carbon fibers and carbon black) constitutes an extremely efficient composite material (S-CNT) for lithiation/de-lithiation reactions. The presence of MWCNT is decisive, and the cathode performs better for higher levels of MWCNT in the mixture[29] (Figure 3.21).

---

28 The LiPS solid electrolyte is a glass-ceramic of composition $78Li_2S$-$22P_2S_5$ (7822 gc) and with conductivity of over 10^{-3} S/cm at 25°C. It is chemically stable with respect to sulfur and lithium.

29 The S-CNT cathode material is a ternary mixture of MWCNT, sulfur and LiPS electrolyte. These materials are obtained using a relatively complex experimental protocol, including grinding and solubilization phases, leading to the formation of pastes, followed by vacuum evaporation and hot drying. The different S-CNT(i) sulfur composites (i = 1, 2 and 3) correspond to increasing proportions of MWCNT.

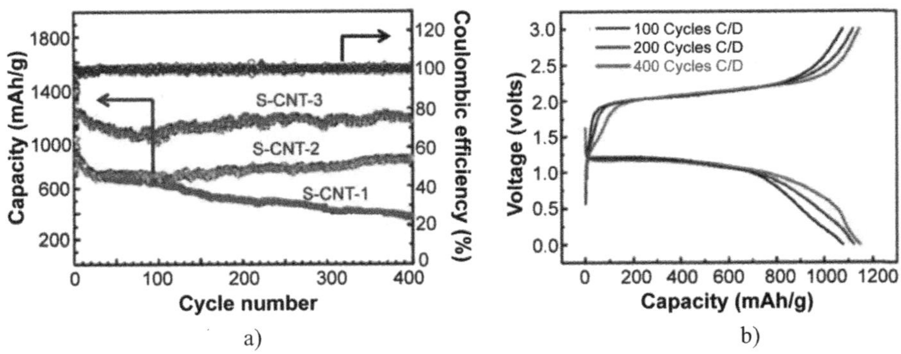

Figure 3.21. *Electrochemical characteristics of all-solid state Li(In)//LiPS//S-CNT batteries. a) Variation of specific capacity (mAh/g) of batteries made with three distinct cathode materials, S-CNT-1,2,3, as a function of the number of C/D cycles. b) Discharge (D, bottom curves) and charge (C, top curves) at 0176 mA/Cm^2 for a Li-S battery with S-CNT-3 cathode and Li(In) anode. Adapted from [ZHA 18c]*

Three particular formulations of sulfur composite cathodes, S-CNT-1, S-CNT-2 and S-CNT-3, corresponding to increasing CNT concentrations, incorporated in an all-solid state battery of the type [Li(In)//LiPS//LiPS/(S-CNT-i)] were tested over long periods of lithiation-de-lithiation cycles.

The best performance was obtained with the S-CNT-3 cathode, which contains the highest amount of CNT. Its initial capacity (1209 mAh/g) decreases slightly to 1141 mAh/g after 400 C/D cycles at the rate of 0.176 mA/cm^2 (i.e., approx. 0.1 C). Applying a higher velocity of 0.25 C (0.44 mA/cm^2), the specific charge of the battery is still 834 mAh/g after 1000 C/D cycles. According to the authors, this result is much better than previous values obtained for all-solid state batteries in terms of both durability and capacity, and demonstrates the important role played by CNTs. The porosity of these composites results in increased ionic conduction of Li^+ ions inside the cathode material; furthermore, the presence of CNTs improves electronic connectivity between sulfur particles.

Note, however, that the voltage available during the discharge reaction is not very high (1.2 V), resulting in specific energy of between 800 and 600 mWh/g; this corresponds to the threshold value expected of new-generation batteries[30]. The

30 Using an indium anode to insert the lithium also contributes to shifting the Li^+ discharge potential to more positive potentials, reducing the voltage between the two electrodes accordingly.

discharge rates are also low, preventing these batteries from being used to provide the high operating powers required by electromechanical devices. Finally, although CNT is a particularly efficient material for use in cathodes, the high cost of the material is problematic.

In conclusion, the main issue hindering the development of Li-S batteries with liquid electrolytes concerns loss of sulfur mass: the solubility of Li polysulfides leads to their extraction from the cathode and causes a rapid decrease in capacitance during cycling. Adding materials to complex these polysulfides is one means of limiting sulfur loss. Encapsulating sulfur in a membrane to slow down the migration of polysulfides in the electrolyte by causing their adsorption is a good solution, and one which is still being researched. Similar effects are also obtained by mixing sulfur with carbon-based nanomaterials, functionalized in various ways, obtained from natural products and from porous carbon biomass derivatives [YUA 19]. The introduction of a sulfur-doped carbon-based intermediary layer in Li/S batteries is another suggested solution; this has the additional effect of reducing polysulfide migration in the electrolyte solution [YAN 16]. All-solid state Li/S batteries, using solid electrolytes, are not subject to the same problems, and are also safer. On the other hand, cell impedance is considerably increased, making it difficult for these batteries to operate at high powers. This is one of the reasons so much research is still being carried out on the subject of liquid electrolytes, with the aim of fully eradicating the lithium polysulfide dissolution phenomenon. Another alternative, Li/Air batteries, is also the subject of active research.

3.2.3.7.2. Li/Air batteries

From a theoretical standpoint, Li/Air (Li/O_2) batteries present a number of advantages compared to Li/S batteries. For the same theoretical specific capacity of 1672 mAh/g for the sulfur and oxygen cathodes[31], the theoretical operating voltage of a Li/O_2 battery is higher than that of a Li/S battery (2.93 V against Li/Li^+ compared to 2.15 V for Li/S). This results in a theoretical specific energy of around 3500 Wh/kg, much higher than that of a Li/S battery (2515 Wh/kg). Furthermore, air or oxygen is an attractive option for use as cathode material because of its natural availability and better ecological credentials than sulfur.

In practice, the operation of Li/Air batteries is just as complex as that of a Li/S battery, and the technology is still not at a level to permit effective use, despite the number of papers already published on the subject. There are many problems

---

31 This specific capacity of 1,672 mAh/g is calculated with respect to the mass of O_2. Some authors calculate it in relation to the mass of the additives alone, which can lead to much higher values.

inherent to the operation of these batteries, notably in relation to the cathode in an aprotic electrolyte medium [PAR 18, ZHA 18b, CAI 19, WAN 19a].

Both the charge and discharge reactions in Li/Air batteries involve the formation and destruction of lithium oxide, Li_2O_2. In the discharge reaction, the lithium anode releases Li^+ in the solution according to the reaction $2Li \rightarrow 2Li^+ + 2e^-$; the released electrons are transferred to oxygen, which is reduced and reacts with the Li^+ ions to give Li_2O_2, according to the reaction $O_2 + 2Li^+ + 2e^- \rightarrow Li_2O_2$. Insoluble lithium dioxide, Li_2O_2, is deposited at the cathode and, during the reverse charge reaction, its oxidation releases Li^+ ions back into the solution. These ions are reduced to Li metal at the anode. Oxygen is simultaneously produced in the solution. The insolubility of Li_2O_2 and its electrical insulating properties constitute a first difficulty for cathode operations. The cathode, generally made of a porous mixture of carbon-based materials and catalysts, rapidly loses its efficiency due to the progressive obstruction of its pores by Li_2O_2. This leads to a high overvoltage (more than 4 V against Li^+/Li) in the charging reaction, corresponding to the release of O_2, and results in parasitic reactions of the electrolyte, responsible for a rapid drop in capacity.

A first improvement was made by Peng *et al.* in 2012 [PEN 12], using dimethylsulfoxide as the solvent in the place of a more "conventional" ether oxide such as dimethoxyethane (DME). Furthermore, replacing the carbon-based materials with a porous gold electrode was seen to speed up the Li_2O_2 decomposition reaction (rates 10 times those observed with a carbon cathode); the charge potential measured with respect to the Li^+/Li couple remained stable, and did not exceed 4 V after around 100 C/D cycles. Most of the parasitic reactions were thus eliminated, and a high specific capacity was retained [PEN 12]. Based on this discovery, a whole series of new carbon-free catalysts, based on transition metals (Co, Mn, Ti, etc.) in oxide form or on noble metals, has since been tested.

Another means of reducing the oxidation overpotential is to add catalysts to the electrolyte solution, which act as redox mediators (RM) for Li_2O_2 oxidation (Figure 3.22).

A wide variety of organic, biological and inorganic RMs [PAR 18, WAN 19a] have been tested; tetrathiafulvalene (TTF) and 2,5-di-tert-butyl-1,4-benzoquinone (DBBQ) are two examples of typical catalysts, used for catalytic oxidation of Li_2O_2 and for O_2 reduction, respectively.

Unfortunately, although these RM catalysts significantly lower the charge potential corresponding to Li_2O_2 oxidation, they also react with the lithium anode due to their diffusion in solution; furthermore, they also contribute to deterioration of cell characteristics through a series of parasitic reactions.

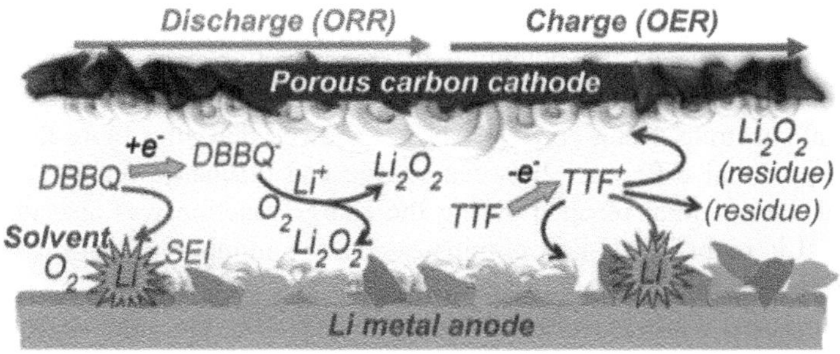

Figure 3.22. *Action of two redox catalysts DBBQ and TTF, in a Li/O_2 battery with a metallic Li anode and a porous C cathode in the absence of a separator*

COMMENT ON FIGURE 3.22.– *DBBQ and TTF intervene during the formation and decomposition, respectively, of Li_2O_2 (toroidal particles, shown in yellow, deposited on the cathode). DBBQ in the reduced state catalyzes O_2 reduction (ORR) and Li_2O_2 formation during the discharge reaction. Conversely, TTF catalyzes the oxidation and decomposition of Li_2O_2 during the charge reaction, through its oxidized form TTF^+ (OER). Note that both redox catalysts also react with lithium, causing a degradation of the specific capacity of the battery. Adapted from the Supporting Information of [QIA 18b].*

One solution to this new problem of RM diffusion toward the anode is to place a separator between the lithium electrode and the cathode, with sufficiently small pores to confine the RM to the cathode compartment. Qiao *et al.* [QIA 18b] proposed a solution of this type, covering a traditional membrane separator (glass fiber (GF), Celgard membrane) with an MOF (see Chapter 2), which, due to its nanoporous structure, considerably slows RM diffusion toward the lithium electrode; it thus contributes to maintaining a constant, low charge overvoltage, which has the effect of increasing the lifespan of the cell (Figure 3.23).

Using the HKUST-1 MOF alongside the redox mediators tetrathiafulvalene (TTF) and 2,5-di-tert-butyl-1,4-benzoquinone (DBBQ), the authors observed a lower rate of RM diffusion toward the anode compartment, resulting in a significant decrease in the charge potential. According to the authors, this improvement is explained by the small size of the MOF nanopores, smaller than the TTF and DBBQ molecules: thus, the RMs are unable to diffuse toward the lithium anode, but Li^+ ions are still able to travel between the two electrodes[32].

---

32 HKUST-1 is an MOF marketed by BASF under the name Basolite® C-300. It is the condensation product of Cu^{2+} and benzene-1,3,5-tricarboxylate (BTC), with the global formula $Cu_3(BTC)_2$. It can be obtained in the form of a membrane of around 10 µm thickness.

The use of inorganic catalysts such as lithium iodide has also resulted in significant reductions in the potential difference between charge and discharge. Using a titanium oxide black cathode, TiO_{2-x} (conductive), deposited on a porous nickel metal substrate, Kang *et al.* [KAN 17] showed that after more than 300 C/D cycles the potential difference between charge and discharge remains low (under 0.4 V) and the potential reached during the charge reaction was also under 3.5 V against Li^+/Li; in practice, this eliminates a large number of parasitic reactions corresponding to decomposition reactions in the electrolytic medium. The specific capacity measured with respect to the mass of TiO_2 is in the order of 1000 mAh/g, and remains constant throughout the 300 C/D cycles.

Li/O_2 electrochemical cells present a number of advantages compared to Li/S batteries. However, one major problem relates to the maintenance of regular oxygen diffusion inside the cathode. This requires controlled permeability of the cathode, which can only be effective if Li_2O_2 can be formed and dissolved in a reversible way, so as to prevent the obstruction of the pores inside the cathode. The choice of redox catalysts inert to the lithium electrode, or of separators establishing a diffusion barrier to prevent the RMs from reaching the lithium, could allow batteries of this type to operate in an optimal manner.

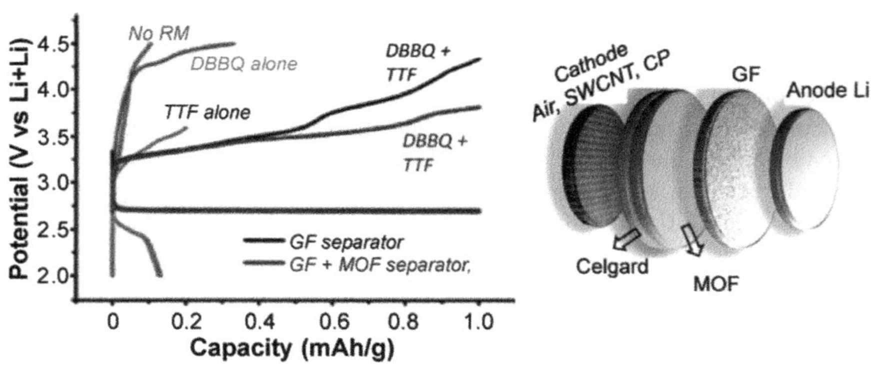

a) b)

Figure 3.23. *Galvanostatic charge and discharge (i = 1A/g) curves between 2 V and 4.5 V in a button cell, in the presence of different MR and for a charge limit of 1 mAh/g*

COMMENT ON FIGURE 3.23.– *a) C/D curves obtained using a conventional separator (glass fiber-GF) and using a GF separator coupled with an MOF. Note that in the case of the RM mixture (TTF + DBBQ) the charge overvoltage in the presence of the MOF is around 0.5 V lower than that of the cell without MOF. b) Exploded view of the button cell with three membranes attached: a Celgard membrane + MOF + GF. The cathode is made up of MWCNTs mixed with carbon paper sheets (CP). Adapted from [QIA 18b].*

3.2.3.8. *Sodium-ion batteries*

Sodium-ion batteries (SIBs) present an advantage compared with LIBs in that they are made from an abundant and low-cost material: unlike lithium, there is no likelihood of future supply problems. This supply issue has been a major motivating factor for research since the 2010s, particularly in light of predicted future demand for large-scale energy storage, which will involve extremely large quantities of cathodic and anodic materials. Figure 3.24 shows the proportions of certain strategic materials contained in the Earth's crust and is particularly revealing; it shows the extent to which the scarcity of lithium, and consequently its cost, could rapidly become prohibitive if lithium technology was to become widespread for storing photovoltaic and wind-generated energy.

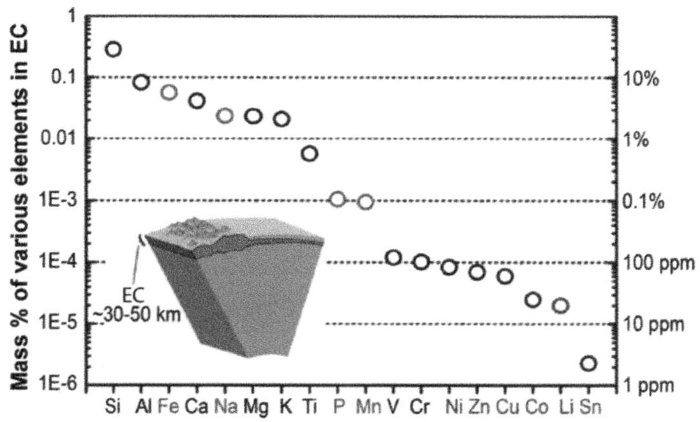

Figure 3.24. *Natural abundance of certain strategic elements used in battery production in the Earth's crust (EC). Note the lithium-sodium ratio of around 1:1000. Adapted from [YAB 14]*

The problems encountered with respect to sodium-ion batteries are very different from those encountered in LIBs. Sodium atoms are much larger than lithium atoms, meaning that sodium cannot be intercalated in graphite in the same way as Li^+ ions. Moreover, the very high reactivity of Na metal toward organic solvents makes it more difficult to use as an anode. Unlike lithium, which is protected by an SEI layer, sodium metal corrodes continuously when in contact with a non-aqueous electrolyte, and a protective layer is only obtained with certain fluorinated salts ($NaPF_6$) and mixtures of carbonated solvents such as ethylene propylene carbonates (EC-PC) [PON 12]. Another difference compared to lithium is that the Na^+/Na redox potential is lower than that of lithium (–2.71 V against a standard hydrogen electrode (SHE), compared to –3.01 V for lithium); in all cases, whatever the choice of anode and cathode materials, this leads to much lower specific energies. A wide variety of materials for sodium ion insertion and a number of alternative organic solvents have been proposed and are described in detail by Yabuuchi *et al.* [YAB 14].

The most common anode material currently used is hard carbon, HC, for which specific capacities between 200 and 300 mAh/g have been obtained by means of Na^+ ion intercalation. The choice of the solvent is decisive for achieving high capacities and for attaining good lifespans in terms of C/D cycles. A recent publication shows that, contrary to carbonate solvents, the use of TEGDME (tetraethylene glycol dimethyl ether) alongside a preliminary treatment of the HC with lithium makes it possible to reach a specific capacity of around 220 mAh/g, twice that obtained in carbonate medium. Moreover, around 85% of the specific capacity is retained after 1000 C/D cycles, carried out at rates of over 4 C (1 A/g) [XIA 18b].

HCs can be assimilated to graphene fragments with a turbostratic structure, corresponding to strongly interlinked disordered graphitic domains [IRI 15, XIA 19b] (Figure 3.25). They are obtained by pyrolyzing natural carbon compounds, and are commercially available. The pyrolysis temperature depends on the choice of the precursor, and is around 1000°C.

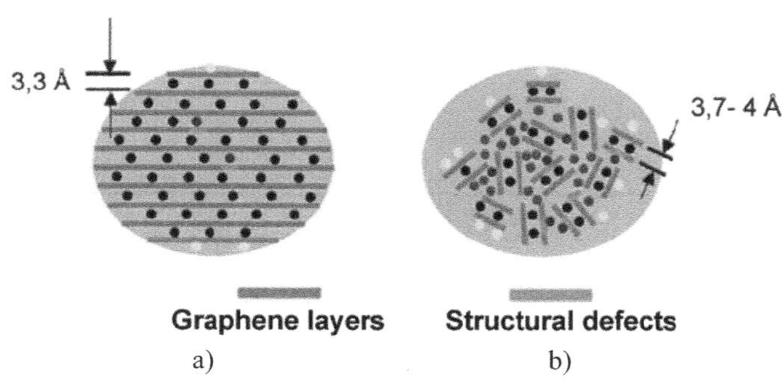

Figure 3.25. *Comparison of ion insertion modes in graphite and turbostratic graphene*

COMMENT ON FIGURE 3.25.– *Ion insertion in graphite (a) and in turbostratic graphene (b). The black dots represent intercalation sites between graphene sheets; the yellow and blue dots are adsorption sites onto graphene sheets and inside pores, respectively. The purple dots are ion adsorption sites on structural defects (shown in green). Adapted from [XIA 19b].*

Recent research has also been carried out into other options for anode materials, based on the conversion reactions of elements such as phosphorus. With "red" phosphorus, which can be converted into Na_3P (giving a theoretical capacity of 2596 mAh/g), specific capacities in excess of 2000 mAh/g can be obtained from the first charging reaction; however, these drop rapidly after the first few C/D cycles, falling to values of around 300 mAh/g after around 10 cycles [MAR 18].

Similarly, extensive work has been carried out on identifying suitable inorganic cathode materials for sodium or potassium intercalation. Various compounds based on transition metal oxides, polyanions (phosphates, pyrophosphates and fluorophosphates), oxychlorides and NASICONs have been tested, alongside a large number of organic compounds. Sun and coworkers provide a detailed description of the performance of these different classes of materials [HWA 17]. Among the different inorganic material options, NASICON $Na_3V_2(PO_4)_3$ coated with a thin layer of carbon, around 5 nm thick (NVP/C), was found to be remarkably stable during extended C/D cycles (beyond 10000 cycles) (Figure 3.26).

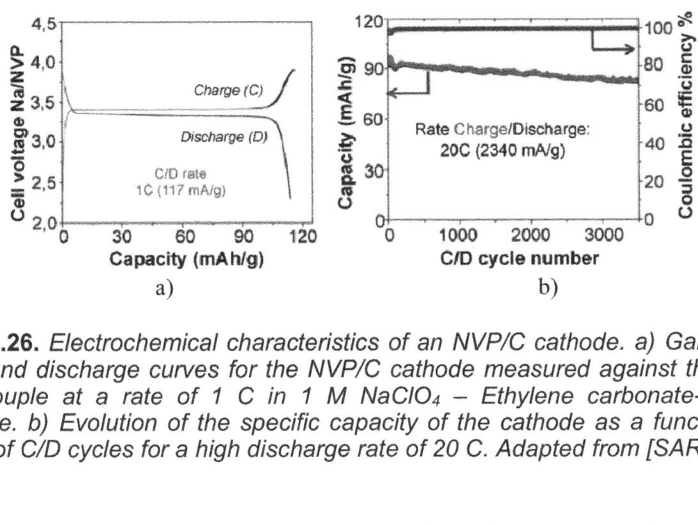

Figure 3.26. *Electrochemical characteristics of an NVP/C cathode. a) Galvanostatic charge and discharge curves for the NVP/C cathode measured against the Na+/Na redox couple at a rate of 1 C in 1 M $NaClO_4$ – Ethylene carbonate-Propylene carbonate. b) Evolution of the specific capacity of the cathode as a function of the number of C/D cycles for a high discharge rate of 20 C. Adapted from [SAR 13]*

Specific energies of the order of 400 mWh/g against a Na metal anode can be obtained with this cathode at low C/D rates [SAR 13]. The material is also able to withstand extremely high discharge rates of 40 C; in this case, it still retains almost 50% of its initial capacity after 30000 C/D cycles (around 100 mAh/g). Furthermore, the C/D potential measured with respect to the Na^+/Na couple remains almost constant, between 3.3 and 3.4 V, for a large part of the charge and discharge periods. According to the authors, these exceptional electrochemical characteristics make NVP/C a good potential material for use in producing sodium batteries for large-scale static storage purposes.

Since these discoveries, a large amount of work has been carried out in this field and several promising inorganic materials with high capacities have been proposed; these are well documented in the literature [YAB 14, FAN 18b]. Research is primarily focused on the development of new cathode materials. NVP-type materials continue to be the subject of improvements, as shown very recently by Zhang *et al.* [ZHA 21], who propose a fluorinated derivative of NVP doped with potassium (K-doped $Na_3V_2(PO_4)_2F_3$), which also exhibits very good performance in terms of specific energy and cycling.

Although organic compounds have received less attention than mineral materials, they are also considered to be potential candidates for energy storage applications and for producing organic sodium-ion batteries (OSIB). Several families of redox compounds, insoluble in organic solvents, may be used as cathode or anode materials, depending on the value of their oxidation-reduction potential E° (high E°, cathode; low E°, anode). These operate through reduction or oxidation, followed, respectively, by the insertion of an alkaline cation (Na^+, K^+ or Li^+) or by the insertion of the anion associated with the Na^+ cation. Finally, some materials can be bipolar, characterized by two distinct redox potentials, meaning that they can be used either as anodic or cathodic materials (Figure 3.27).

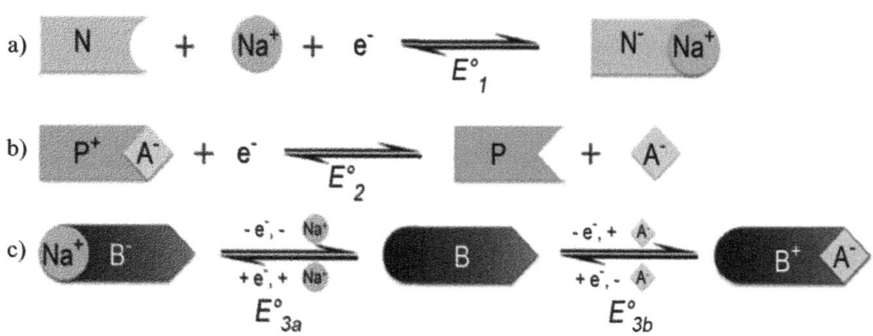

Figure 3.27. *Organic materials used in sodium batteries*

COMMENT ON FIGURE 3.27.– *a) n-type organic compound (N), characterized by a redox potential $E°_1$ close to that of Na^+/Na, suitable for use as an anode. b) p-type organic compound (P), with a redox potential $E°_2$ which is much higher than that of Na^+/Na: this makes it suitable for use as a cathode, with complexation of the counter-ion A^- associated with P^+. c) Bipolar compound B: it can be reduced or oxidized, respectively, at the two redox potentials $E°_{3a}$ and $E°_{3b}$ such that $E°_{3a} < E°_{3b}$. This means that the material may be used as an anode or as a cathode. Adapted from [XU 18].*

Organic compounds are generally cost-effective. The absence of heavy metals, their lightness and flexibility, and the possibility of controlling the redox properties by specific functionalization all combine to make these materials promising candidates for electrochemical storage applications [XU 18].

The largest group of organic compounds is that of carbonyl derivatives with carboxyl, anhydride, quinonyl and imide groups, considered to be n-type anode materials. The reversible single-electron reduction of the carbonyl group, with the formation of an anion, is compensated by the addition of a Na^+, Li^+ or K^+ ion; this is similar to the way Li^+ ions are inserted into carbon-based anode material. Amino compounds, such as polyaniline derivatives, and all conductive polymers (polypyrrole, polythiophene and polyphenylene) constitute p-type cathode materials; after oxidation and the reversible formation of a cation, these are neutralized by the addition of an anion (ClO_4-, PF_6-). Bipolar compounds, which possess both n- and p-type functional groups, can be reduced sequentially from their cationic oxidation state. The application of a decreasing discharge potential thus results in progressive transformation of the material from the cationic to the anionic state, transitioning through a neutral state corresponding to the open circuit potential of the initial compound.

Sakaushi *et al.* [SAK 13] used a porous graphitic-type bipolar material, belonging to the covalent triazine frameworks (CTF) family, discovered in the 2000s by Kuhn *et al.* [KUH 08]. Obtained by condensing para-dicyanobenzene at 400°C in the presence of $ZnCl_2$, the material consists of a covalent chain of benzene and triazine rings arranged in a honeycomb pattern. The presence of triazine (electron acceptor) and benzene (electron donor) nuclei gives the material a bipolar character (Figure 3.28a). However, unlike graphitic structures, this material has low conductivity, of the order of 4.09×10^{-6} S cm^{-1} at 25°C in the neutral state.

After initial p-doping, which transforms the compound into a cation, its progressive reduction toward the anionic form involves p-de-doping between 4 and 2.8 V, followed by n-doping between 2.8 and 1.3 V; the 2.8 V potential corresponds to the neutral form of the material[33]. The equivalent specific capacity for

33 Electrodes are made by mixing 70% of the active cathode material with 20% carbon black and 10% binder (carboxy cellulose). The electrode thus contains 1.5 mg/cm^2 of cathodic charge, for a cathode thickness of approximately 100 µm. The specific capacity is calculated on the basis of the dicyanobenzene trimer (M = 387 g), which gives a theoretical capacity of 69.3 mAh/g for one electron exchanged per trimeric unit. The specific capacities of 55 mAh/g and 185 mAh/g obtained for the p and n discharges, respectively, correspond to electronic exchanges equivalent to 0.79 and 2.65 electrons. According to the authors, this difference between the two doping modes of the material could be due to the fact that the p-doping

p-de-doping (between 4.8 and 2.8 V) is 55 mAh/g; it is much higher (185 mAh/g) for the n-doping region (between 2.8 and 1.3 V). Overall, a specific capacity of 240 mAh/g is thus obtained for a very slow discharge rate of 0.01 A/g. The compound is able to withstand a very large number of C/D cycles without noticeable degradation: for over 7000 C/D cycles at low discharge rates (0.01 A/g) in a propylene carbonate/ 1 M $NaClO_4$ medium, a capacity retention of about 80% was achieved with a coulombic efficiency of 100% between charge and discharge. Finally, after a few C/D cycles performed at very high rates (10 A/g, i.e., over 40 C), the system recovered its initial capacity for lower rates of 0.1 A/g, illustrating the high chemical stability of the compound (Figure 3.28b).

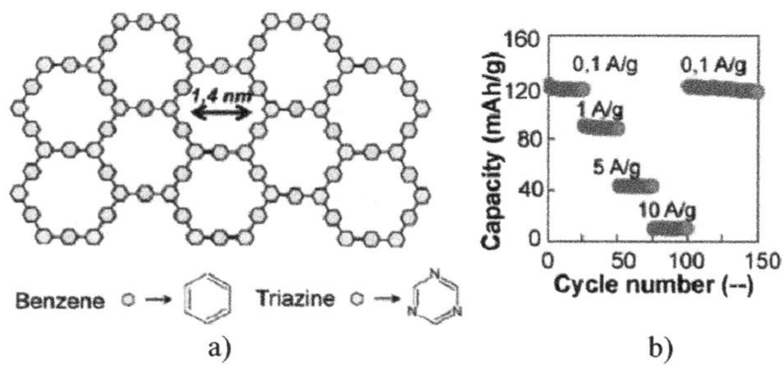

Figure 3.28. *Porous graphitic-type bipolar material consisting of a 2D network of benzene and triazine rings*

COMMENT ON FIGURE 3.28.– *a) Structure and average size of the nanopores in the material. b) Variation of the specific capacity of the bipolar material with the number of cycles and for very high discharge rates. Note that after a series of C/D cycles at 10 A/g (~40 C), the material returns to its initial capacitance of around 120 mAh/g when the C/D rate is brought back to 0.1 A/g (~0.4 C). Adapted from [SAK 13].*

The examples discussed in this section provide an illustration of the large number and wide variety of solutions which exist for the design and creation of sodium-ion batteries for high-capacity static storage. Although the energies and capacities obtained using these technologies are slightly lower than those of LIBs, the natural abundance of sodium compared to lithium presents a clear advantage, justifying current research efforts in this area. Compared to Na/S battery technology, which works with sulfur and sodium in the molten state (300°C), sodium-ion batteries also present a safety advantage, while providing specific energies of the same order of magnitude [SAK 13].

corresponds to a surface reaction (capacitive type), while the n-doping corresponds to a faradic process due to the reduction of the material, followed by the insertion of Na^+.

3.2.3.9. *Thin-film microbatteries*

The emergence of connected objects (the Internet of Things, IoT) over the past decade went hand-in-hand with an upsurge in the use of various electronic sensors and micro-electromechanical systems (MEMS) (medical implants, wireless RF receiver-transmitters, self-powered integrated circuits, hearing aids, RFID tags, "smart" watches, quartz oscillators, etc.), characterized by low operating powers ranging from 1 mW to 10 μW [SAL 09]. The subsequent need for small, independent energy sources has given rise to research into new storage systems, capable of delivering micro-quantities of energy, of the order of 100 μAh/cm^2 (~300 μWh/cm^2 for an average voltage of 3 V), compatible with the microsystems cited above. One recent solution, thin-film solid state microbatteries (TSSMBs), integrates a 3D architecture in a thickness of between one and a few tens of microns.

The first all-solid state thin-film LIBs, using phosphorus oxynitride (LiPON) as a solid electrolyte, were developed in the late 1990s at Oak Bridge National Laboratory (the United States) by Dudney and coworkers [BAT 00]. This type of battery is made with a planar structure, with a total thickness (including the protective layer) not exceeding around 15 μm. Batteries are made by vacuum vaporization and sputtering techniques, successively depositing a thin layer of an anode material (Li or various tin, zinc and silicon-based alloys) on different solid supports, followed by the electrolyte (LiPON, thickness ~2 μm) and the cathode material ($LiCoO_2$ or $LiMn_2O_4$, of the same order of magnitude 2.5 μm). Considering the thin-film structure of this type of battery, capacities are evaluated in μAh/cm^{2}μm, but these values can be easily converted into mAh/g[34].

Considerable progress has been made since these first works were published, notably concerning the architecture of the deposits: the aim is to increase the effective surface of deposits as much as possible while maintaining the same footprint. This implies 3D structuring of the film. One key approach consists of creating a regular network of cylindrical pores in a silicon sheet then covering the walls with different layers to make up the microbattery (MB). Using this technique, for cylindrical microholes of 3 μm in diameter spaced 6 μm apart (center to center) in a silicon sheet and with a depth of 12 μm, the equivalent surface area of the electrode materials deposited on the walls is around 4 times higher than that of the

34 In the case of a 2.5 μm thick $LiCoO_2$ cathode, the measured capacity for a surface of 1 cm^2 is 160 μAh, from which we deduce a capacity of 132 mAh/g taking account of a $LiCoO_2$ density of 5.03 g/cm^3. This value is close to the theoretical value of a $LiCoO_2$ cathode (137 mAh/g) for which only 0.5 Li is involved in the discharge. Considering that the average discharge voltage is 3.9 V, the average energy stored is 624 μWh (i.e., 2.25 Joules) for a rate of 0.1 mA/cm^2 and an average power of 390 μW. This power is more than sufficient for quartz oscillators (0.1 μW), smart watches (1 μW) or RFID tags (10 μW).

surface footprint; for a hole depth of 30 μm, the equivalent value is 10 times higher and corresponds to an area enhancement factor (AEF)[35] of 10.

The diagrams in Figure 3.29 show a planar structure and a 3D architecture of cylindrical pores in silicon. For the same footprint, the 3D structure provides both higher capacitance and higher power. Only one of these properties can be obtained using a planar structure: a thick film provides high capacitance but low power, while a thin film provides high power, but low capacitance.

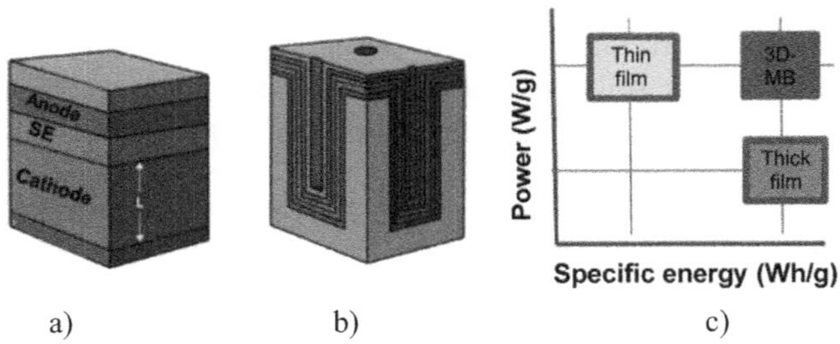

Figure 3.29. *Planar and three-dimensional structures of thin film all-solid microbatteries*

COMMENT ON FIGURE 3.29.– *a) Planar structure (SE: solid electrolyte, the two outer layers covering the anode and cathode are the current collectors of the anode and cathode). b) Three-dimensional structure (same color codes as for (a), corresponding to coating the walls of the cylindrical pores with the different constituent materials of the microbattery. c) The 3D MB structure (red square) potentially has the same specific power as a thin film (gray square) and the same specific energy as a thick film (blue square). a) and b) Adapted from [PEA 18].*

Pearce *et al.* [PEA 18] recently described the fabrication of all-solid state microbatteries, obtained by ionic machining of silicon, followed by evaporation and condensation under vacuum of the different component materials of an LIB. The technology used is based on the ALD (atomic layer deposition) method; compared to traditional sputtering techniques, this approach has the advantage of producing highly homogeneous deposits, whatever the geometry of the substrate surface (Figure 3.30).

35 Adopting a three-dimensional structure for the different layers has the effect of increasing the effective area while maintaining the same surface footprint. The ratio between the effective area A and the surface footprint A_f is defined by an increase factor AEF such that AEF = A/A_f.

Perfectly calibrated cylindrical holes are formed by deep reactive ionic etching (DRIE) of the silicon. The various thin layers making up an all-solid state battery are then deposited, successively, by ALD (Figure 3.30, step 1). These layers include a ruthenium collector, a layer of V_2O_5 as a cathode material (transformed into LiV_2O_5 after electrochemical reduction in a $LiClO_4$ solution), a solid electrolyte film (LiPON-Li_2PO_2N) and an SnN_x film as the anode, coated with TiN as a current collector electrode. A circular copper layer (diameter 1 mm) is then deposited by evaporation through a mask (step 2) in order to electrically connect the microbatteries arranged under the copper disk. Finally, the microbatteries are isolated from the rest of the structure by ionic stripping of the TiN and SnNx layers which surround the disk using Ar^+ ions (step 3). The voltage and current are measured between the copper and ruthenium layers.

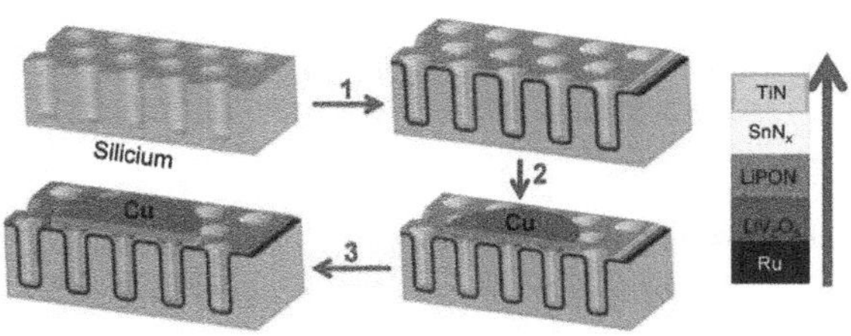

Figure 3.30. *Steps in the creation of [Ru//LiV_2O_5//LiPON//SnNx//TiN] microbatteries from a silicon sheet, structured by a network of cylindrical micropores, 3 μm in diameter and with 6 μm spacing (center-to-center)*

COMMENT ON FIGURE 3.30.– *Step 1: ALD of different layers on the walls of the micropores: Ru (40 nm), V_2O_5 (70 nm) transformed into LiV_2O_5 by electrolysis, LiPON (50 nm), SnNx (10 nm) and TiN (25 nm). Step 2: vapor deposition of the copper disk, followed by ion etching of the TiN and SnNx layers (step 3). Adapted from [PEA 18].*

Although the choice of materials in this case was not optimal in terms of capacity and voltage, the technology involved is particularly interesting, showing that microbatteries may be implanted directly into electronic circuit using techniques borrowed from the semiconductor industry. Pearce *et al.* clearly showed that the specific capacities and powers of this type of battery can be significantly increased by the use of increasingly deep cylindrical holes, resulting in greater surface areas with an AEF of between 4 and 10 (Figure 3.31).

These results show that the use of extremely powerful ion etching techniques to produce perfectly calibrated microstructures paves the way for all-solid state microbatteries, the capacity and power of which can be improved by modifying the architecture. At present, calibrated arrays of cylindrical holes and ALD deposition seem to offer the best means of producing microbatteries for direct implantation on integrated circuit platforms.

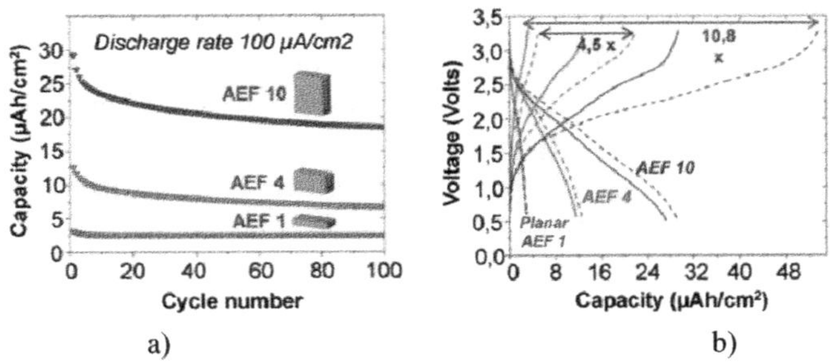

Figure 3.31. *Electrochemical characteristics of three microbatteries [Ru// LiV_2O_5 // LiPON // SnNx // TiN] with planar (AEF 1) and three-dimensional (AEF 4 and 10) structures*

COMMENT ON FIGURE 3.31.– *a) Evolution of discharge capacities as a function of cycle number: note the proportionality of the capacity values to the effective surfaces (1:4:10). b) Galvanostatic charge and discharge curves for batteries with various AEFs, obtained at a rate of 100 µA/cm². The dotted and solid lines show the first and second galvanostatic C/D curves performed between 3.2 and 0.5 V. Note the continuous, and approximately linear, voltage variation throughout the discharge process. This corresponds to a surface redox behavior, comparable to a capacitive-type process. Adapted from [PEA 18].*

3.2.4. *Supercapacitors*

Helmholtz first described the double electrochemical layer which is characteristic of positively or negatively polarized metal electrodes over a century ago. Ions are adsorbed from the solution onto the surface of each electrode, with cations adsorbed on the negative electrode and anions on the positive electrode. The capacitance of the double layer on each electrode is obtained using the classical formula $C = \varepsilon_r \varepsilon_0 A/d$, where ε_r is the dielectric constant of the medium, ε_0 is the dielectric constant in a vacuum, A is the active area of the material on which the ions are adsorbed and *d* is the distance that separates the ions from the electrode surface.

Compared to a planar capacitor with a solid insulating dielectric (alumina, plastic insulator, ceramic, etc.) for which the average thickness *d* is around 100 nm (Figure 3.35a), a double-layer electrochemical capacitor has a spacing *d* between the positive and negative charges which is measured on an atomic scale, and which is less than 1 nm. This is the reason why a capacitor with plastic insulation of average thickness 100 nm and with an average dielectric constant ε_r of 3 has a capacitance of 78 nF/cm^2, whereas for an electrochemical double-layer capacitor in an aqueous electrolyte the capacitance is around 20 μF/cm^2 [BAR 01], that is, over 250 times that of a solid capacitor. The use of activated carbons or any other porous material with a high specific surface area, often exceeding 1000 m^2/g, makes it possible to achieve even higher capacitances of the order of 100 F/g in an aqueous medium [FER 08b]. Thus, an electrode with an apparent surface area of 1 cm^2 containing a charge of 10 mg of activated carbons will have a capacitance of 1 F/cm^2, 50000 times that of a single double-layer capacitance. Electrochemical supercapacitors are generally made up of an assembly of activated carbon microparticles with liquid electrolyte in the gaps; this amounts to introducing numerous nano-sized anfractuosities into the electrode, considerably increasing its active surface and thus its capacitance (Figure 3.32b).

Depending on the type of activated carbon, this results in specific powers between 20 and 10^4 W/kg and specific energies between 2 and 15 Wh/kg. These characteristics are very different from those of rechargeable batteries, for which the specific powers and specific energies vary between 10 and 10^3 W/kg and 20–180 Wh/kg, respectively.

Another way to increase the capacitance of an electrode is to add a reversible redox system to the surface. This increases charge transfers to the electrodes by means of a fast faradic exchange, and thus increases the global capacitance of the device, known in this case as a pseudo-capacitor. Systems of this type have been created using metal oxides such as RuO_2, Fe_3O_4 and MnO_2, and with n- and p-type conducting polymers such as poly (3-(4-fluorophenyl)thiophene), which was first proposed in the 1990s [RUD 94]. Another solution is to use graphene as an electrode substrate, onto which a variety of rapid redox systems can be adsorbed. The very high specific surface area of graphene (2650 m^2/g) and its quasi-metallic conductivity make it particularly suitable for use in supercapacitors and pseudo-capacitors with a high power output in excess of 10 kW/kg: these values are 10 times higher than those of most current rechargeable batteries.

The energy, given by the relationship E = ½ CV2, also depends on the potential that can be applied across the capacitor. Unlike solid dielectric insulator capacitors, which can withstand high voltages (lower than breakdown voltages, that is, several thousand volts/cm), an electrochemical capacitor can only withstand voltages which fall below the decomposition voltages of the electrolyte. In an aqueous medium, this voltage is less than 1 V, and in a non-aqueous medium it is around 2–3 V. Depending on the type of activated carbon, this results in power densities of between 20 and 4000 W/kg and energy densities of between 2 and 15 Wh/kg [FER 08b].

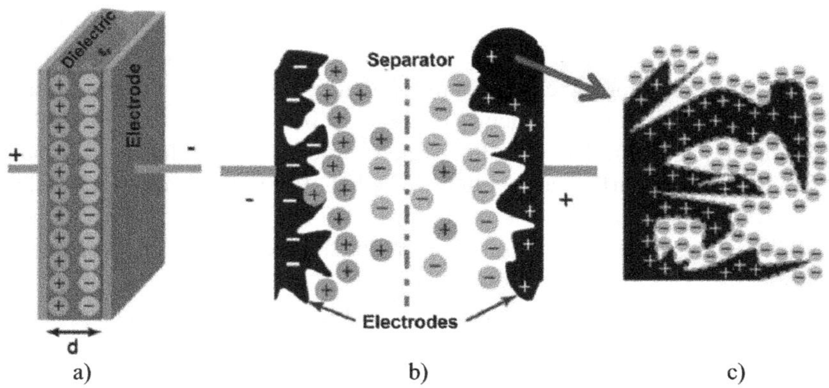

Figure 3.32. *Comparison of different capacitor types*

COMMENT ON FIGURE 3.32.– *a) Solid dielectric insulator capacitor with dielectric constant ε_r. b) Electrochemical capacitor with an electrolytic double layer and electrodes made of an assembly of carbon microparticles. Note that the overall capacitance of the cell is equivalent to two capacitors in series, one corresponding to the positive electrode and the other to the negative electrode. c) Schematic representation of the interior of an electrode (positive electrode b, in black) showing the anfractuosities inside the electrode which create a significant increase in the ion adsorption surface. Adapted from [KIM 15].*

3.2.4.1. *Influence of film porosity on supercapacitor values*

The size of the pores in the electrode material plays a determining role in the value of a capacitor. Simon and Gogotsi [SIM 08] have shown that the capacitance is maximal when the size of the anions and cations (desolvated) is equal to that of the pores in the electrode material. This is a particularly important result, and has been verified in the case of porous carbon materials in contact with an EMI-TFSI (ethyl-methyl-imidazolium-bis(trifluoromethane-sulfonyl)imide) ionic liquid[36]. In the case where the pores and ions are of similar sizes, the capacitor can be modeled as an electric wire placed in the axis of a cylinder. When the pore dimensions largely exceed those of the ions, a planar plate model applies. The capacitance values obtained in the second case, which are encountered in the majority of current systems, are significantly lower (Figure 3.33).

36 Carbon-based materials, of which the porosity can be precisely calibrated, are obtained by decomposing titanium carbide using chlorine, according to the reaction $TiC + 2Cl_2 \rightarrow C + TiCl_4$. Average pore size can be adjusted with a range 0.6–1.1 nm by varying the reaction temperature between 400 and 1000°C [LAR 08].

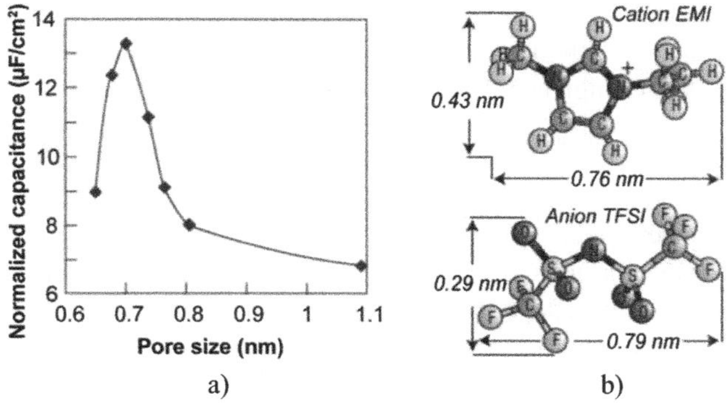

Figure 3.33. *Influence of the pore size of an electrode material on its capacitance value. a) Variation of the surface capacitance (normalized) of carbon-based materials with different porosities and in contact with EMI-TFSI ionic liquid. b) Dimensional molecular models of EMI (cation) and TFSI (anion) ions. Adapted from [SIM 08]*

The capacitances obtained with these carbon-based materials are of the order of 160 F/g and 85 F/cm^3, much higher than those obtained with activated carbons, which are of the order of 100 F/g and 50 F/cm^3 under the same conditions using EMI-TFSI ionic liquid.

3.2.4.2. *Graphene as a base material for supercapacitors*

As we have seen, controlling the porosity of electrode materials is crucial in order to obtain high capacitances; extensive work has been carried out into simple means of producing efficient materials for supercapacitor fabrication, given that supercapacitors are likely to be incorporated into many portable and flexible devices over the coming years.

Graphene and CNTs have been widely used in supercapacitors. Compared to the activated carbons used commercially, they present an advantage of excellent electronic conductivity; *a priori*, this will allow a considerable increase in C/D rates, resulting in higher powers. Various electrode materials using graphene, notably the reduced form of its oxide (r-GO), have been developed and described in the literature [KIM 15, SHA 15, DON 16, KE 16, GUO 17a].

One example of a product developed from r-GO is that of r-GO films, a porous material which is stable over time. Various techniques have been used to produce these films for use in supercapacitors. A main difficulty encountered with this type of material is the strong tendency for the sheets to aggregate with each other via a π-π stacking effect. The result is a loss of porosity over time, with a consequent deterioration of the capacitive properties of the material. The freeze-casting process[37] is a novel technique used to overcome this type of problem. Kaner and coworkers used this method to develop time-stable, high-porosity r-GO thick films (r-GO-P), which give high capacitance values [SHA 16] (Figure 3.34).

Films made in this way are characterized by low series resistances (under 1 Ω); furthermore, inter-pore communication allows ions to access the entire surface of the film. With a conventional r-GO film, in which the films are agglomerated together, it is much harder for ions to penetrate into the core of the material, and this decreases the specific capacitance. Using a porous r-GO film (r-GO-P), a mass specific capacitance of 284 F/g is thus obtained in an aqueous acid medium (1 M H_2SO_4) and for a charge rate of 1 A/g. At a charge rate of 500 A/g, this value is still 174 F/g (giving a retention rate of 61%). The mass specific capacitance of a non-porous r-GO electrode decreases from 181 to 50 F/g under the same conditions, representing a retention rate of only 28%. r-GO-P also offers higher stability, evaluated over long periods of C/D cycles between 0 and 1 V. Capacitance retention is over 97% after 10000 C/D cycles, compared to 86% for standard r-GO films. However, the volume densities are reversed for the two types of materials: around 45 F/cm^3 for r-GO-P, compared to 172 F/cm^3 for standard r-GO. This is due to differences in the compactness and porosity of the two materials, with densities of around 0.95 g/cm^3 for r-GO and 0.16 g/cm^3 for r-GO-P [SHA 16].

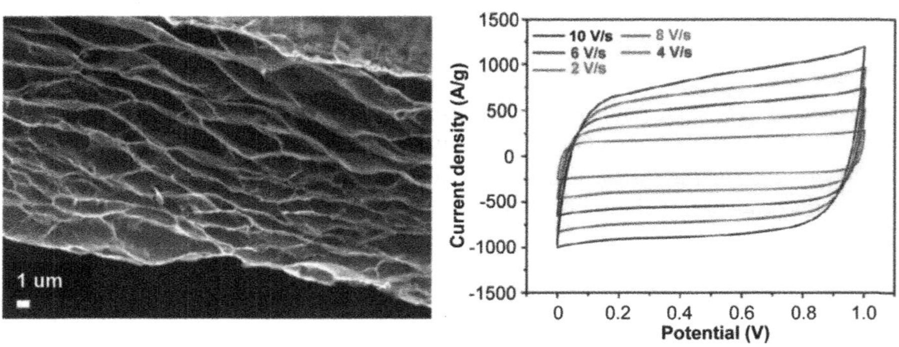

a) b)

Figure 3.34. *Structural and capacitive characteristics of r-GO-P films*

37 There are several steps involved in obtaining the porous film. A suspension of graphene oxide nanosheets in water is partially reduced with ascorbic acid, resulting in an r-GO micro-gel. This is then filtered to obtain a thick film (between 10 and 40 μm). The freezing of the film by liquid nitrogen, followed by sublimation evaporation of the water, results in a porous r-GO film (r-GO-P).

COMMENT ON FIGURE 3.34.– *a) SEM image of an r-GO-P film in cross-section, showing the honeycomb structure of the material. b) Cyclic voltammograms of an r-GO-P film as a function of the scan rate in acidic medium. Note the quasi-rectangular shape of the voltammograms, characteristic of a capacitive behavior, with the capacitance C deduced from the relation i = C(dV/dt). Adapted from [SHA 16].*

The specific powers and energies of r-GO-P, measured for a charge of 0.2 mg/cm^2, are also much higher than those of r-GO films: 282 kW/kg and 9.9 Wh/kg, respectively, compared to 70.8 kW/kg and 6.3 Wh/kg for standard r-GO lamellar film. Note, however, that these specific power values decrease sharply as a function of film thickness, indicating that C/D rates become much slower as the thickness of the film is increased.

Due to their excellent mechanical properties, these films have been used to create flexible capacitive devices, using a gel made up of a PVA (polyvinyl alcohol) sheet impregnated with PO_4H_3 as the electrolyte. Other supercapacitor systems have been designed for use in portable and flexible devices, taking the form of carbon-based fibers or papers, and are also described in the articles cited above.

3.2.5. *Pseudocapacitors*

The capacitive effect of supercapacitors is due exclusively to the formation of a double layer between the ions of the solution adsorbed on the surface of the electrode and the surface charges of the electrode material, the reason for their alternative name of EDLCs. Pseudo-capacitors differ from this model by the additional presence of a Red/Ox couple adsorbed on the surface, capable of rapidly transferring charges when the electrode potential reaches that of the oxidation-reduction reaction, Ox + ne$^-$ $\leftrightarrow$ Red.

Pseudocapacitors offer higher specific energy compared to supercapacitors, but maintain fast charge and discharge times on the condition that the redox couple is adsorbed on the surface in the form of an ultra-thin film. The aim for pseudocapacitors is, essentially, to produce new energy storage systems with the ability to deliver a high-intensity electric current, thus high power, but which also have the capacity to store higher specific energy than supercapacitors. In short, pseudocapacitors constitute a "half-way house" between rechargeable batteries and supercapacitor technology, with the capacity to deliver specific energies of between 10 and 100 Wh/kg.

At present, pseudocapacitors are generally produced using transition metal oxides; a detailed description is given by Simon, Dunn and coworkers [AUG 14].

One of the first materials, studied by Trasatti in the 1970s, is ruthenium oxide (RuO_2). Adsorbed as a thin film, RuO_2 behaves like a capacitor, but with oxidation-reduction reactions, leading to two distinct charges: q_o (outer), resulting from a surface reaction, and q_i (inner), corresponding to a deeper penetration of the material, implying two distinct time constants and earning the system its "pseudo-" qualification [ARD 90]. The capacitance, evaluated in farads per gram, is very high and can reach values of the order of 1500 F/g [AUG 14].

The prohibitive cost of ruthenium has led to the development of a range of other metal oxides. Among these, oxides of Mn (Mn_3O_4), Fe (Fe_3O_4) and Co (Co_3O_4) have been extensively evaluated as materials for use in pseudocapacitors, alongside mixed crystalline oxides such as $MnFe_2O_4$ and $NiCo_2O_4$. Again, detailed descriptions can be found in Augustyn *et al.*'s [AUG 14] widely cited article.

Carbon-oxide composites have also been considered, combining carbon nanomaterials (graphene and CNT) with metal oxides or other redox systems in various composite structures. Graphene, in the form of r-GO, features in a very large number of composites, many of which have been used in supercapacitors or pseudocapacitors, recently described in the literature [GUO 17a].

Sathiya *et al.* [SAT 11] developed a hybrid system with a pseudocapacitor made from multi-walled CNTs, covered by a thin layer of vanadium oxide (V_2O_5), for which a precise evaluation of the faradic and capacitive contributions of the charge is available (Figure 3.35).

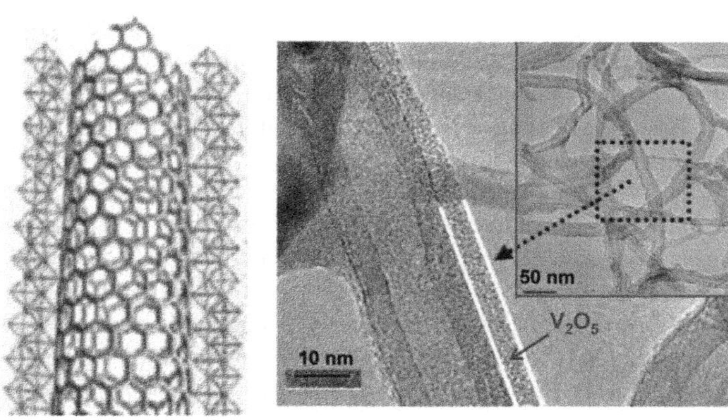

a) b)

Figure 3.35. *Pseudocapacitor made from a V_2O_5-CNT hybrid material. a) Multi-walled CNT coated with an ultrathin layer of V_2O_5. b) TEM image of a CNT covered by the V_2O_5 layer. Adapted from [SAT 11]*

By applying a linear variation of the voltage as a function of time to the electrode, it is possible to distinguish between the part of the current resulting from a surface reaction due to ionic species adsorbed on the surface of the electrode and the part of the current resulting from a volume faradic reaction, corresponding, for example, to the insertion of ions in the material. In the first case, the absence of diffusion leads to a linear variation of current intensity as a function of scan rate v ($i_c = a_1v$); in the second case, the current intensity is often controlled by an ion diffusion process, which leads to a variation of the current intensity proportional to the square root of the velocity ($i_D = a_2v^{1/2}$). The total intensity for a potential V is thus of the form $i(V) = i(v) = a_1v + a_2v^{1/2}$. a_1 and a_2 can be determined graphically by plotting $i/v^{1/2}$ as a function of $v^{1/2}$, giving straight lines of slope a_1 and y-intercept a_2, from which it is possible to deduce the capacitive and diffusional components of the current [SAT 11].

The cell used by the authors is comparable to that used in a rechargeable LIB, with lithium metal as the anode and the V_2O_5-CNT composite as the cathode, with V_2O_5 grafted onto the walls of the CNT in such a way as to form a layer of about 4–5 nm thickness[38].

Determining the capacitive and faradic contributions, we see that as the V_2O_5 content of the composite increases beyond 25%, the surface redox reactions on the CNTs become less and less important, corresponding to a decrease in capacitive contribution. Li^+ ion insertion into the V_2O_5 remains essentially constant above 25%, at 260–270 mAh/g; this implies the insertion of around 1.7 Li per mole of V_2O_5 (Figure 3.36).

For the example cited, the maximum specific capacity is around 800 mAh/g, of which the surface redox component accounts for 67% and the volume insertion reaction accounts for 33%. This proportion is reversed to 30–70% when the capacitive component becomes weaker compared to the insertion reaction.

The variable capacitive effect determines the shape of the galvanostatic discharge curves. In the case of a pseudocapacitor (ultra-thin redox layer), and for a constant current discharge rate, the decrease in potential is almost linear as a function of time, whereas in the case of a rechargeable battery, the potential variation remains almost constant before dropping off abruptly as the charge tends toward 0 (Figure 3.36a)

38 Vanadium oxy-trihydroxide [$OV(OH)_3$] is grafted onto CNTs which have been functionalized by hydroxyl groups. After several dehydration-condensation steps a thin layer of V_2O_5 is anchored on the surface of the CNTs.

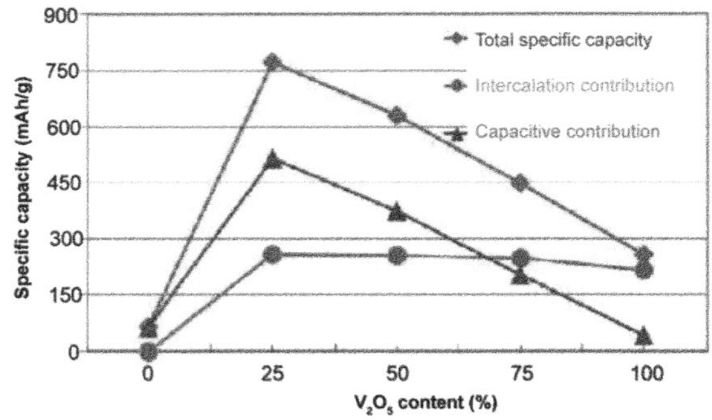

Figure 3.36. *Faradic and capacitive contributions to the specific capacity of the V_2O_5-CNT pseudocapacitor (mAh/g) as a function of the V_2O_5 content of the composite*

COMMENT ON FIGURE 3.36.– *Values calculated from cyclic voltammetric curves, obtained at a rate of 0.5 mV/s and in 1 M $LiClO_4$/EC (ethylene carbonate)-DMC (dimethyl carbonate) medium with a Li counter electrode (anode) associated with V_2O_5-CNT (cathode). Adapted from [SAT 11].*

Thus, for a given electrode charge, the available power may be considered as a function of the discharge rate, which itself depends on the structure of the redox layers adsorbed on the electrode. $LiCoO_2$, which is widely used as a cathode material in LIBs, can also exhibit capacitive behavior when the electrode is made up of very small crystallites, favorable to surface reactions (Figure 3.37a).

This may explain why some LIBs can be assimilated to pseudocapacitors with high power and low energy or, conversely, to batteries with high energy and medium power; this can be seen from a Ragonne diagram representing and comparing a number of battery and supercapacitor systems (Figure 3.37b).

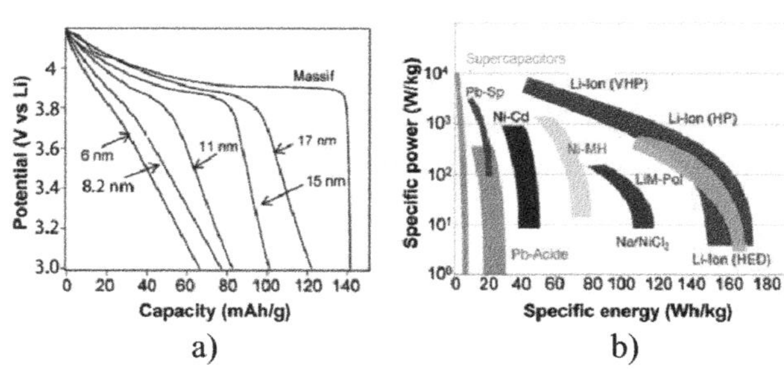

Figure 3.37. *Galvanostatic discharge characteristics of $LiCoO_2$ nanocrystallites of different sizes and Ragonne diagram of different batteries*

COMMENT ON FIGURE 3.37.– *a) For very small crystallites (6 and 8.2 nm), the discharge curves observed are those of a pseudocapacitor. The discharge rate is 1 C (0.14 A/g). b) Ragonne diagram for different electrochemical systems. Note the very wide range of properties observed in the case of LIBs, from very high specific powers (VHP) and low energy densities to high energy densities (HED), but low power. a) Adapted from [OKU 07]. b) Adapted from [IEA 11].*

3.3. Conclusions on energy storage

Energy storage is a complex issue, requiring a whole range of different solutions according to the objectives and the type of application considered. Very different constraints are encountered in different situations, from storing large quantities of energy supplied by photovoltaic or wind fields to powering all-electric vehicles or portable objects. In the first case, weight and volume limitations are of secondary importance; in electric vehicles, however, these elements are key. High powers and capacities are crucial for attaining high autonomy, requiring batteries with specific energies beyond 500 Wh/kg; furthermore, the weight of these systems after assembly should not exceed 300 kg. In the case of portable objects where high energies are not required, LIBs are particularly suitable, and have been developed and used as a reliable solution for these purposes since the 1990s. Graphite anodes and lithium insertion cathodes, mostly including transition metal oxides (Ni, Co and Mn), currently form the basis of these batteries. One new challenge is to produce all-solid state batteries with performance levels comparable to those of liquid electrolyte systems. Efforts in this area are justified by potential safety gains, including elimination of the flammability risk inherent in liquid electrolytes.

However, the increasing demand for high-energy, high-capacitance batteries for electric vehicles poses a problem in terms of supply: the production cost of lithium-based systems is likely to increase considerably over the coming years due to a dwindling, finite lithium supply. For this reason, research into new battery types, using sodium anodes and sulfur or oxygen cathodes, is crucial. Many problems remain to be solved in this area, via the development and use of new materials.

Supercapacitors also have a wide range of applications in cases where high energies are required over a short time period. The main challenge in this respect lies in improving energy density; carbon-based and organic nanomaterials are key to these efforts. Many innovations have been put forward using nanotechnologies, but have yet to be commercialized. The next step is to test the reliability of these new systems over long periods of use, and to evaluate their feasibility for large-scale, cost-effective production.

4

Energy Conversion

Today, electricity represents less than a quarter of the primary energy consumed in the world (see section 3.1). It is mostly produced by the combustion of fossil chemical compounds (natural gas, oil, coal, etc.) which, unfortunately, generate large quantities of carbon dioxide, the main cause of global warming. The main non-polluting, renewable energy sources, such as wind, hydro and photovoltaic, currently represent only a small proportion of total electrical energy production; rapid development of these technologies is crucial in enabling a transition toward totally decarbonized energy production by the end of this century. One key element of this development concerns energy storage systems, which must be sufficiently efficient to respond to the intrinsically intermittent nature of wind and solar power. Reducing electricity use is another important means of reducing CO_2 emissions. Urban and residential lighting accounts for a significant proportion of total consumption, and can be reduced by systematically replacing incandescent or electric discharge lighting with more efficient light-emitting diodes (LEDs). A final important area of research concerns the use of hydrogen as an energy source: hydrogen is easy to store and isolate in chemical form, and can then be restituted in the form of electrical energy in a fuel cell by means of water splitting (via electrolysis or photolysis). Three main forms of energy conversion – photovoltaics, electroluminescence and electrolysis – will be described in this chapter. The efficiency of these transformation processes depends on the nature of the materials involved; as we will see, the majority of recent material innovations are the result of the introduction of new, composite nanostructures.

For a color version of all the figures in this chapter, see: www.iste.co.uk/lacaze/nanotechnology.zip.

4.1. Photovoltaics

An inexhaustible source of natural power, the sun provides an average energy estimated at 1.22×10^{17} joules (or 122000 TW) to the Earth's surface every second. Predicated primary energy consumption for 2050 is not expected to exceed 2.3×10^{13} J per second (23 TW), that is, over 5300 times less than the energy provided by the Sun – a source expected to last for over 5 billion years[1]. The transformation of light energy into electrical energy is not a new idea; in 1839, Edmond Becquerel was the first to observe an increase in electrical current under illumination between electrodes immersed in an electrolyte. Much later, in 1954, Chapin, Fuller and Pearson, at Bell Laboratories in the United States, observed a notable photovoltaic effect (energy yield of about 6%) using a silicon *p-n* junction.

Silicon-based photovoltaics have been progressing steadily ever since: silicon is the main material used in domestic solar panels for housing and for photovoltaic farms. Progress is required mainly in terms of manufacturing technologies, with the aim of lowering industrial production costs. These costs remain high essentially because of the energy needed to transform silica into silicon. For this reason, new processes based on the use of new mineral, organic or hybrid active materials are constantly being put forward. Thin-film "second-generation" devices, using amorphous silicon (a-Si), CIGS (copper-indium-gallium-selenium) or CdTe (cadmium-telluride) semiconductor (SC) alloys, are currently being developed and may be used in specific applications. The energy yields of CIGS and CdTe are higher than those of amorphous silicon, but they use materials which are rare (Te) or toxic (Cd). High-efficiency cells have also been designed using GaAs, and are mainly used for aerospace equipment, where performance is key and cost is a lesser consideration.

Organic semiconductor materials offer another pathway for investigation. They are particularly numerous and can be obtained using techniques which require less energy than metallurgical transformations. Long limited by energy yields not exceeding 10%, recent advances have enabled organic materials to perform at levels above this threshold, meaning that the technology is now viable as an alternative to poly-Si cells. The last, and most recent, approach concerns the use of perovskite photovoltaic cells, organo-mineral hybrid materials that are proving to be extremely efficient. Since their discovery in 2009, perovskites have been used in an increasing

1 Assuming average illumination of 12 hours per day and a 30% loss in radiation due to crossing the atmosphere, considering an average inclination of the rays with respect to the equator which reduces the initial AM0 (Air Mass 0) radiation of 1,336 W/m^2 to an average value of 684 W/m^2, we obtain an average effective radiation of 240 W/m^2. Given the total surface of the Earth (510 million km^2), this represents an average power of more than 122000 TW ($1.22*10^{17}$ W).

range of devices, achieving energy yields approaching those of monocrystalline silicon. However, their long-term reliability still needs to be proved, and significant progress is needed before they can be marketed as a genuine alternative to monocrystalline silicon.

4.1.1. *General principles of the photovoltaic process*

A semiconductor of band gap E_g, irradiated by light of frequency ν such that hν > E_g, produces an exciton corresponding to the formation of a hole and an electron, located, respectively and after thermalization, in the top of the valence band (VB) and the bottom of the conduction band (CB) (Figure 4.1a).

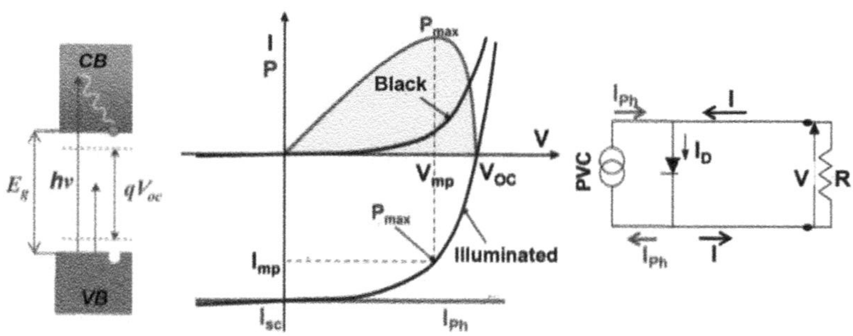

Figure 4.1. *Operating principle of a photovoltaic cell*

COMMENT ON FIGURE 4.1.– *a) Production of an electron-hole pair in a semiconductor of band gap E_g by irradiation of the SC with photons of energy hν > E_g. The hole (white dot) is at the top of the valence band (VB); the electron (ochre dot) is, after thermalization, located at the bottom of the conduction band (CB). The open circuit voltage (V_{OC}) is such that qV_{OC} is slightly less than E_g. b) Curves I = f(V) and P = f(V) of a p-n junction in the dark and under illumination. $P_{max} = I_{mp} \times V_{mp}$ is the maximum power point, defining the optimal operating conditions of the photovoltaic cell. c) Circuit equivalent to a photovoltaic cell connected to a load resistor R and assimilated to a current generator (PVC, photovoltaic cell) flowing in a p-n junction. a) Adapted from [POL 12]. b) Adapted from [JÄG 14].*

The recombination of the electron and the hole inside the SC can be prevented by choosing a *p-n* junction consisting of the assembly of two *n*- and *p*-doped semiconductor layers, which has the effect of creating an electric field at the interface. It is this permanent electric field which is at the origin of the diode effect of the *p-n* junction and means that the current of the diode can be effective only for a particular direction of polarization. In the dark, the current intensity I_D is given by the relation

$$I_D = I_{D,S}\,[\exp(qV/kT) - 1]^2,$$

where V is the voltage applied to the *p-n* junction, q the charge of the electron, k the Boltzmann constant and T the temperature. I_{DS} is the saturation current, which is opposite to the direct current and of very low intensity. The diode is said to pass or block, respectively, when the polarization voltage applied to the SC *p* is direct (voltage V applied to the positive SC *p*) or reverse (negative voltage V).

Under illumination, a photocurrent I_{Ph} is produced, which is constant and of opposite sign to the diode current. The total current I is thus given by the relation

$$I = I_{D,S}\,[\exp(qV/kT) - 1] - I_{Ph}.$$

From this formula, the open circuit voltage V_{OC} (I = 0) of the photovoltaic cell can be deduced as a function of I_{DS} and I_{Ph}:

$$V_{OC} = (kT/q)\,[1 + \mathrm{Log}(I_{Ph}/I_{D,S})]$$

At zero current, the V_{OC} is maximum, while at zero voltage (V = 0), the current is maximum and is known as the short circuit current, I_{SC} (Figure 3.30b). From the curves I = f(V), a fill factor (FF) and power conversion efficiency (PCE) can be defined such that

$$FF = I_{mp}\,V_{mp}/I_{sc}\,V_{oc}$$

$$PCE\,(\%) = (P_{mp}/P_{in}) \times 100 = (I_{sc} \times V_{oc} \times FF/P_{in}) \times 100,$$

---

2 An *n-doped* semiconductor (SC) corresponds to an intrinsic SC (*i-SC*) such as silicon (of valence IV), in which a fraction of the Si atoms has been replaced by atoms of valence V (phosphorus, arsenic). This insertion has the effect of transforming As into As^+ and releasing an electron, thus increasing the concentration of free electrons in the SC, while the whole remains neutral. In the same way, a *p*-doped SC is obtained by substituting some silicon atoms with valence III elements (boron) which are transformed into B^- and create holes (positive charges) in the SC, corresponding to a decrease in the concentration of free electrons in the SC. When the two SCs, *n* and *p*, are connected, a space charge is created, resulting from the diffusion of the electrons of the SC *n* toward the SC *p*. This results in the creation of a thin, double ionic layer at the interface along with a permanent electric field, directed from the positive charges toward the negative charges, located on the separation surfaces of the *n* and *p* SCs, respectively.

where P_{mp} is the maximum power obtained for I_{mp} and V_{mp}, and P_{in} is the power of the incoming light flux.

It is then possible to estimate the maximum values of current which can be obtained, starting from the spectrum of solar energy received on the Earth's surface, which extends from 280 to 4000 nm and which corresponds to a photon flux with an average power of 100 mW/cm^2 (Figure 4.2a)[3]. Thus, considering that the absorption of a photon results in an electron and a hole, used to produce a current, the maximum value of the intensity for a semiconductor characterized by a particular value of E_g can be deduced (Figure 4.2b). For a silicon photovoltaic cell, with an E_g of around 1.2 eV, the number of photons absorbed between 280 and 1000 nm (1000 nm corresponds to the E_g of silicon) represents 53% of the total flux radiated by the Sun, from which a maximum current of 33.9 mA/cm^2 and a maximum energy efficiency of around 30% are deduced; this latter figure is known as the Shockley-Queisser (SQ) limit.

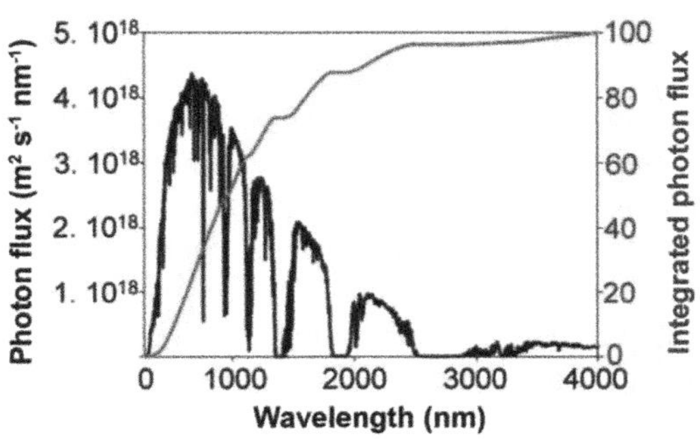

a)

λ(nm)	% photons	J(mA/cm2)
500	8	5.1
600	17.3	11.1
650	22.4	14.3
700	27.6	17.6
800	37.3	23.8
900	46.7	29.8
1000	53.0	33.9
1250	68.7	43.9
1500	75.0	47.9

b)

Figure 4.2. *Photon flux of the solar spectrum and maximum values of photovoltaic currents as a function of E_g (evaluated in nm; 1,240 nm is equivalent to 1 eV)*

COMMENT ON FIGURE 4.2.– *a) Photon flux as a function of wavelength (black curve) and integrated photon flux (red curve) from 280 to 4000 nm. This solar spectrum is typical for AM 1.5 irradiance. The hollows observed in the spectrum correspond to absorption by the main atmospheric gases (O_2, O_3, H_2O). b) Maximum theoretical values of photovoltaic currents for different semiconductors with E_g between 2.48 eV*

3 The solar energy outside the Earth's atmosphere is 136.6 mW/cm^2 and corresponds to an irradiation noted AM 0; after crossing the atmospheric layer, it is only around 100 mW/cm^2 at the equator (irradiation AM 1). For the northern latitudes of Europe and North America, the energy losses due to the atmospheric absorption are greater and correspond to an irradiation noted AM 1.5, due to an inclination of the Sun at 45° above the horizon (1.5 = 1/cos45) [BUN 07].

(500 nm equivalent) and 0.83 eV (1500 nm equivalent). Inset in red: the position of crystalline Si, for which E_g ~1.2 eV (1000 nm equivalent). Adapted from [BUN 07].

4.1.2. *Photovoltaic technologies*

4.1.2.1. *Current photovoltaic cells*

Ninety percent of the current photovoltaic market is occupied by silicon cells, with an energy yield (and cost) dependent on the quality of the silicon used. Monocrystalline silicon (c-Si, accounting for around 35% of the market in 2014), with an E_g of 1.12 eV, gives the highest energy yields (around 22% for large modules). Unfortunately, production is dependent on the expensive Czochralski method of ingot formation, in which silicon is produced from silica at very high temperatures (1700°C) and then cut into wafers. Polycrystalline silicon, also known as poly-Si (55% of the market share), with an equivalent E_g of 1.7 eV, is obtained using a simpler metallurgical process and is consequently cheaper than c-Si; however, its energy yields are lower (18% for large modules) [JEA 15]. Another disadvantage of silicon is its low light absorbency, which requires the use of thicknesses of around 100 μm, increasing the cost of the photovoltaic panels. This is one of the main reasons for the development of the thin-film photovoltaic cell industry, representing the remaining 10% of the photovoltaic market. This technology uses SCs characterized by higher light absorption coefficients than crystalline Si, such as hydrogenated amorphous silicon and certain multinary combinations like gallium arsenide (GaAs, a combination of elements from columns III and V of the periodic table), Cd telluride (CdTe, combination of II-VI elements), copper indium diselenide (CIS, combination of elements I-III-VI) or CIGS, in which part of the indium of CIS is replaced by gallium. Their energy yields are over 20% in the case of simple *p-n* junctions; the highest yields are obtained with GaAs (29% in laboratory conditions and 24% on a module scale). However, its high cost limits its use to specific applications, for example, in spacecraft. The detailed characteristics of these different systems (yields, V_{OC}, I_{SC}, FF, etc.) along with manufacturer names can be obtained by consulting the tables periodically published by Green *et al.* [GRE 20].

The processes used to manufacture these materials are less energy-consuming than those for mono- and polycrystalline Si, and the thicknesses produced are significantly lower (from 1 to a few μm). On the other hand, the conversion yields are much lower than those of c-Si. Hydrogenated amorphous Si (a-Si:H), obtained by plasma-enhanced chemical vapor deposition of SiH_4 (PE-CVD) at temperatures of around 200°C, is a relatively cheap product, but its energy yields are low, ranging from 5% to 7% in module form[4]. Other materials such as CdTe, CIS and CISG are also produced by vapor deposition. In the case of CdTe, the thickness of the active layers is around 5 µm and yields are between 9% and 17%, with an average value of 13% in module form. Similar values are obtained for CIGS cells, for which the energy yields are between 11% and 18%, with material thicknesses of around 1.5–3.5 µm. All of these systems are produced commercially by different companies, and have undergone considerable developments in terms of manufacturing, always with the aim of obtaining the best possible balance of cost and efficiency (Figure 4.3).

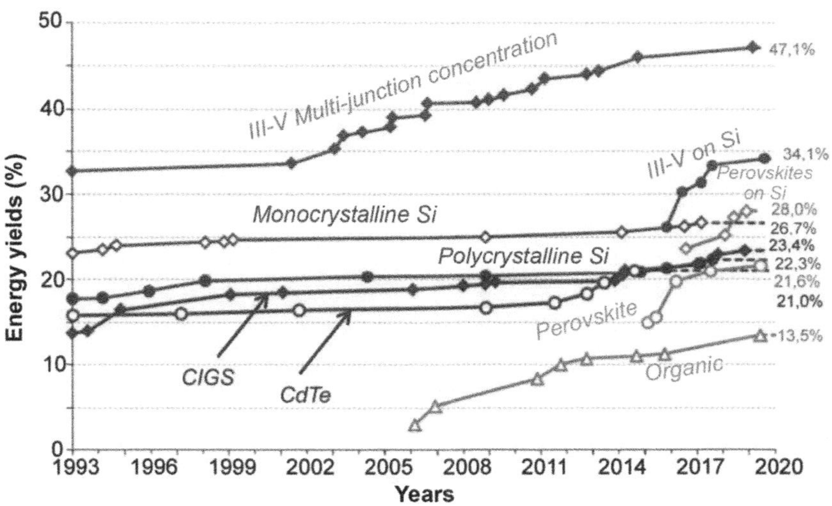

Figure 4.3. *Evolution of energy yields of major photovoltaic technologies over time (laboratory conditions). Adapted from [FRA 20]*

This evolution, the result of successive innovations stretching over a period from the 1960s to the present day, will not be described in detail here. Advances and technological innovations in inorganic photovoltaic cells using silicon or the aforementioned materials are described in a recent collaborative paper published by specialists from 20 different laboratories [RAM 16]. Similarly, recent reports by the ITRPV (International Technology Roadmap for Photovoltaic) describe the prospects for improving manufacturing processes and the objectives of the silicon industry for

4 A 300 nm thick a-Si:H film absorbs 90% of the solar spectrum between 280 and 730 nm (730 nm is equivalent to an E_g of 1.7 eV), and can be used to make flexible and lightweight photovoltaic cells. However, the market share of amorphous Si photovoltaic cells is very small, around 2%; they account for 22% of the thin-film cell market [JEA 15].

the period up to 2028. These documents notably recommend the development of 72-cell modules, delivering 390 Wp (Watt peak) with polycrystalline Si, and 415 Wp with *p-type* monocrystalline Si [INT 18]. Figure 4.4 shows a typical model of a single cell, a photovoltaic module and of a solar panel made up of three photovoltaic modules.

The model shown below consists of 36 photovoltaic cells, arranged in four strings of nine cells. Depending on the way the cells are connected to each other – for example, as series of two strings, connected in parallel – the output voltage and the output current of the module will be 18 times and two times that of a single cell, respectively[5].

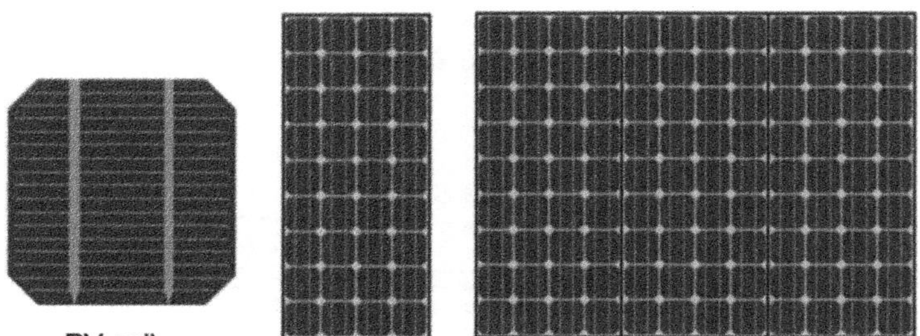

Figure 4.4. *Front view of a photovoltaic cell, module and solar panel assemblies*

COMMENT ON FIGURE 4.4.– *The cell has a surface of around 140 cm² in size; the two wide vertical strips (in gray) are the metallic conduction lines that collect the current supplied by the whole cell, distributed along the thinner, horizontal metal lines. From [JÄG 14].*

In the following section, we shall describe emerging technologies involving various nanotechnologies or nanomaterials. The new technologies, corresponding to dyed or dye-free organic photovoltaic (OPV) cells, or, more recently, using perovskites, constitute significant advances. The latest cells have certified yields that are close to those of the dominant silicon-based technology. As we will see, each

5 For example, with solar cells of about 140 cm^2, each delivering an I_{SC} current of 5 A with a V_{OC} voltage of 0.6 V, a photovoltaic module with 36 solar cells in which two strings of 18 cells (connected in series) are connected in parallel will be characterized by a V_{OC} of 10.8 V and a short-circuit current of 10 A, that is, a maximum power of 108 W. The same theoretical power is also obtained in the case of an all-series connection (21.6 V × 5 A).

form presents its own advantages and disadvantages, making them suitable for different fields of application.

4.1.2.2. *Organic photovoltaic cells*

Research in this field is much more recent than that concerning the development of silicon cells. Cost issues are a major motivation, along with the search for flexible systems for use in various portable electronic devices, deposited, for example, on plastic, paper or textile supports. The first organic cell was produced by Tang at the Eastman Kodak laboratories in New York in the 1980s. This invention had a high impact, despite low conversion efficiency (<1%). It triggered an upsurge in interest, leading active research still ongoing today. Huge progress has been made in recent years: the 10% threshold was first breached in 2016, and 15% yields were obtained in 2019.

Tang's proposed system is very simple. It consists of a heterojunction between two organic materials: an electron donor and an electron acceptor (the equivalent of a *p-n* junction), arranged in ultra-thin layers between two ITO (indium tin oxide) and Ag electrodes (Figure 4.5).

For the system to function, a photoactive material must be present to absorb photons and produce excitons (electron-hole couples) which then diffuse toward the donor-acceptor (D-A) interface. Charge separation occurs at the interface due to the presence of an electric field, resulting from differences between the energy levels of the two materials. Materials are selected using two criteria: (1) the energy difference between the HOMO (highest occupied molecular orbital) and the LUMO (lowest unoccupied molecular orbital) of the donor must be small enough to absorb a maximum of photons from the solar spectrum; (2) the energy level of the donor LUMO must be higher than that of the acceptor LUMO, so that the transfer of electrons from the donor LUMO can take place without an energy barrier. Electrons and holes are collected by a silver electrode and a transparent ITO electrode, respectively. These also differ in their fermi levels, and condition the injection of holes by ITO and the collection of electrons by Ag.

The low energy efficiency of the cell is largely due to the short exciton diffusion distances (under ~10 nm). This results in a large number of electron-hole recombinations well before the D-A separation surface is reached, leading to low electrical current yields. It also constitutes a key difference between this technology and that of inorganic SCs, in which exciton dissociation takes place immediately and leads directly to a charge separation, followed by the injection of the electron into the CB and injection of the positive charge (hole) into the VB. The other disadvantage of this system compared to inorganic SCs lies in its low charge mobilities and conductivities, along with the difficulty inherent in balancing the conductivities of the electrons and holes in the acceptor and donor materials.

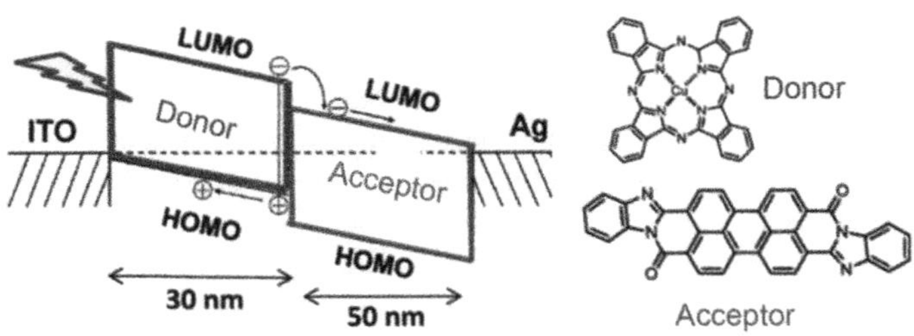

Figure 4.5. *Simplified energy diagram of an organic photovoltaic cell consisting of a heterojunction between an electron donor material (copper phthalocyanine) and an electron acceptor material (perylene-3,4:9,10-bis(benzimidazole) – PTCBI)*

COMMENT ON FIGURE 4.5.– *Note that LUMO (lowest unoccupied molecular orbital) and HOMO (highest occupied molecular orbital) are equivalent, respectively, to the CB and VB of an SC. The alignment of the fermi levels of the ITO and Ag electrodes corresponds to the short-circuit current of the cell. Adapted from [TAN 86].*

Tang's first findings [TAN 86] provided the trigger for a large number of subsequent studies. A first improvement in yield was obtained by the choice of other *n* and *p* organic materials; in the 1990s, still greater progress was made by Sariciftci *et al.* [SAR 92], Ma *et al.* [MA 05] and Halls *et al.* [HAL 95][6], who replaced the bilayer heterojunction with a bulk heterojunction (BHJ), obtained by finely mixing acceptor (A) and donor (D) materials.

In this type of structure, the donor and acceptor are finely mixed, meaning that the D-A interface is deployed throughout the full volume of the material: this means that excitons are generated at a distance from the interface which is compatible with their diffusion range (around 10 nm). This increases the rate of conversion of excitons into separate electrons and holes, which then diffuse within the respective A and D domains and are collected by the electrodes deposited on either side of the organic layer (Figure 4.6).

---

6 Volumetric heterojunctions have been produced by mixing poly[2-methoxy-5-(2-ethyl-hexyloxy)-1,4-phenylene-vinylene] (MEH-PPV), a *p*-type electron-donor polymer, with fullerenes (*n*-type electron acceptors), or by using interpenetrating arrays of *p*- and *n*-polymers (MEH-PPV and CN-PPV).

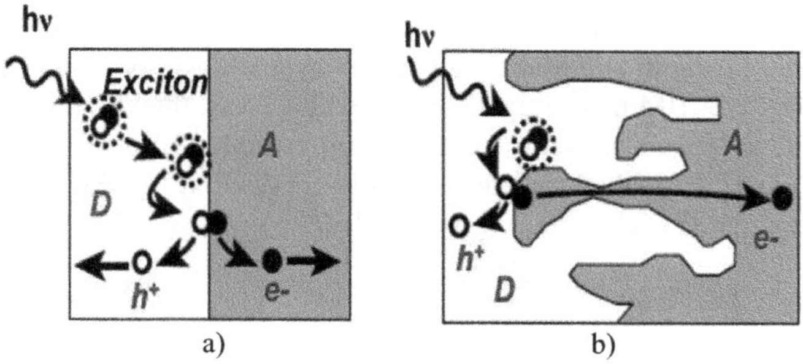

Figure 4.6. *Exciton dissociation mechanisms in different types of heterojunctions*

COMMENT ON FIGURE 4.6.– *a) Planar bilayer heterojunction between an electron donor (D) and acceptor (A). b) Bulk heterojunction (BHJ). In the first case (a), an exciton formed at a distance from the interface greater than its diffusion length cannot reach the interface and does not contribute to the production of an electric current. In the second case (b), the greater proximity of excitons to the donor-acceptor interface statistically leads to more charge separations and thus more current. Adapted from [PEU 03].*

This volume heterojunction principle, alongside new organic materials with higher electron and hole conductivities and mobilities, resulted in significant progress. In 2009, Heeger and coworkers [PAR 09] reported a conversion record of 6% using a polymeric SC composed of thiophene, carbazole and thiadiazole groups (PCDTBT) as the *p*-type material and a C_{70} fullerene derivative (phenyl C_{70} butyric methyl ester-$PC_{71}BM$) as the *n*-type material. Although the result was considered important at the time, the operating characteristics were modest: the short-circuit current I_{SC} was only 10.9 mA/cm^2, with an open-circuit voltage (V_{OC}) of 0.88 V and an FF of 0.66 at an illumination of 100 mW/cm^2 [PAR 09]. The authors highlighted the fact that the degree of nanostructuring of the active layer is decisive for obtaining a high charge separation rate from the excitons. This implies specific proportions between A and D, and the choice of a suitable solvent to optimize the homogeneity of the solid A-D thin film obtained after solvent evaporation[7].

7 The best charge separation efficiency (IQE – internal quantum efficiency), close to 100%, is obtained for a 1:4 mixture of PCDTBT and $PC_{71}BM$, solubilized in dichlorobenzene. A nanostructured photoactive layer is obtained after deposition of a small amount of the solution containing the PCDTBT and $PC_{71}BM$ mixture on the ITO/PEDOT electrode, evaporation of the solvent and then annealing of the formed layer at low temperature.

Compared to Tang's bilayer junction, the BHJ cell described by [PAR 09] has two additional layers in addition to the active layer (thickness 80 nm) formed from the PCDTBT : $PC_{71}BM$ mixture: a titanium oxide (TiOx, thickness 10 nm) electron-conducting layer in contact with the aluminum cathode (10 nm), and a conductive polymer (PEDOT:PSS, thickness 40 nm) hole-conducting layer in contact with the anode (transparent ITO electrode) (Figure 4.7).

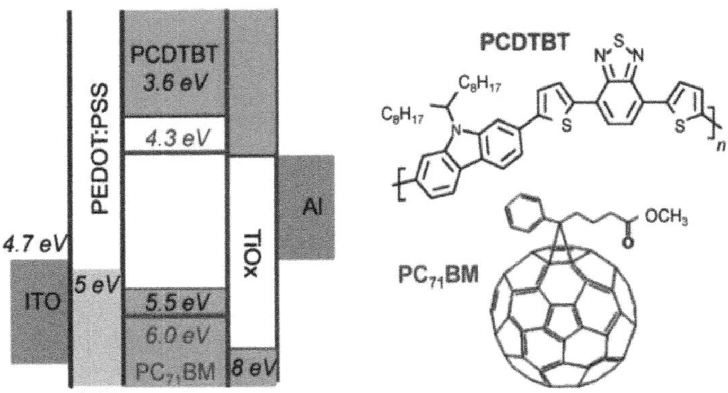

Figure 4.7. *Energy diagram of the BHJ photovoltaic cell with 6% efficiency. The active material consists of a composite layer (80 nm) of PCDTBT (donor, green) and $PC_{71}BM$ (acceptor, HOMO and LUMO in blue)*

COMMENT ON FIGURE 4.7.– *Note the very good relative arrangement of the characteristic energy levels of each of the constituent materials of the cell (negative with respect to the vacuum level 0). The LUMO of PCDTBT (–3.6 eV), located above the energy level of the LUMO of $PC_{71}BM$ (–4.3 eV), itself aligned with the bottom of the CB of TiOx and the fermi level of aluminum, favor the flow of electronic charges toward the aluminum cathode. Conversely, the HOMO of PCDTBT (–5.5 eV) which contains the positive charges (holes) from exciton dissociation is favorably positioned with respect to the fermi levels of PEDOT:PSS (–5 eV) and ITO (–4.7 eV). This arrangement favors the flow of positive charges and electrons to the ITO and Al electrodes, respectively. (All energies are measured against the 0 energy of vacuum and are negative.) Adapted from [PAR 09].*

Optimal operation of the cell is ensured by careful choice of the HOMO and LUMO energy levels of the donor and acceptor materials. The need to block the diffusion of holes toward the Al cathode justifies the addition of a thin film material (10 nm) such as titanium oxide (TiO_2), characterized by a very high ionization potential (7.2 eV); in addition to this hole-blocking function, it also acts as an optical spacer and prevents the formation of parasitic interferences. Similarly, the deposition of a conductive layer of poly (3,4-ethylene dioxythiophene): (polystyrene sulfonate), (PEDOT:PSS, 40 nm), on the transparent ITO electrode improves the electrical contact between the active layer and the ITO anode.

With the exception of aluminum, which is deposited by vacuum evaporation in the final stage, all other compounds (PEDOT:PSS, PCDTBT:PC70BM, TiOx) are deposited on the ITO by a wet process in successive layers, followed by solvent evaporation and air annealing, the latter performed at low temperatures, in the range of 120–140°C; this represents a clear gain in terms of production costs compared to silicon.

Building on these first, promising results, extensive research has been carried out in the aim of improving OPV function. Approaches include varying the nature of the donor compounds (polymeric and molecular), while, most often, retaining fullerene derivatives (C_{70} and derivatives) as the acceptor materials. Similarly, particular attention has been paid to improving the homogeneity of the nanostructuring of the active BHJ layers, by the choice of specific solvents and additives [HEE 14]. Nevertheless, despite all these improvements in photovoltaic technology, the energy yields of BHJ cells remained below the 10% threshold until 2014 [GUO 14, DIA 15, WAD 19, YIN 20].

The reason for using fullerenes in OPVs is that they are strong electron acceptors and that they can be used in the formation of nanoaggregates, perfectly suited for fabricating BHJs with a donor material. However, the very high symmetry of fullerenes' molecular orbitals means that they absorb virtually nothing in the UV-visible spectrum. They simply conduct the electron current resulting from the ionization of the *p*-donor. This is the main reason for replacing C_{60} with $PC_{71}BM$, which has a symmetry that allows a higher number of permissible optical transitions. Despite this progress, absorption remains very low in the visible UV range, and electron mobility is lower (around 10^{-3} cm^2/Vs, on average) than that of symmetrical fullerenes such as C_{60} [ZHA 18d]. Another disadvantage of the use of fullerenes is their tendency to aggregate and form microdomains, hampering the exciton dissociation mechanism, over time [WAD 19].

As the use of fullerenes such as $PC_{71}BM$ did not increase yields beyond 10%, other molecular acceptor compounds have been considered for producing more efficient BHJs in terms of optical absorption, while offering higher conductivities and mobilities with respect to electrons. A series of fullerene-free compounds of ternary structure A-D-A (where A and D are acceptor and donor, respectively), known as non-fullerene acceptors (NFA), has been synthesized and tested in BHJ photovoltaic cells [YU 14] (Figure 4.8).

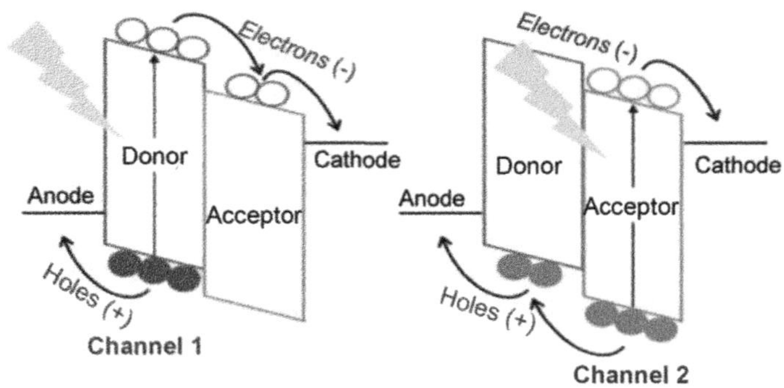

a) b)

Figure 4.8. *Photogeneration of electrons and holes in a non-fullerene BHJ OPV. Charges induced in the (a) donor (channel 1) and (b) acceptor (channel 2). Adapted from [WAD 19]*

The alternation of the acceptor (A) and donor (D) groups, which are unsaturated and conjugated, results in an overall electron acceptor compound with an intense absorption spectrum; depending on the choice of A and D, this may cover a spectral region distinct from that of the donor in the BHJ. Thus, instead of having the single light absorption channel resulting from the presence of the donor in fullerene-based BHJs with fullerenes, additional contributions are obtained in certain NFAs, via a second absorption channel which is characteristic of the NFA absorption spectrum [QIU 17].

Some of these NFAs have proven to be particularly successful, achieving and exceeding the 10% energy yield threshold as early as 2016. Fluorination of the compounds used in BHJs has also been shown to be an effective way to increase energy yields. The Hou group, in China, showed that after fluorination of the donor and acceptor compounds, the energy yield of OPV cells could be increased from 11% to more than 13%, using the fluorinated polymer PBDB-T-SF as the donor and IT-4F as the NFA for the BHJ [ZHA 17f] (Figure 4.9)[8].

8 Fluorination has the effect of simultaneously lowering the HOMO and LUMO energy levels of the compounds and improving charge transport due to better crystallinity of the materials resulting from F...H and S...F interactions. It also increases the polarization of the molecules and lowers the coulombic potential between positive and negative charges, and sometimes leads to stronger light absorption [ZHA 17f].

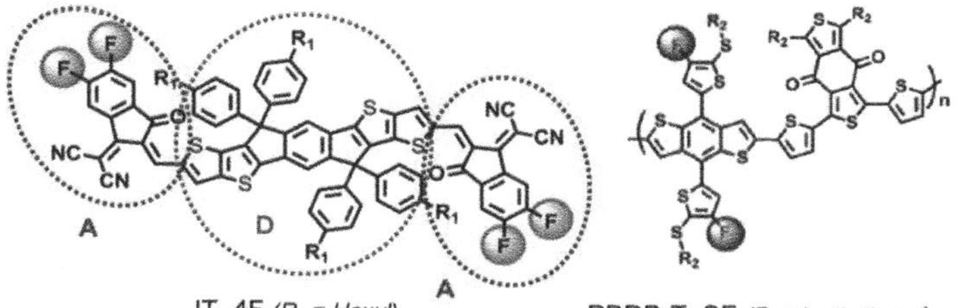

Figure 4.9. *Fluorinated acceptor (IT-4F) and donor (PBDB-T-SF polymer) materials used in the fabrication of BHJ, obtained after fluorination of ITIC (IT-4F compound without fluorine atoms) and PBDB-T (PBDB-T-SF without fluorine) compounds. Adapted from [ZHA 17f]*

Recently, the same group further improved the operation of these OPVs by replacing the PBDB-T-SF donor with a new PTO2 donor (Figure 4.10c), leading to a further increase in energy yields from 13% to 14.7% [YAO 19]. The authors give two main explanations for these high yields. First, the two compounds have complementary absorption domains (Figure 4.9b), resulting in a significant broadening of the absorption spectrum; second, the formation of a charge-transfer complex between the two compounds induces a strong local electrostatic field, favoring the dissociation of excitons and increasing the charge separation efficiency. Theoretical calculations for the BHJs [PTO2:PBDB-T-SF] and [PTO2:PC71BM] show a stronger electrostatic-type interaction between donor and acceptor for [PTO2:PBDB-T-SF], resulting in a strong intermolecular electric field between PTO2 and PBDB-T-SF.

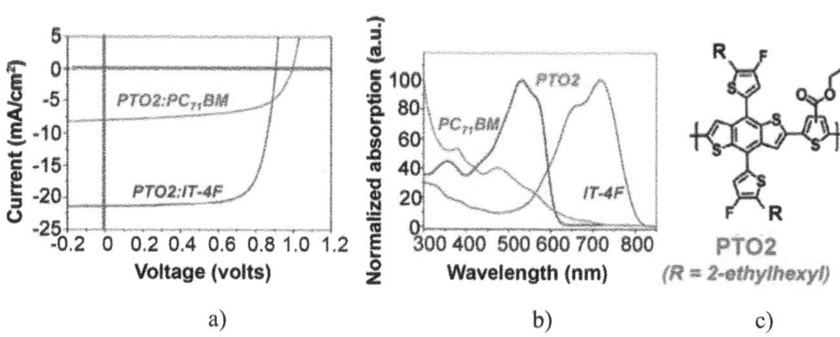

Figure 4.10. *Performance of photovoltaic cells made with two different BHJs containing the same PTO2 donor, but differing in the nature of the acceptor: one with Fullerene (PTO2:PC$_{71}$BM), the other using IT-4F as an NFA (PTO2:IT-4F)*

COMMENT ON FIGURE 4.10.– *a) Current-voltage characteristics of photovoltaic cells. b) Normalized absorption spectra of the donor compound PTO2 and the acceptor compounds IT-4F (NFA) and $PC_{71}BM$. Note the good complementarity of the IT-4F and PTO2 spectra. c) Structure of the donor polymer PTO2. Adapted from [YAO 19].*

Experimental results confirm this idea of a higher charge separation in the first case, resulting in a high short-circuit current of 21.5 mA/cm^2, an open-circuit V_{OC} voltage of 0.91 V, a fill factor FF of 0.75 and a PCE of 14.7%. In the case of the BHJ [PTO2:$PC_{71}BM$], the short circuit current is much lower (8.1 mA/cm^2), as is the fill factor FF (0.62), giving an energy efficiency of only 5% despite a high V_{OC} voltage of almost 1 V (Figure 4.10a).

Recently, a theoretical study was carried out by a group of researchers from six different laboratories with the aim of defining structural criteria likely to optimize the value of the open circuit potential of cells. Unlike inorganic cells (Si type), the open circuit potential of OPVs is much lower than the E_g value of the material, due to non-radiative energy losses produced during charge separation [QIA 18a]. The authors cite two main conditions which must be respected in order to minimize the energy loss: (1) the donor and acceptor materials must be characterized by energy levels which are close to each other; and (2) the photo-active material selected for the BHJ must have a low E_g and very high photoluminescent efficiency.

Many works now describe high energy yields of well over 15% obtained by an increasingly diversified choice of donor and acceptor materials. Improved nanostructuring of donor and acceptor phases still presents an important challenge, requiring innovation in BHJ design and manufacture. In many cases, binary D-A combinations have given way to ternary and quaternary combinations of materials, aimed at improving the nanostructure of the active layers, a crucial factor in obtaining high yields. The addition of certain additives to the active layer has proved to be beneficial in charge separation, leading either to an improvement of the V_{OC} or to an increase in the short-circuit current. The use of buffer layers between the photo-active layer and the electrodes, improving the transport of positive and negative charges, has also contributed to increased yields, while the arrangement of the layers in "tandem" OPV cells contributes to increased energy efficiency due to the addition of the V_{OC}s. Meng *et al.* [MEN 18] achieved yields of over 17%, bringing OPV cells closer to other thin-film photovoltaic systems; however, the construction of these cells is more complex (Figure 4.11).

A tandem cell consists of two OPV cells. The front OPV cell (BHJ-f), which receives the light, contains an active layer consisting of a mixture of the PBDB-T donor and the F-M acceptor; it is in contact with the transparent ITO electrode via an electron conductor (PFN-Br) and the ZnO semiconductor, which corresponds to a reverse OPV setup (electron flow toward the ITO) [WAN 16b]. This front OPV cell is connected via a PEDOT:PSS conductor to the rear OPV cell (BHJ-r), consisting of the ternary mixture [$PC_{71}BM$: O6T-4F: PTB7-Th]. The front OPV absorbs in the region from 300 to 750 nm, and the second OPV absorbs in a complementary region from 750 to 1050 nm. The connection of the two OPVs thus results into the addition of two V_{OC}s; this is the reason for the high overall V_{oc} of 1.62 V and for the high efficiency of cells of this type.

The final point to address concerning the future development of OPV cells is their lifetime which, at present, struggles to compete with that of silicon photovoltaic cells, currently guaranteed to operate over periods in excess of 25 years. They are not currently suitable for use in large-dimensional panels intended for static storage of solar energy. At this time, the operation of these materials has not been tested over periods of more than 1 year; nevertheless, current results suggest an operating lifetime of around 5 years, and this may be sufficient for use in low-power electronic devices, notably connected portable objects (IoT – Internet of Things) [YUN 18].

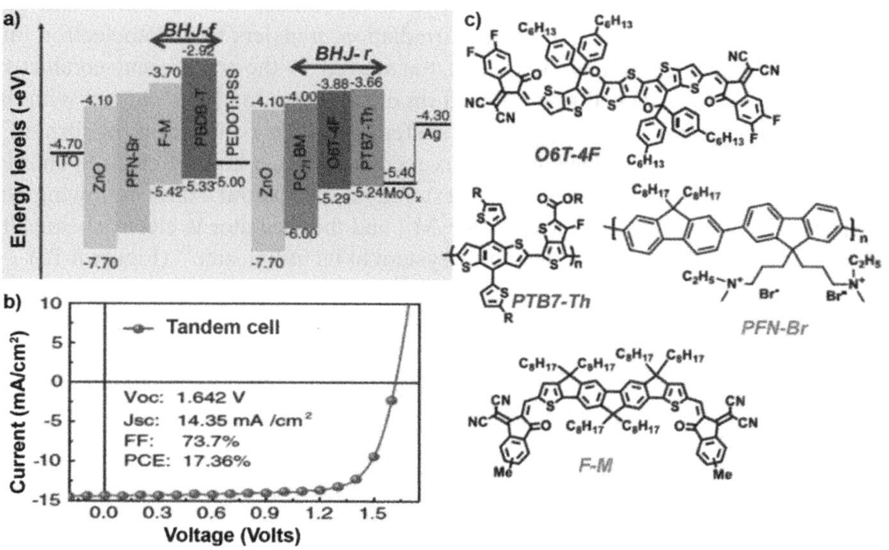

Figure 4.11. *Characteristics of a tandem OPV cell consisting of two BHJs with complementary optical absorption domains (300–700 nm for the front cell BHJ-f and 750–1050 nm for the rear cell BHJ-r)*

COMMENT ON FIGURE 4.11.– *a) Energy levels of the different compounds in the tandem OPV cell. b) Current-voltage curve for the tandem cell. c) Structures of the constituent materials of the two donor-acceptor BHJs. Adapted from [MEN 18].*

One of the major advantages of OPV cells compared to Si cells and other inorganic materials is the very high levels of light absorption, which allow the use of very thin active layers (thickness <100 nm), permitting their use in flexible devices. Another notable advantage of OPV cells is the almost infinite number of options for customizing their opto-electronic properties, via the functionalization of various organic compounds. Finally, the fact that organic layers are manufactured using the wet-coating process means that they may be easily produced on an industrial scale and at low cost, using an R2R (roll-to-roll) type printing process. It is also important to note that these systems are particularly efficient in indoor environments, a further factor in support of their use in portable objects[9].

4.1.2.3. *Dye-sensitized solar cells*

The concept of dye-sensitized solar cells (DSSCs) was first invented by O'Regan and Grätzel in the very early 1990s [ORE 91], based on the operation of a photoelectrochemical cell in a liquid electrolyte medium. The innovative aspect of the cell lies in the use of an organic dye S in contact with a wide band gap semiconductor (TiO_2), which, after light irradiation, transfers the photoelectron into the CB of the semiconductor. It is then transferred to the transparent conductive electrode, of ITO or FTO (fluorine-doped tin oxide) type, in ohmic contact with the semiconductor. Following this electron transfer, the dye is oxidized in the form of a radical cation $S^{\bullet+}$. The presence of a redox mediator (M), a kind of electron shuttle, in the solution allows the dye to be restored to its neutral state, following the electron exchange process $S^{\bullet+} + M \rightarrow S + M^{+}$, and the mediator is electrochemically reduced at a cathode, restoring the whole system to its initial state[10] (Figure 4.12).

The energy produced during this photo-electrochemical process is a function of the current intensity and voltage of the electrochemical cell. The current is proportional to the electron flow, itself proportional to the total amount of dye, adsorbed in monolayer on the TiO_2 NPs; the voltage at the terminals of the cell is

---

9 The CEA-LITEN and the Japanese Toyobo company recently announced that they had created organic photovoltaic cells with the ability to reach conversion efficiencies of 25% when placed indoors (*PV Magazine*, March 23, 2020).

10 The electrolyte medium (redox mediator) is tetrapropyl ammonium iodide (5 M) in the presence of iodine (0.04 M), solubilized in a mixture of ethylene carbonate and acetonitrile. The dye is the ruthenium complex shown in Figure 4.12c.

equal to the potential difference between the fermi level of the semiconductor (E_F) and the redox potential of the mediator M/M^+.

The other original element in this process lies in the use of a mesoporous layer of titanium dioxide (TiO_2), which is around 10 μm thick, transparent and semi-conductive, and consists of an assembly of nanoparticles (diameter 15 nm) on which the dye is adsorbed [ORE 91]. The large equivalent surface provided by the TiO_2 NPs (around 2000 times the footprint) means that a large quantity of dye is exposed to the light, explaining the high energy yields obtained by these systems; from the outset, DSSCs reached and exceeded the 7% threshold under simulated AM 1.5 radiation, and 12% in diffuse light.

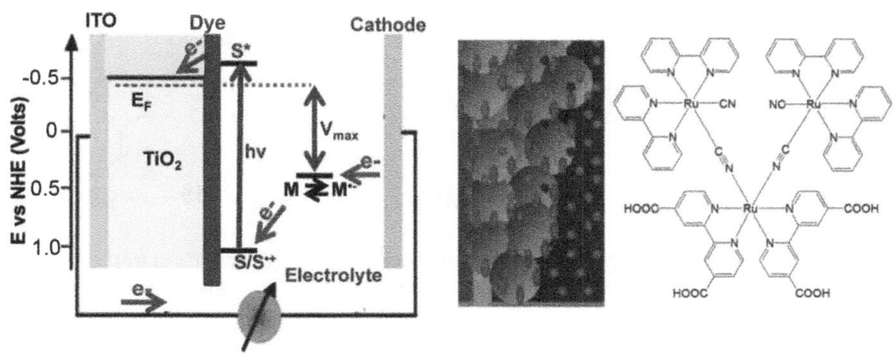

a) b) c)

Figure 4.12. *Dye-sensitized photovoltaic cell*

COMMENT ON FIGURE 4.12.– *a) Operating principle of the electrochemical photovoltaic cell. The arrows in blue indicate the direction of electron flow after excitation of the dye S to S*, followed by the transfer of the photoelectron into the semiconductor CB and the two cascade reduction reactions between $S^{\bullet +}$, $M^{\bullet +}$, M and the cathode, corresponding to the regeneration of the system to its initial state. The redox potentials of the dye (~ 1 V) and the mediator (~0.4 V) are evaluated relative to a normal H_2 electrode (NHE). In this case, the redox mediator is the iodine/iodide couple (I_3^-/I^-). Note a significant loss of energy of the order of 0.6 eV, in addition to the electron-hole recombination reactions, caused by the dye regeneration reactions. b) Schematic representation of the mesoporous layer of TiO_2 nanoparticles on which the dye molecules (in red) are adsorbed. The green and blue dots represent the electrolyte. c) Trimeric ruthenium complex used as S dye. Adapted from [GRÄ 09].*

One reason for these high yields is that the photoelectrons are transferred to the semiconductor with a time constant of between 10^{-13} s and 10^{-11} s. This transfer rate is much higher than those corresponding to the recombination reactions of the photoelectron with $S^{\bullet+}$ and I_3^-, added to that of the dye's return to the ground state, with time constants of 10^{-4} s, 10^{-2} s and 10^{-8} s, respectively, evaluated for the most efficient ruthenium dyes such as N719 [HAG 10, FAB 11] (Figure 4.13).

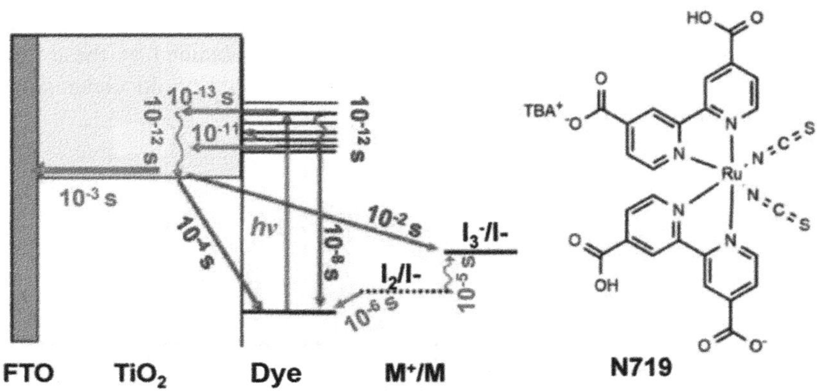

Figure 4.13. *Orders of magnitude of the time constants corresponding to the different electronic exchange reactions between the TiO_2 mesoporous layer, the N 719 dye adsorbed on the TiO_2 NPs and the redox mediator I_3^-/I^- (M) in a DSSC*

COMMENT ON FIGURE 4.13.– *The arrows in red correspond to the electron recombination and deactivation reactions that contribute to reducing the overall efficiency of the photovoltaic effect. Adapted from [HAG 10].*

However, the rate of diffusion of electrons in TiO_2, and thus their transfer to the photoanode, is much slower (time constant of around 10^{-3} s). It is made slower by the presence of traps distributed at different energy levels below the fermi level of TiO_2. For this reason, the recombination reactions of the electrons with the dye or the mediator, which have rates comparable to that of the diffusion of the electrons in the TiO_2 layer, will contribute to decreasing the overall conversion efficiency of photons into electrons, with a consequent effect on energy yields.

In terms of reliability, the long-term performance of the system is very good: placed under visible light ($\lambda > 400$ nm) for a two-month period, the conversion efficiency decreased by only 10%, corresponding to over 5 million $S^{\bullet+}/S$ redox cycles without significant degradation. A series of dyes corresponding to different ruthenium complexes has since been synthesized to absorb the widest possible section of the light spectrum, from visible to near IR (400 nm $< \lambda <$ 900 nm).

Photocurrents of above 20 mA/cm^2 have been obtained, with open circuit voltages between 0.7 and 0.8 V and certified energy yields exceeding 10% [NAZ 01].

Even in the early stages, when the yields of all-OPV cells did not exceed 2%, these new devices attracted significant interest. The process was commercialized very quickly, resulting in the production of colored, electricity-generating facade panels, with an attractive decorative effect, for use in the building industry (Figure 4.14). These commercial panels used all-organic panchromatic dyes in the place of ruthenium: a much cheaper, but equally effective, alternative.

a) b)

Figure 4.14. *Examples of multicolored solar panels inserted in building facades. a) Swiss Tech Convention Center in Switzerland (panels by Solaronix). b) The Toyota dream house in Japan (panels manufactured by the Aisin-Seikin subsidiary, outlined in red)*

At the time of writing, energy yields under AM 1.5 illumination (100 mW/cm^2) have not progressed much beyond 12%–14%, despite a great deal of work aimed at improving the various components involved in dye-sensitized liquid electrolyte photovoltaic cells. The synthesis of non-ruthenium dyes has been a particular focus of research efforts, essentially due to the high cost of ruthenium: alternatives include cobalt complexes, zinc porphyrins, phthalocyanines, quantum dots and, more recently, high-absorption all-organic dyes with a [donor-pi-acceptor] structure in which the donor and acceptor groups are in conjugation via an unsaturated spacer with π double bonds [HAR 12, YEL 13, JOL 14]. The conversion efficiencies of DSSCs using most of these dyes are lower than those obtained with Ru complexes, and only a few can really compete with ruthenium; their structure and behavior are described in detail in the literature [HAG 10, HAR 12, YE 15a, CAR 18].

The semiconductor/electrolyte interface is another important area of research, with the main objective of eliminating electron recombination reactions between the SC and the redox mediator (electrolyte). Attempts have been made to improve operations via nano-structuring of the photoanodes and the use of CNT (carbon nanotube) or graphene; however, the manufacturing process for these solutions is much more complex, resulting in increased cost. Nevertheless, one advantage of these new materials is that they can be used to create flexible photovoltaic cells on plastic supports [YAN 13c].

Other teams have worked on replacing the I_3^-/I^- redox mediator with less corrosive compounds. Grätzel's group proposed the use of cobalt complexes as early as 2011, achieving yields in excess of 12% [YEL 11][11]. This improvement (see Figure 4.11) is largely due to a significant increase in the V_{OC} (close to 1 V), resulting from the fact that the redox potential of the mediator is much more positive than that of I_3^-/I^- and very slightly lower than that of the dye (Zn-YD2-o-C8 porphyrin).

The described process involves the use of liquid electrolytes, but the current trend is toward all-solid systems for obvious engineering reasons. An all-solid solution was developed a few years after the liquid electrolyte cell, replacing the redox mediator by a solid organic layer which conducts positive charges, or holes (HTL – hole transport layer).

Using this new approach, the holes originating in the $S^{\bullet+}$ radical cation migrate to the cathode; the same process could be described by saying that the electrons injected by the cathode migrate by stages, via a hopping mechanism, into the HTL layer, reducing the $S^{\bullet+}$ radical cation and regenerating the dye S, following an "electron pump" model.

The first results obtained with all-solid dye cells were disappointing, to say the least, and these cells are far from being able to compete with liquid electrolyte alternatives. The first all-solid-state cell proposed by Grätzel's group in 1998 attained a woefully low conversion efficiency of less than 1%. The cell included a

11 The mediator is a cobalt (II/III) complex coordinated with three bipyridyl groups of formula $Co^{II/III}(bipyr)_3$. It is associated with the zinc porphyrin dye YD2-o-C8 of type π. The porphyrin ring forms the π conjugate bridge; it is substituted by two $C_6H_3–(O–C_8H_{17})_2$ groups and an anilino group $N[C_6H_4–(C_6H_{13})]_2$, each of which has an electron donating effect. It can also be substituted by the acceptor group $–C≡C–C_6H_4COOH$, which is strongly adsorbed on TiO_2 via COOH and contributes to increasing photoelectron transfer to the semiconductor. Long alkoxyl and alkyl chains prove beneficial in slowing down electron recombination reactions, and thus improving conversion efficiency.

mesoporous layer of TiO_2 into which a hole-transporting material (spiro-OMeTAD) was inserted, intended to replace the previous I_2/I_3^- redox mediator [BAC 98][12].

This very low yield, due in part to the low conductivity of the hole transporter, can also be explained by the difficulty of achieving good levels of interstitial insertion of the organic semiconductor into the mesoporous TiO_2 NP layer. Many solutions have been put forward to improve infiltration into the TiO_2 layer; other authors have suggested alternative hole carriers as replacements for spiro-OMeTAD, notably conducting polymers (derived from PEDOT). Yields of around 6% were first achieved in 2010 using a PEDOT derivative and improved modes of insertion into the mesoporous TiO_2 layer, via *in situ* polymerization from the monomer which, unlike the polymer, is easier to insert into the TiO_2 nanopores [LIU 10b].

Two years later, the Kanatzidis group took an important step by using an important milestone using an inorganic *p*-type semiconductor ($CsSnI_3$), easily processable in solution, with very high positive charge mobility in the solid state (585 cm^2/Vs) [CHU 12]. Associated with a dye such as N719, high conversion yields are obtained by doping $CsSnI_3$ with fluorine and SnF_2. The best results were obtained using a $CsSnI_{2.95}F_{0.05}$ compound doped with 5% SnF_2. After dissolution in solvent and injection into the mesoporous layer, optimal insertion into the TiO_2 nanopores was attained, resulting in conversion yields in excess of 10%.

Paradoxically, despite a higher light absorption range than N719 dye (significant absorption up to 789 nm in IR, whereas for N719, absorption stops at 677 nm), the energy yield observed with $CsSnI_{2.95}F_{0.05}$ alone is very low, below 1%. The use of the dye is essential to attain high yields. According to the authors, the improvement of yields is due to a synergistic effect of light absorption between the dye and the semiconductor, in conjunction with improved charge flow. It should also be noted that the structure of $CsSnI_3$ corresponds to that of a perovskite with a low energy band gap (1.3 eV); this family of compounds has recently proven to be extremely efficient for producing photovoltaic cells with efficiencies in excess of 15%, as we shall see later.

More recently, a series of copper complexes with more strongly positive redox potentials than the I_3^-/I^- couple, combined with dyes with good energy compatibility with the mediator, has been used to reach yields of around 11% under AM 1.5 illumination. This improvement essentially results from a significant increase in

12 The photovoltaic cell was constructed by adsorption of the dye $[Ru(II)L2(SCN)_2]$, similar to N719, in which the ligand L is 4,4′-dicarboxyl-2,2′-bipyridyl, onto the surface of TiO_2 NPs (mesoporous layer of 4 μm thickness). This first deposit was followed by that of the spiro-OMeTAD with a spirannic bifluorene structure [2,2′,7,7′-tetrakis(*N,N* di-p-methoxyphenyl-amine) 9,9′-spirobifluorene].

open circuit potential [CAO 17]. The achievement of efficiencies in the 10%–11% range using all-solid DSSCs constitutes a significant advance. Durability tests carried out over periods of a few 100 hours have shown promising results, but the reliability of the materials over periods of several years still needs to be determined.

In conclusion, DSSCs using liquid or all-solid electrolytes offer energy conversion yields of between 10% and 12% at best, depending on preparation techniques, and seem to have reached a ceiling at this level. They are in competition with all-organic BHJ cells, long characterized by low yields of around 8%–10%, but which have crossed the 14%–15% yield threshold in recent years; they are just as simple to implement, and no more expensive. However, DSSCs have a notable advantage over other systems in that they are transparent and can therefore be illuminated on both sides; their operation in diffused light is very good, which allows them to work under domestic interior lighting. This property makes them particularly attractive for self-charging connected objects, for which traditional energy sources (batteries or cabling systems) can be replaced by DSSCs coupled with storage devices (micro-batteries or super-capacitors) [YUN 18].

4.1.2.4. *Perovskite solar cells*

The development of perovskite solar cells (PSCs) represents a paradigm shift, the result of replacing dyes in electrochemical and all-solid-state DSSC cells with a perovskite semiconductor having high light absorbing capacity. In 2009, Kojima *et al.* [KOJ 09] were the first to suggest replacing the ruthenium dyes in liquid electrolyte DSCC cells with the hybrid perovskite $PbI_3NH_3CH_3$. However, the combination of this perovskite with a liquid electrolyte and a Br^-/Br_2 redox mediator was not particularly successful, with yields (between 3% and 4%) well below those obtained using the same perovskite inserted into an all-solid cell.

Nevertheless, this discovery of the hybrid perovskite resulting from an association of lead iodide (PbI_2) with methylammonium iodide (CH_3NH_3I), leading to the three-dimensional (3D) crystal structure $PbI_3NH_3CH_3$ characteristic of a perovskite (MAX_3) (Figure 4.15), triggered exponential development in this new area, finally resulting in the removal of sensitizing dyes and redox mediators used in DSSCs.

In just a few years, efficiencies have increased from 3.8% to over 17% [ZHO 14]. According to Snaith [SNA 13], this represents the dawning of a new era in low-cost, high-efficiency solar cell design. Activity in this field has taken off in recent years, involving many laboratories around the world; results continue to improve in terms of both energy efficiency and system reliability.

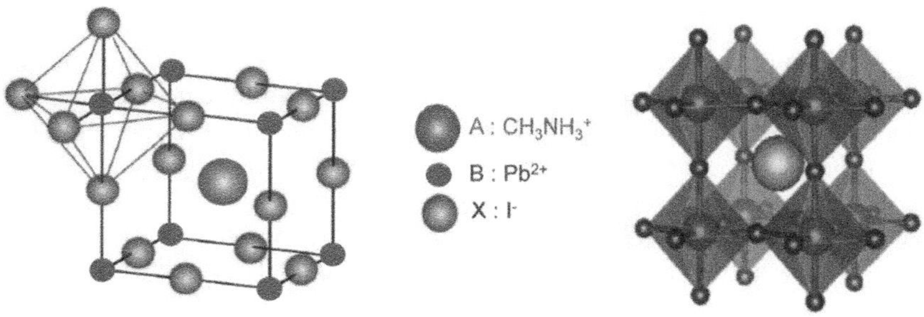

Figure 4.15. *Structure of the perovskite $PbI_3CH_3NH_3$ (ABX3) showing the insertion of the ammonium cation into a crystal lattice and within the crystal. Adapted from [GRÄ 17b]*

There are several reasons for this level of interest:

– $PbI_3NH_3CH_3$ is a semiconductor with a band gap very close to that of crystalline silicon (1.5 eV). Unlike Si, however, absorption corresponds to a direct transition between the VB and the CB, giving the perovskite a very high absorption coefficient α of 8×10^4 cm^{-1} at 550 nm; for silicon, α is only 6×10^3 cm^{-1}. Moreover, its absorption range is very wide, from 300 to 800 nm; under AM 1.5 solar illumination, this results in an average EQE (external quantum efficiency)[13] of 80%–90%;

– $PbI_3NH_3CH_3$ has a high dielectric constant. The electron-hole binding energy in the exciton is thus very low (0.05 eV), leading (as in the case of crystalline Si) to a total dissociation of the exciton into electron and positive charge (hole), followed by their direct injection into the CV and BV, respectively, of the semiconductor;

– electrons and hole mobilities are very high: 7.5–30 cm^2/Vs for electrons and 12.5–60 cm^2/Vs for holes, corresponding to an increase of 1000–10000 times compared to conventional organic semiconductors. Moreover, the diffusion lengths of the charge carriers are high, of around 1 μm, making it hard for traps to capture the electronic charges.

All of these elements combine to make $PbI_3NH_3CH_3$ both a low-gap semiconductor and a highly absorbent dye, resulting in easy exciton dissociation; this explains the exceptionally high conversion efficiencies. An analysis of the phenomena involved and of the methods used to produce these different materials can be found in [JUN 15] and [JEN 19].

13 EQE (external quantum efficiency) is the ratio between the produced current and the absorbed photon flux, expressed as a percentage.

Different types perovskite cell structures have been considered. The perovskite is often placed in direct contact with an electron conductor (generally TiO_2 or ZnO in the form of compact or mesoporous layers), in contact with the cathode, and a hole conductor (usually spiro-OMeTAD), in contact with an anode. In some cases, the electron conductor may be replaced by Al_2O_3 (Figure 4.16).

A typical planar thin-film structure with high conversion yields, ranging from 17% to 19%, was produced by the Zhou group in 2014. They created a perovskite containing two distinct halides (I– and Cl–), of formula $CH_3NH_3PbI_{3-x}Cl_x$, characterized by a band gap of 1.55 eV (Figure 4.17).

According to the authors, the high yields they obtained are the result of the other materials used alongside the perovskite; modifications were made to the surface of these materials to adapt their energetic characteristics to those of the perovskite, thus minimizing charge recombination reactions and optimizing charge extraction by the electrodes.

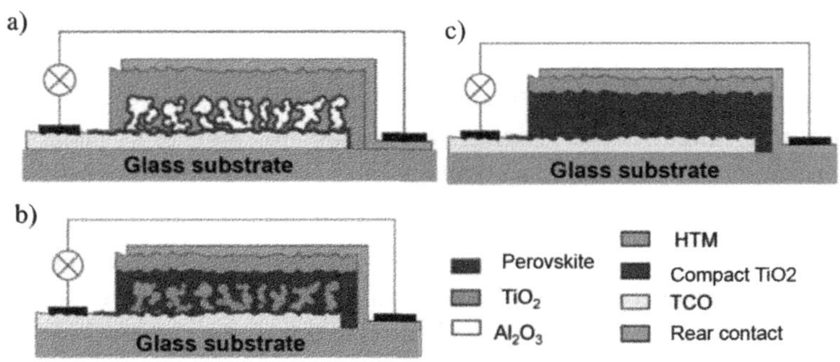

Figure 4.16. *Main architectures used in perovskite cells (PSC)*

COMMENT ON FIGURE 4.16.– *a) and b) are characterized by the presence of a mesoporous layer of* TiO_2 *or alumina (*Al_2O_3*) on which the perovskite is deposited, covered by the hole transfer material (HTM). The mesoporous layers are in contact with a compact layer of* TiO_2*, deposited on a transparent conducting oxide (TCO) electrode. The whole structure is covered in a metallic deposit constituting the back contact of the cell. c) Planar structure of PSC with a planar and parallel arrangement of the different materials, deposited successively in the following order: TCO, compact thin* TiO_2 *layer, perovskite, HTM and metallic contact. Adapted from [YUN 18].*

The work function of the ITO electrode is lowered from 4.6 to 4.0 eV by depositing an ethoxylated polyethylene imine (PEIE) film on its surface, making it compatible with the work function of TiO_2 (the electron transport material), and facilitating the flow of electrons from the CB of the perovskite (–3.75 eV) toward the ITO electrode. The conductivities of TiO_2 and spiro-OMeTAD (the hole transfer material – HTM) are enhanced by yttrium (Y) and lithium and cobalt salt doping, respectively, giving these compounds substantially equivalent conductivities (around 2×10^{-5} S/cm). This equivalence is particularly important for the operation of the PSC, balancing the flow of positive charges and electrons to both electrodes (Figure 4.17).

There are two drawbacks to PSCs containing perovskites of the $PbI_3CH_3NH_3$ type. The first concerns the use of lead, which raises environmental issues, although effective lead recycling systems now exist, particularly in the field of batteries. The second concerns their low stability over time, due to their sensitivity to humidity, which causes degradation of the perovskite [WAN 16a].

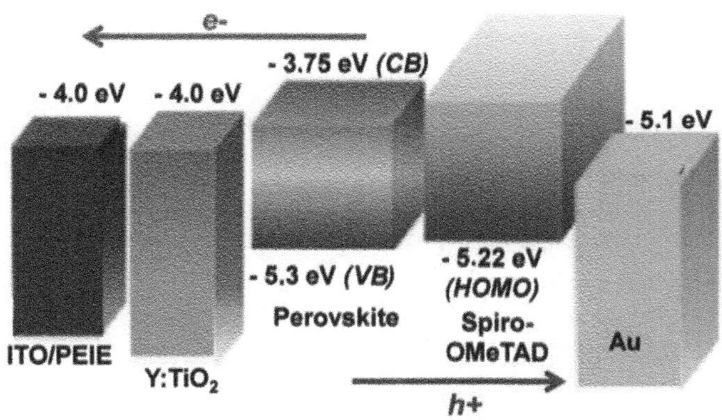

Figure 4.17. *Planar structure of a $CH_3NH_3PbI_{3-x}Cl_x$ perovskite solar cell*

COMMENT ON FIGURE 4.17.– *The values –4.0 eV, –4.0 eV and –5.1 eV correspond to the work functions of ITO/PEIE (150 nm), Y-TiO_2 (40 nm) and gold (100 nm), respectively. The perovskite $CH_3NH_3PbI_{3-x}Cl_x$ is characterized by the energy levels of its valence (VB, –5.3 eV) and conduction (CB, –3.75 eV) bands. Note the favorable alignment of the energy levels through which electrons (CB of perovskite, Y-TiO_2 and ITO/PEIE) and holes (h+) transit from the VB of perovskite to the HOMO of spiro-OMeTAD (200 nm) and gold. The entire cell is of sub-micron thickness; the individual Y-TiO2, perovskite and spiro-OMeTAD layers are successively wet deposited on the ITO electrode and subjected to annealing in ambient air then at low temperature. Finally, the gold electrode is deposited by vacuum evaporation. Adapted from [ZHO 14].*

Significant improvements have been made in recent years by replacing methylammonium (MA) with various organic and inorganic cations. An energy yield of 21% was notably obtained by the Grätzel and Hagfeldt groups by replacing MA with a ternary mixture of cations consisting of MA, formamidinium ($H_2NCH = NH_2^+$, FA) and cesium (Cs) [SAL 16]. Out of all the different $Cs_x(MA_{0,17}FA_{0,83})_{(100-x)}Pb(I_{0,83}Br_{0,17})_3$ perovskites, the one containing 5% Cs (Cs_5M, x = 5) was found to be the most stable, retaining a yield of 18% after 250 hours of operation in a nitrogen atmosphere at room temperature and with AM 1.5 illumination. This represents a considerable improvement compared to the Cs-free perovskite (Cs_0M, x = 0), for which the yield drops to below 8% after 100 hours of operation (Figure 4.18).

More recently, Grätzel [GRÄ 17b] has produced perovskite cells with a large surface area (10×10 cm^2) that remain stable over a period of more than 1 year. This result is significant as the process may be easily adapted for large-scale production, and that it gives energy yields of around 14% with a standard mesoporous cell architecture.

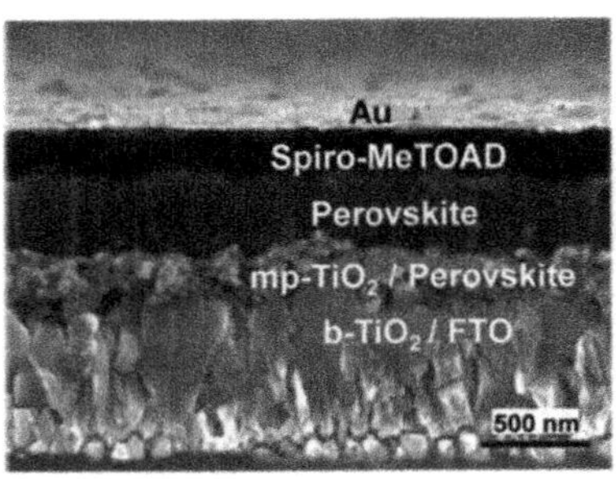

a)

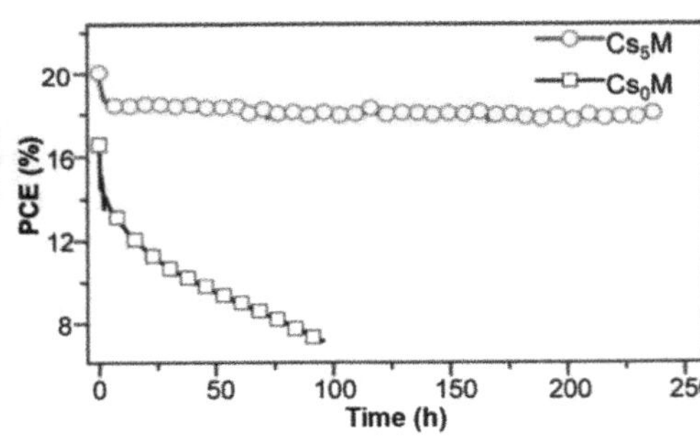

b)

Figure 4.18. *Hybrid perovskite containing a ternary mixture of MA, FA and Cs cations*

COMMENT ON FIGURE 4.18.– *a) Cross-section of the PSC showing the different constituent layers of the cell. Note that the perovskite is both embedded in the mesoporous TiO_2 layer (mp-TiO_2) and arranged in a compact layer of about 500 nm above the mp-TiO_2. The compact TiO_2 layer (b-TiO2) deposited on the FTO electrode has the effect of blocking the diffusion of holes to the FTO. b) Behavior of perovskites with and without Cs (Cs_5M and Cs_0M, respectively) over time. a) Adapted from [GRÄ 17b]. b) Adapted from [SAL 16].*

According to the authors, this stability results from the introduction of a small quantity of 2D perovskites into the cell, mixed with the 3D perovskites; 2D perovskites are known to exhibit greater chemical inertia with respect to humidity than their 3D counterparts. Ammonium iodide of amino valeric acid, $HOOC(CH_2)_4NH_3I$ (AVAI), larger in size than CH_3NH_3I, crystallizes precisely with PbI_2 in the form of a 2D perovskite corresponding to the formula $[HOOC(CH_2)_4NH_3]_2PbI_4$. According to the authors, the application of different solutions of 2D and 3D perovskites following a precise protocol creates a type of structural gradient throughout the whole thickness of the composite perovskite, resulting in a significant gain in stability. After more than 10000 hours of operation, under AM 1.5 G illumination, the yield losses remain negligible; this is one of the best results to date in terms of stability, and it suggests further possibilities for improvement by varying the materials used in the composition of perovskites. Efficiencies of 14% are obtained with a single cell, and around 12% with 10×10 cm^2 modules.

The same group recently made a notable advance by subjecting δ-$FAPbI_3$ perovskite to heat treatment (100°C) in the presence of methyl ammonium thiocyanate in the vapor state, resulting in its transformation into a very stable crystalline phase (α-$FAPbI_3$). The performances observed over a period of 500 hours are remarkably constant and far exceed the photovoltaic characteristics of untreated perovskites [LU 20][14].

Tandem assemblies of perovskites have also been fabricated with the aim of capturing a wider portion of the solar spectrum. The principle, described for OPV cells (see section 4.1.2.2), consists of coupling two cells in series: one with a small gap, to capture the IR part of the solar spectrum, and the other with a large gap to collect the UV-visible part of the spectrum. Tandem cells of this type have been created by assembling two perovskites, one with a low gap (1.25 eV,) with structure $[(FA\text{-}SnI_3)_{0.6}(MA\text{-}PbI_3)_{0.4}$ or $(FA\text{-}SnI_3)_{0.6}(MA\text{-}PbI_3)_{0.4}]$ modified by guanidinium thiocyanate-GA-SCN at 7%, and one with a wider gap (1.63 eV), of formula $Cs_{0.05}FA_{0.8}MA_{0.15}PbI_{2.55}Br_{0.45}$). Coupling these wide-gap and low-gap perovskites in series results in a cell which absorbs from 300 to 800 nm (light hitting the top cell) and from 700 to 1000 nm (bottom cell receiving light filtered through the top cell). The respective efficiencies of the two cells are 18.2% and 6.8%; a record overall efficiency of 25% was achieved with a 4T-tandem setup [TON 19][15].

---

14 An ITO/SnO_2/α-$FAPbI_3$/spiro-MeOTAD/Au type cell is characterized by a PCE of 23.1%, a J_{SC} short-circuit current density of 24.4 mA/cm^2, a V_{OC} of 1.165 V and an FF of 81.3%.

15 2T and 4T cells correspond to two and four contact assemblies of a tandem cell. The 2T arrangement allows the output voltages of the two cells to be added together, but the output current is dictated by the cell with the lowest energy output. In the 4T four-contact arrangement, the voltages and currents of each cell are taken separately and summed in an external circuit.

The current trend is toward the development of tandem cells with a low-gap (1.2 eV) silicon cell – a tried and tested technology – coupled with a large-gap PSC. An efficiency of 23.6% was recently achieved with a 2T tandem cell of 1 cm^2 by coupling a silicon cell with a PbI_2 perovskite cell containing formamidinium (FA) and cesium cations, known to increase the stability of perovskites; the perovskite presents an E_g of 1.63 V for the composition $Cs_{0.17}FA_{0.83}Pb(Br_{0.17}I_{0.83})_3$ (CsFA). The combination of this perovskite with the silicon cell allows a large absorption spectrum (from UV to near IR) to be covered. However, a certain number of precautions must be taken in order to protect the organic compounds used in manufacturing a tandem cell; for example, intermediate inorganic layers are deposited using ALD (atomic layer deposition) techniques, which are operational at low temperatures (around 120°C), avoiding degradation of the organic compounds [BUS 17].

The way in which PSCs may develop and their commercial prospects both remain to be seen. Small cells have reached levels above 22% over short periods of time, meaning that they now have the capacity to compete with amorphous silicon cells, CIGS and CdTe based thin-film cells [LI 18d]. Many material innovations have been put forward with the aim of increasing the stability and energy yields of perovskites and their energy yields [JEN 19]. However, energy yields fall considerably (to around 12%–15%) for cells larger than 1 cm^2. Nevertheless, the fact that PSCs can be produced using wet process technology is an advantage; many teams have developed procedures to create large surface area PSCs (over 100 cm^2) using a variety of coating techniques, and encapsulation processes may be used to increase their longevity beyond 1200 hours [MAT 16]. Rapid drying processes using flash IR annealing, which do not alter the properties of perovskites, were recently developed by Hagfeldt and coworkers [SAN 20]. This IR curing technique is a good alternative to thermal annealing; tested on a perovskite cell made of a ternary mixture of formamidinium (FA), cesium (Cs) and guanidinium (GA) cations, with structure [FTO-mpTiO_2/perovskite (FA:Cs:GA)/spiro-OMeTAD], it resulted in stable operation with 20% yield over a period of more than 1 month. This result is comparable to that of cells prepared using standard thermal annealing techniques.

In conclusion, PSCs are a promising new technology; recent scientific literature is full of announcements of record yields, bringing PSCs into genuine competition with inorganic cells (poly-Si, CdTe and CIGS). They can be fabricated using a wet process and the low cost R2R type printing technology can be envisaged in this case: an important advantage in terms of large-scale production. Unfortunately, however, PSCs contain lead, which is problematic in environmental terms, and their stability over very long operating periods has yet to be confirmed. To the best of our knowledge, no hybrid perovskite has yet achieved operating times in excess of 1 or 2 years, clearly insufficient for immediate commercialization. This is also one of the main reasons why they are not suitable for static storage of solar energy at the present time; for these purposes, yield losses must remain below 10% over operating periods of 25 years. Nevertheless, PSCs may soon be used for other applications, for example, as autonomous energy sources in connected objects, where operating periods of 5–10 years are less problematic [KAN 19b].

4.2. Electroluminescence, lighting and display

Electroluminescence is a well-known phenomenon, first discovered in the early 20th century. In contrast with the photovoltaic process, electroluminescence results from the application of a current to a *p-n* junction. The first LEDs were made using mineral materials; later, the approach was extended to organic materials, resulting in the production of thin-film OLEDs (organic light-emitting diodes). OLED televisions are one of the more recent products which have been brought to market using this technology.

From an energy perspective, LEDs constitute an ideal SSL (solid state lighting) system, far surpassing traditional incandescent lighting systems using filaments or fluorescent tubes. Consuming far less energy and lasting much longer than previous lighting technologies, current LEDs have attained energy yields very close to their theoretical limits, and much higher than those of incandescent light bulbs and fluorescent tubes. LEDs are currently more expensive to produce than incandescent lamps; however, production costs should continue to fall, and given that their useful lifetime – in excess of 50000 hours – far exceeds that of traditional bulbs, they should become increasingly competitive in commercial terms.

The widespread replacement of older lighting devices with LEDs represents a significant energy saving; a study by the US Department of Energy (DoE) estimates that the cumulative energy savings brought about by a complete phased introduction of SSL in the United States. from 2017 to 2035 would represent around 78 quads[16], or, in financial terms, $890 billion [DOE 19].

Energy usage is also critical in the field of display, concerning all connected objects and connected devices with screens. Traditional inorganic diodes have been gradually replaced by OLEDs, which have the advantage of being able to provide a wide variety of colors; they consume less energy, and, unlike inorganic light-emitting diodes (ILEDs), are easy to use in large display panels and for ambient lighting.

16 The *quad* is a unit used by the US DoE for energy assessment and is equivalent to 1.055×10^{18} joules (just under 300 billion kWh).

The following section concerns recent improvements in LED and OLED design, and gives an overview of a new concept in LED technology based on the use of QDots.

4.2.1. *Inorganic light-emitting diodes*

The electroluminescence phenomenon was discovered by accident in the early 20th century, in 1907. H.J. Round, working on the rectification phenomenon created by a SiC crystal subjected to alternating polarization, was the first to observe a light emission[17]. Given the slow progress of knowledge concerning solids and semiconductors, and the relatively late adoption of energy band theory in the 1940s, significant developments in the field of diodes did not occur until the first bipolar transistors emerged in the late 1940s. The feasibility of LEDs was established in the 1950s–1960s, using alloys of elements from columns III-V in the periodic table. A detailed history of lighting through the ages, from candles up to LEDs, is given by Weisbuch [WEI 18] in a recent article.

The main devices used to produce light emissions are made of inorganic semiconductors assembled in a *p-n* junction, or in a double junction with a photoactive layer. These devices were invented well before the 2000s, but have undergone continuous development ever since; recently, advances in the manufacture of nanocrystals have given rise to a new generation of LEDs based on the use of QDots (QLEDs). Other LED devices (such as OLEDs) have been developed using organic semiconductors.

4.2.1.1. *Electroluminescence in a p-n homojunction or a p-i-n double heterojunction*

Shortly after the invention of the transistor in the Bell Labs in 1947, it became apparent that the *p-n* junction between two *n*- and *p*-doped semiconductors, then a double hetero junction, subjected to direct polarization, could give rise to photon emissions (Figure 4.19).

In the first case, emissions result from the recombination of electrons and holes injected on both sides of the contact zone (depletion region) between the two semiconductors *p* and *n*; in the second case, it is the accumulation, then recombination, of electrons and holes in a semiconductor with a smaller gap which causes photons to be emitted, with a much higher efficiency than that found in the homojunction. Another advantage of the double heterojunction (DH) is the ability to

---

17 H.J. Round was an engineer with the Marconi Company working on radio sets and AC rectifier materials.

modulate the emission wavelength according to the band gap E_g of the active layer; in the case of a homojunction, the emitted wavelength is determined by the E_g of SCs *n* and *p* alone.

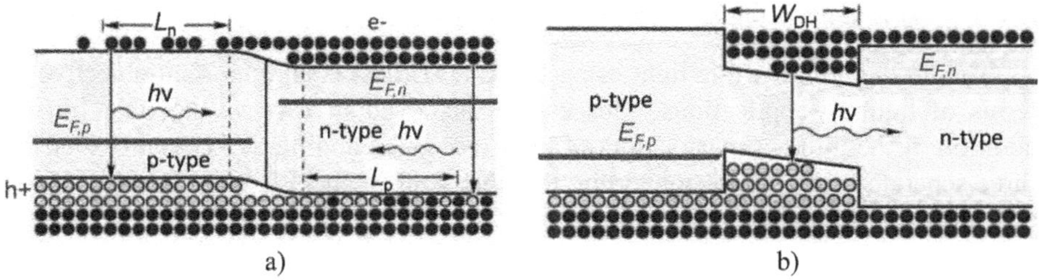

Figure 4.19. *(a) p-n homojunction; (b) a DH double heterojunction*

COMMENT ON FIGURE 4.19.– *a) Under a direct polarization $V = E_{Fn} - E_{Fp}$ (where E_{Fn} and E_{Fp} are the Fermi levels of SCs n and p), electrons and holes (h+) diffuse to the opposite semiconductor, over respective diffusion distances Ln and Lp; their recombination with opposite sign charges results in the emission of hv photons such that hv ≈ E_g (band gap). b) The DH is an assembly of two SCs, n and p, separated by an intrinsic SC with a thickness of around 100 nm and a gap smaller than that of SCs n and p. This constitutes the active layer in which the electrons and holes diffuse and are confined prior to recombining and emitting photons; the energy of these photons is equal to the E_g of the active layer. Adapted from [SCH 06].*

Significant progress has been made since the first discovery of LEDs, which long pre-dates nanotechnology. The first commercial LEDs appeared in the late 1960s in Hewlett-Packard calculator displays, using LEDs emitting dark red light with GaAsP semiconductors. This light emission was very low; by the early 1970s, the energy yields of "red" diodes had still not exceeded 1 lumen per watt, compared to efficiencies of around 15 lm/W for the incandescent bulbs available at the time [CRA 92][18]. A wide range of different semiconductor materials were later developed

18 The lumen (lm) defines a flux of light energy and corresponds to an intensity of 1 candela (cd) contained in an angle of 1 steradian (sr) or 1 lm = 1cd × 1sr. The candela is the intensity radiated in a given direction by a source emitting monochromatic radiation at 555 nm and with an energy intensity of 1/683 watt/sr in that direction (definition established by the International Commission on Illumination (*Commission Internationale de l'Eclairage*, CIE), 1979). For example, a candle emits a light energy of around 1 cd/sr , and a light source that emits a light intensity of 1 cd in a solid angle of 4π radiates an energy flux of 12.57 lm. 555 nm corresponds to a green color, to which the eye is most sensitive.

and tested, resulting in improved-efficiency red LEDs; in 1990, levels of around 10 lm/W were achieved using AlGaAs semiconductors and, more recently, values of 150 lm/W have been attained using AlInGaP, in the case of DH structures [TSA 14].

4.2.1.2. *Blue light emission and white light production*

The production of white light using LEDs (WLEDs) is an important objective in terms of lighting applications. This can be achieved in two ways: either by the addition of the three primary colors of light red, green and blue (RGB) provided by three separate LEDs, or by converting the blue light of an LED using green and red or yellow phosphors[19].

Access to reliable, durable blue-emitting diodes is thus crucial, and the development of these diodes was a major focus of research for over 30 years. The problem was finally solved in the mid-1990s by Nakamura, Akasaki and Amano, earning them the 2014 Nobel Prize for Physics. Earlier, in the 1980s, it was noted that two SCs, GaN and ZnSe, were identified as sources of blue emissions. ZnSe can be easily deposited on a GaAs crystal support, but the many defects produced during the manufacturing process result in degraded emissions; satisfactory results were never achieved using ZnSe diodes. GaN posed a different problem. Akasaki and Amano resolved the issues inherent in n- and p-doping GaN (necessary to create homojunction LEDs); the final step was to find a way to deposit a low-gap SC, such as InGaN, in order to create a DH of the n-GaN/InGaN/p-GaN type. However, the solution to this problem was a long time in the making, and for many years; the GaN-based approach was considered to be a dead end by most of the scientific community. Nakamura, however, persisted, developing a chemical technique for vaporizing organometallic compounds (MOCVD – metal-organic chemical vapor deposition), with the capacity to produce very pure GaN deposits of gallium nitride; this can be doped with indium (In). This ground-breaking discovery[20] rapidly opened the way to the creation of the first DH with a very thin (20 nm) active layer

---

19 White light can be obtained from a blue-emitting LED via a conversion process, by depositing a phosphorescent material (phosphor) in the diode that emits yellow light when excited by blue light. The addition of blue and yellow light is perceived by the eye as white light; the corresponding LED is designated PC-WLED (phosphor-converted white LED). A white light of greater purity can be obtained by using two phosphors R and G, which is equivalent to adding the three primary colors B, R and G. It should be noted that blue light has harmful effects on health (mood disorders, circadian disturbance, insomnia and fatigue) and efforts have recently been made to produce new WLEDs that minimize the blue light component [HAR 21].

20 Nakamura describes the approach that led him to achieve dislocation-free GaN deposits in the presentation which he gave to the Nobel Committee in December 2014 [NAK 14].

of GaInN, arranged between layers of n-GaN and p-GaN [NAK 93]. The blue emissions obtained are very pure, characterized by an emission peak that can be modulated between 400 and 450 nm by varying the In composition in the $In_xGa_{1-x}N$ alloy. Initial yields were very low, but increased rapidly as a result of successive improvements in the deposition techniques used for the different layers; the invention of quantum well DHs (QWDH) was also decisive. In this final case, the thickness of the active film is reduced to a few atomic layers, that is, a few nm, influencing the band gap and thus the color of the emission [NAK 95].

Gallium nitride continues to be the material of choice for lighting devices emitting white light. Numerous improvements in preparation techniques have increased the energy efficiency of LEDs, which now far exceeds that of incandescent lamps and fluorescent tubes. White light LED lamps with energy yields of 300 LMW (300 lumen/watt) were first offered commercially by Cree in 2014; this represents a considerable improvement on the initial values obtained in the 1990s. The cost of LEDs has also fallen considerably since the 1960s; over the period to 2010, the average cost of LEDs (red and white) fell by a factor of 10 every 10 years, while their efficiency increased by a factor of 20 over the same time intervals (Figure 4.20).

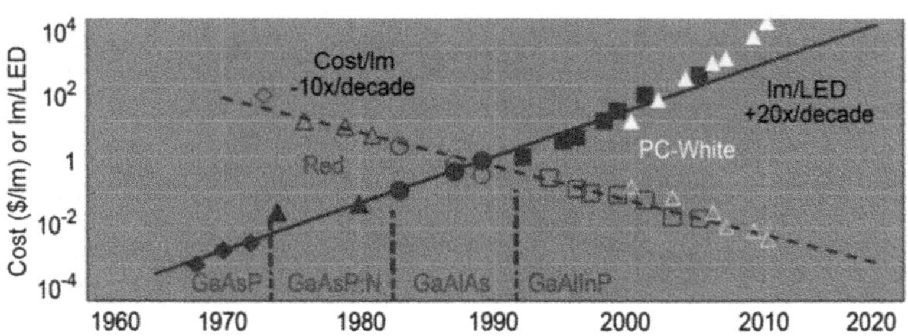

Figure 4.20. *Evolution of costs ($/lm) and lighting power (lm/LED) of red- and white-light (PC-White) LED lamps between the 1960s and 2010, including four generations of red-light diodes: GaAsP, GaAsP:N, GaAlS and GaAlInP. Adapted from [TSA 14]*

4.2.1.3. *Problems of LEDs*

White light producing LEDs, using of GaN and GaInN SCs, are currently made up of an active layer with several quantum wells (QWs), which are simply DHs with an extremely thin GaInN active layer (around 2–3 nm). Unfortunately, two important phenomena which occur at high current densities (above 10 A/cm^2) hinder their commercial use [CHO 13]: a significant weakening of the internal quantum efficiency (IQE), and a decrease in efficiency when the emissivity shifts from blue to longer wavelengths (green), a phenomenon known as the green gap.

These phenomena are discussed in detail in the literature, and it is broadly accepted that these yield losses mainly result from non-radiative relaxation processes that can occur within and outside the active region of the LED [CHO 13]. There are two non-radiative relaxation processes that take place inside the QW: an Auger recombination between electrons and holes, leading to "hot" electrons of higher energy than the one at the bottom of the GaN CB (Figure 4.21d), and deactivation of an electron and a hole toward a structural defect in the InGaN following an SRH (Shockley-Reed-Hall*)* mechanism (Figure 4.21b). Electrons may also be lost via an overflow effect above the CB of the p-GaN layer, followed by recombination outside of the QW (Figure 4.21a).

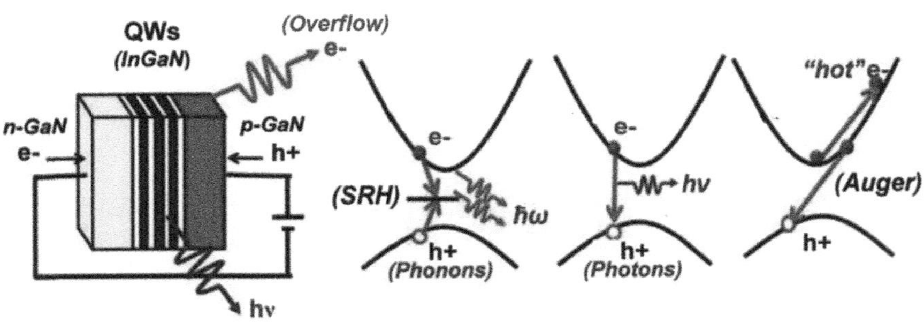

a) b) c) d)

Figure 4.21. *Different phenomena involved in the injection of electrons (e-) and holes (h+) into a multiple quantum well (MQW) LED*

COMMENT ON FIGURE 4.21.– *a) LED cell with four quantum wells (InGaN, in white) surrounded by GaN layers (in purple). The emitted electrons correspond to an overflow effect occurring at very high current densities. b), c) and d) Electron and hole recombination reactions inside the InGaN layers. b) Non-radiative recombination to a trap with phonon emission (ħω). c) Radiative recombination with photon emission (hv). d) Non-radiative Auger recombination with high-energy electron emission (hot electron). Adapted from [WEI 20].*

The high intensities of the electric excitation current of the LED, along with the rise in temperature resulting in thermo-ionic electron emission, have the effect of causing electrons to be ejected and to migrate out of the QW; this contributes to decreasing photon emission yields[21]. The lower mobility of holes compared to electrons can also weaken radiative recombination.

Various means have been suggested to avoid this loss of photon emission efficiency at high current densities (several hundred amperes per cm^2). One way to mitigate loss is to increase the number of QWs; another approach is to insert an EBL (electron blocking layer) (AlGaN) into the p-GaN region, which helps to keep electrons confined to the QWs and thus increase overall photoemission efficiency. However, the optimal number of QWs is still limited to below 10 [WEI 20].

Another way of preventing a loss in emissions at high current densities is to replace planar SC layers with nanostructured layers. The Bhattacharya group used this technique with arrays of nanorods (NR), each consisting of an n-GaN/InGaN/p-GaN assembly [GUO 10]. The NRs are aligned and deposited vertically on a silicon support and are fabricated using the MBE (molecular beam epitaxy) technique[22]. Due to the high surface to volume ratio, and despite the crystal mismatches between Si, GaN and InGaN, NRs have very few dislocations, which, as we have seen previously, cause a drop in emission efficiency at very high current densities.

Very dense arrays of GaN NRs ($1–2 \times 10^{11}/cm^2$), 2 μm in length and with diameters between 10 and 50 nm, are thus deposited on a crystalline silicon substrate. Remarkably, it is possible to modulate the In content in the middle part of the NR (InGaN) experimentally, either toward a concentration corresponding to a determined color (green, yellow or red), or according to a concentration gradient to produce a white light emission [GUO 10] (Figure 4.22).

---

21 The size differences between In, Ga and N lead to a high concentration of dislocations in the InGaN active layer. This results in the creation of traps in the band gap of the SC, and for most SCs, results in low radiative emission rates; InGaN, however, continues to emit in the blue with high yields. Different interpretations of this phenomenon can be found in the literature [CHO 13, REN 15a, WEI 15].

22 This physical technique consists of evaporating materials contained in effusion cells under ultra-vacuum, resulting in the formation of "molecular jets", atomic or molecular beams of very small diameters. These are directed toward and condensed onto a crystalline substrate maintained at a temperature lower than that of the material in the vapor state, making it possible to obtain epitaxial growth of the material with respect to the crystallinity of the support. Molecular jets are a result of the effusion process. They occur when a gas enclosed under pressure in a heated chamber escapes through an orifice with a diameter much smaller than the mean free path of the molecule or atom in a vacuum. This technique is used industrially to produce very pure thin-film deposits. Compared to other methods, it has the advantage of not requiring a catalyst, meaning that high-purity semiconductor deposits can be obtained.

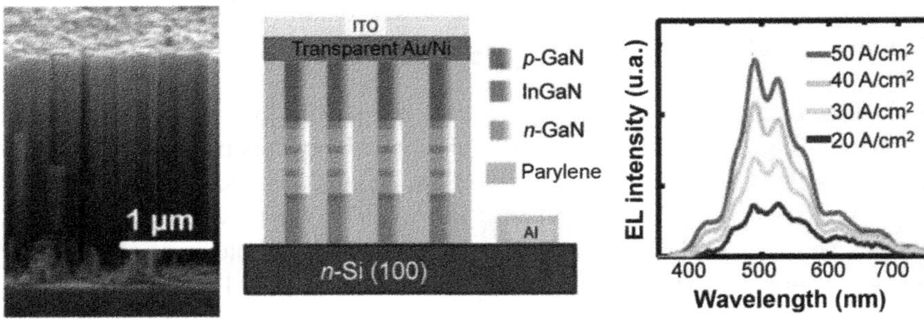

a) b) c)

Figure 4.22. *Electroluminescence produced by a GaN/InGaN NR array*

COMMENT ON FIGURE 4.22.– *a) SEM cross-section image of a GaN NR array deposited on a crystalline n-Si substrate using the molecular beam epitaxy (MBE) technique. (b) Simplified schematic of an NR LED cell with several InGaN nanodisks (≈2 nm thick) separated by n-GaN and p-GaN nanocylinders and embedded in (transparent) parylene. (c) White light emission intensity for different current densities. The white light results from the fact that the concentration of In in GaN has been adjusted according to a concentration gradient which has the effect of extending the emission spectrum from blue (450 nm) to red (650 nm). Note that when the current density increases to very high values (50 A/cm^2), emission intensity continues to grow, in contrast with the situation observed with planar diodes; furthermore, the position of the emission peaks remains unchanged (no green gap). Adapted from [GUO 10].*

A number of teams have adopted this type of nanostructuring approach, some using graphene [LAT 17]. German and Austrian researchers have recently developed an original method for NR array production by applying the RIE (reactive ionic etching) technique to a planar LED structure.

A mask made up of an assembly of silica microspheres was arranged on top of the planar GaN/InGaN/GaN structure, in a way which is equivalent to performing lithography using these silica spheres. After ion etching and various chemical operations, followed by passivating deposition of Al_2O_3 on the surface of the NRs an NR-LED lattice was obtained. The mechanical stresses within the NRs in this lattice are extremely low, eliminating the risk of dislocations and of the radiative electron-hole recombination process dropping off. The results show much more intense light emissions than those observed with planar LEDs; these results are also due to a more efficient mode of extracting light from the diode[23]. While the approach described here has yet to move beyond the experimental stage, it does appear promising for potential future applications [KOU 20].

4.2.2. *Organic light-emitting diodes*

The first OLEDs were produced by Tang and VanSlyke [TAN 87] in the Eastman Kodak laboratories in 1987, following on from their invention of the first OPV cell the previous year. The highly simplified diode comprised a glass substrate on which a conductive layer of ITO is deposited. A first organic hole conductor layer, followed by the fluorescent emitter layer of Alq_3 (8-hydroxyquinoline aluminum), was deposited, along with a second electrode made of a Mg:Ag alloy. Under polarization of around 10 V, the diode emitted green light with a brightness of 1000 cd/m^2 and an EQE of about 1%, comparable to that of ILEDs of the time.

Further innovations followed, leading to OLEDs becoming established as reliable, commercial products; they are currently used in television screens, in ambient lighting and, more generally, electronic devices involving a reading display. The first OLED televisions were produced by Sony and LG as early as 2011. They differ from LCD (liquid-crystal display) televisions in that they have an ultra-thin screen (thickness of about 2 mm), a 180° viewing angle, a fast response time (around 1 microsecond), very high brightness and very pure colors.

The steps leading up to this point have included the discovery of new organic materials, with properties which lend themselves to new optoelectronic mechanisms, improving light intensity, the energy yields and the chromatic quality of colors. Compared to ILEDs, the EQEs of OLEDs are better, and they provide a source of white light (WOLED) that can be deployed across large surfaces, and can supply soft ambient light, which is particularly suitable for indoor lighting.

4.2.2.1. *Typical structure of an OLED*

In its simplest form, an OLED is an electro-emissive device comprising one or two organic pi-conjugate layers sandwiched between two metal electrodes with

23 GaN is a piezoelectric material. Its deposition on a substrate with different crystalline characteristics creates mechanical stress that locally generates an electric field, directed along the directional axis of deposition (c axis of the GaN crystal). This electric field has the effect of attenuating the spatial overlap of the wave functions of holes and electrons, and consequently reduces the probability of radiative recombination of holes and electrons, according to the quantum confined stark effect (QCSE).

different work functions between which a voltage is applied. The cathode, characterized by a low work function, allows electrons to be injected into the organic layer while the anode, characterized by a high work function, allows holes to be injected into the same, or another, organic layer. Charge transport through the layers leads to the formation of excitons (electron-hole pairs linked to the same pi-conjugated molecule), the radiative de-excitation of which results in light emissions.

Current OLEDs have moved on from the original model produced by Tang and Van Slyke, and now include a minimum of five thin organic layers (Figure 4.23). This multilayer OLED comprises a central emitting layer which contains the fluorescent or phosphorescent material (EML – emitting material layer), surrounded by blocking layers on both sides: one for the electrons (EBL – electron blocking layer), the other for holes (HBL – hole blocking layer). Two additional layers are used for the injection and transport of holes (HTL) and electrons (ETL), in contact with the anode and cathode, respectively. This structural arrangement is often referred to as a p-i-n OLED, where i corresponds to the EBL, EML and HBL layers. It allows both the quantum and energy efficiencies of the emission to be optimized. The electron and HTLs are usually *n*- and *p*-doped, respectively[24], in order to increase the overall conductivity of the cell, reducing the voltage required for emission and providing an energy gain. Similarly, the presence of blocking layers ensures a better balance between negative and positive charges and also contributes to improving the quantum efficiency of the emission[25].

In addition to the improved conductivity, these *n*- and *p*-doped layers can also have a favorable effect on the injection of charges through the electrodes.

One disadvantage of this type of multilayer structure is that part of the emitted light is lost through interference and parasitic reflection. It is estimated that, without special optical arrangements on the diode surface, only 20% of the emitted light will

24 Doping organic SCs improves their conductivity and enables them to reach values of the order of 10^{-4} S/cm. The *n*-doping of an organic electron acceptor SC (A) can be achieved by mixing it with an electron donor compound (D), possessing a HOMO with an energy higher than the LUMO of A. Conversely, *p*-doping of a D-donor can be achieved by mixing it with an electron acceptor, with an LUMO characterized by an energy lower than that of the HOMO of D. Many examples of doping of organic SCs are described by Lüssem *et al.* [LÜS 13].

25 Optimal operation of an LED is achieved when the electron and hole currents are of equal intensity. This results from the fact that each electron and each hole injected bind together to form an exciton capable of emitting light. Any electron current in excess of, or, conversely, below the hole current is "lost" for light emission purposes as no exciton will be formed. A large part of the technological solutions implemented in OLEDs relate precisely to balancing these two currents.

be recovered externally. In addition, depending on the nature of the emitting material, two types of emissions are possible. Emitted light may correspond to fluorescence – a radiative deactivation of singlet states ($^1S^*$) of excitons – or to phosphorescence – a radiative deactivation of triplet excitons ($^3T^*$). Given that the statistical distribution between singlet and triplet states of the excitons is 1:3, the external quantum yield EQE of the diode thus differs significantly depending on the emission type. Assuming a yield of 100% for fluorescent emissions Φ_F, phosphorescence Φ_P and charge transformation into singlet ($\varkappa_S$) and triplet ($\varkappa_T$) excitons, the maximum EQE is around 5% in the case of fluorescence and 15% for phosphorescent emissions (Figure 4.24).

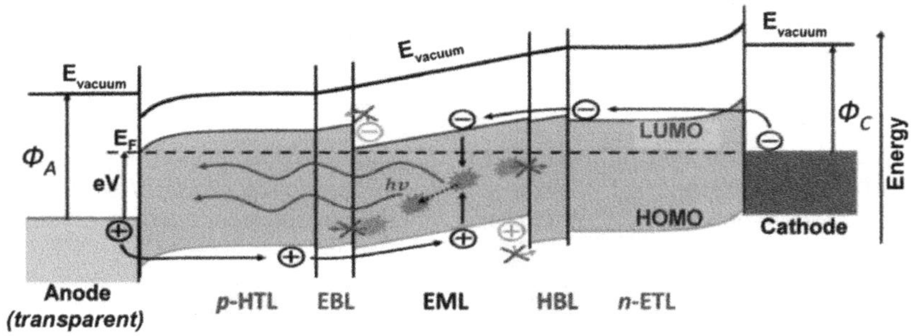

Figure 4.23. *Energy diagram of a p-i-n type OLED. The HTL and ETL layers are p- and n-doped, respectively. The hole blocking (HBL) and electron blocking (EBL) layers generate two energy barriers, preventing holes and electrons from diffusing to the cathode and anode, respectively*

COMMENT ON FIGURE 4.23.– *The blocking layers (EBL and HBL) are around 10 nm thick. The ETL and HTL layer thicknesses vary between 50 and 100 nm, while the emissive layer (EML) is usually around 20 nm thick. Φ_A and Φ_C represent the work function of the anode and cathode, and V is the voltage applied to the diode. The red crosses mark the diffusion boundaries of holes, electrons and excitons. Adapted from [HOF 12].*

As we noted earlier, significant light loss occurs within the diode due to the thin-film structure. However, improvements can be achieved by modifying the surfaces and interfaces. For example, light extraction can be significantly increased by creating periodic nanostructures between the different layers of the OLED. A variety of different technological solutions have been developed and tested in the aim of attaining EQEs close to the theoretical values; a detailed description of these processes is given by Zou *et al.* [ZOU 20].

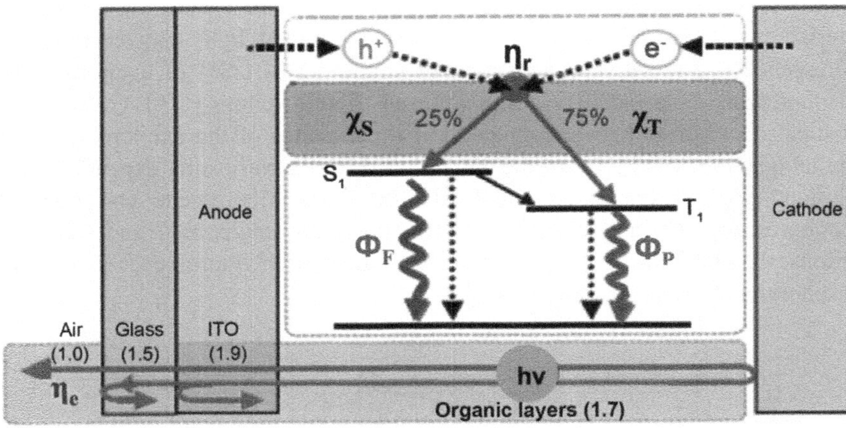

Figure 4.24. *Emitted light hv in an OLED. η_r is the exciton formation yield from the injected charges (h+ and e-); χ_S and χ_T are the respective yields of singlet S_1 and triplet T_1, formation, from which fluorescence or phosphorescence is emitted with respective yields Φ_F and Φ_P*

COMMENT ON FIGURE 4.24.– *The emitted light exits the diode after undergoing a number of reflections and interferences, resulting in a further loss of light η_e of around 20%. The external quantum yield of the extracted light (fluorescence or phosphorescence) from the diode is given by the relation EQE = $\eta_r \chi_S \chi_T \Phi_F \Phi_P \eta_e$. The numbers 1.0, 1.5, 1.9 and 1.7 are the refractive indices of the media through which the light passes, and condition reflection and parasitic interference phenomena. Adapted from [ADA 09].*

The first thin-film OLEDs, based on the fluorescent emissions of a given material, produced very low yields. The first step toward improving yields may take the form of mixing dyes into the emissive material; this technique can also be used to obtain white light emissions. Kido *et al.* [KID 94] used this process to manufacture the first white-emitting OLED (WOLED) in 1994. By mixing the principal blue-emitting fluorescent material known as poly(vinylcarbazole) (PVK), with blue, green and orange fluorescent dyes, the authors obtained an OLED which emits white light with a maximum luminance of 3400 cd/m^2 for a voltage of 14 V; however, this luminance is only 50 cd/m^2 at 10 V and the energy yield is less than 1% (0.83%)[26].

26 For comparison, note that the brightness of a CRT (cathode ray tube) screen is about 100 cd/m^2.

4.2.2.2. *Phosphorescent OLEDs*

As we have seen, the proportion of $^3T^*$ triplet states resulting from hole-electron recombination is three times higher than that of $^1S^*$ singlet excitons. It thus seems logical to use phosphorescent, rather than fluorescent, materials. Unfortunately, most organic phosphorescent compounds are weakly emissive at room temperature, only reaching significant levels at very low temperatures (77 K); moreover, the triplet states are too long-lasting for high efficiencies to be obtained using phosphorescence, due to accumulation and annihilation phenomena in the triplet states, which finally lead to a thermal dissipation of energy.

Baldo *et al.* [BAL 98b, BAL 99] proposed a solution to this problem based on the use of a fluorescent host material doped with a highly emissive phosphorescent material, selected in such a way that the energy of the $^1S^*$ excited state of the fluorophore can be transferred to the $^3T^*$ triplet state of the phosphorescent dopant. This result was obtained using Alq_3 as the fluorophore and platinum octaethylporphyrin (PtOEP) as a phosphorescent dopant[27]. Alq_3 alone emits at 530 nm (green), while PtOEP emits at 650 nm (red). Using PtOEP-doped Alq_3 as the active layer in a diode, the authors observed a double emission of light at 530 nm (weak) and 650 nm (intense) for low doping concentrations (1%). They explain that the red light emission is due to the radiative deactivation of the triplet state of PtOEP, originating in a first conversion between the singlet state of Alq_3 and that of PtOEP, followed by a second conversion from the singlet to triplet state of PtOEP. This red emission increases in intensity as the dopant concentration increases. At a PtOEP concentration level of 4%, the blue emission disappears, and a single emission at 650 nm is observed; maximum intensity is achieved using a concentration of 6%, giving an EQE quantum yield of 4%. At higher concentrations (above 6%), the intensity ceases to increase due to an accumulation effect and the annihilation of the triplet states. While these quantum yield values appear weak in comparison with current performances, they constituted a major advance at the time, and proved that genuine progress could be made using phosphorescent compounds, the validity of which had previously been called into question. They thus paved the way for a second generation of OLEDs.

The use of triplet states for light emission also forms the basis for a new white light production technique and a new form of WOLED . Given that white light is composed of 25% blue and 75% green and red, corresponding to the proportion of excited singlet $^1S^*$ and triplet $^3T^*$ states of the excitons initially formed in a luminescent layer (see Figure 4.24), white light may be produced by an OLED made up of a fluorescent

27 The presence of a heavy metal such as platinum in an organic compound has the effect of facilitating the singlet → triplet conversion reaction (considered to be a "forbidden" transition) and increasing the probability of phosphorescence emission at the expense of fluorescence.

emitting layer (F), producing blue light, juxtaposed with a phosphorescent emitting layer (P), emitting purple light (a mixture of red and green). As the initial excitation takes place from the fluorescent material F, the phosphorescent material can only begin to emit following the conversion of a part of the singlet states of F ($^1S^*_F$) into triplet states of P ($^3T^*_P$). This transformation involves two successive steps[28]: the first corresponds to an energy transfer between $^1S^*_F$ and the ground state of P ($^1S_{P,0}$) and leads to the formation of $^1S^*_P$ following a Förster-type mechanism; next, the intersystem crossing (ISC) transition results in the formation of $^3T^*_P$ (this transition generally has a low probability of occurring, but becomes much more probable when the material includes a heavy metal such as Pt or Ir) (Figure 4.25a).

The second transformation, which again produces $^3T^*_P$ triplet states, corresponds to an exchange reaction occurring between the triplet state of F ($^3T^*_F$) and P according to a Dexter mechanism, and leads to the triplet state of P ($^3T^*_P$). Leo and coworkers [SCH 09] have shown that this collection of triplet states from a fluorophore F is most successful when the energy difference between singlet and triplet states is small, and when the energy of the triplet state of the phosphor P is lower than that of the triplet state of the fluorophore F (Figure 4.25b). IQEs close to 100% can be achieved in this case. In the opposite situation, where the energy of $^3T^*_P$ is greater than that of $^3T^*_F$, some of the triplet states of P are converted to non-radiative triplet states of F, and thus lost in the form of thermal energy (Figure 4.25a).

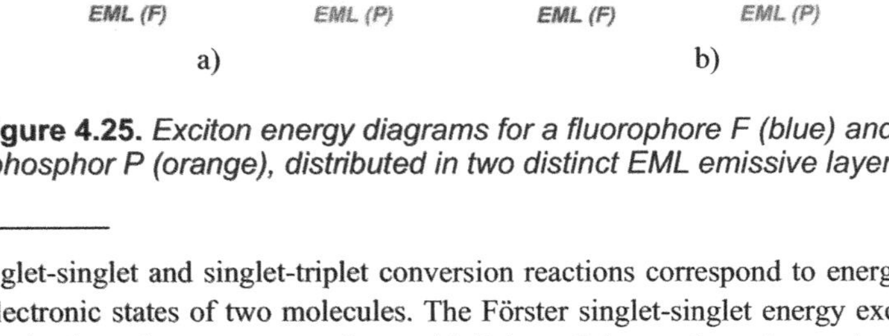

Figure 4.25. *Exciton energy diagrams for a fluorophore F (blue) and a phosphor P (orange), distributed in two distinct EML emissive layers*

---

28 The singlet-singlet and singlet-triplet conversion reactions correspond to energy transfers between electronic states of two molecules. The Förster singlet-singlet energy exchange is a dipolar mechanism that preserves the multiplicity of interacting electronic states and corresponds to the spin exchange reaction between two molecules H and F ($^1S^*_H + {}^1S_{F,0} \rightarrow {}^1S_{H,0} + {}^1S^*_F$). It is effective up to a maximum distance of 4 nm. The Dexter singlet-triplet conversion reaction follows a different mechanism and involves electron transfer between triplet and singlet states, according to the scheme $^3T^*_H + {}^1S_{P,0} \rightarrow {}^1S_{H,0} + {}^3T^*_P$. It can only operate over smaller distances of under 1 nm (electronic states with * correspond to excited states, while an index value $_0$ indicates the ground state).

COMMENT ON FIGURE 4.25.– *a) The fluorophore F is characterized by a triplet state and a singlet state, with energies that are lower and higher, respectively, than those of the triplet and singlet states of the phosphor P. This energy configuration leads to a loss of emission efficiency due to a non-radiative deactivation (n.r.) of the triplet state of F to its ground state. b) In contrast to the previous case, the energy of the triplet state of the fluorophore is higher than that of the triplet state of P, which in this case increases the concentration of the triplets of P, promotes orange emission from P and eliminates non-radiative deactivation of F. Adapted from [SCH 09].*

These energy transfer phenomena between molecular electronic states form the basis for WOLED production. This implies the choice of a blue fluorescent luminophore and green and red phosphorescent compounds, either arranged in distinct emissive layers, or used to dope a host layer, with energy characteristics verifying the criteria described above. Note, however, that phosphorescent emissions degrade at high current intensities and are subject to a roll-off effect. This phenomenon, which is detrimental to the operation of phosphor diodes, is due in part to an accumulation of triplet states in the emissive layer, resulting in different annihilation mechanisms, and in part to an imbalance between the mobilities of positive and negative charges, which also contributes to the non-radiative deactivation of excited states [MUR 13].

Different solutions have been put forward to mitigate this roll-off effect in WOLEDs. For instance, bipolar host materials[29] can be used to balance the flow of electrons and holes; these materials are also characterized by high-energy triplet states, are capable of emitting in the blue and are doped with red and green-emitting substances. High EQEs, in excess of 15%, have been achieved in this case [CHO 10].

4.2.2.3. *P- and E-type delayed fluorescence OLEDs*

In the previous case, singlet states are converted into triplet states and lead to phosphorescence, with an internal quantum efficiency close to 100%. Another new idea is to perform the reverse process by converting the triplet states of an emitting compound into singlet states and to emit fluorescence. The advantage of this approach, compared to the phosphorescence-based system, is that it does not require the use of precious metals such as Ir, Pd, Ru and Pt, which are essential for efficient emissions in the previous case; the roll-off effect at high current densities is also

29 A bipolar compound is characterized by a structure in which an electron-accepting functional group (carbonyl type C = O, sulfoxide type S = O) and an electron-donating group (phenyl-carbazolyl type, phenyl-amino) coexist. Bis-4-(N-carbazolyl)phenyl) phenyl-phosphine oxide (BCPO) is an example of a bipolar compound used as a blue phosphorescent host material in WOLEDs.

reduced, due precisely to an accumulation of triplet states and their non-radiative deactivation.

Depending on the choice of the compound, this process can be carried out in different ways. One mechanism involves merging two $^{3}T^{*}_{1}$ triplet states to produce a triplet state of order m ($^{3}T^{*}_{m}$) which transmits its energy to the $^{1}S_{1}^{*}$ singlet state (with energy lower than $^{3}T^{*}_{m}$). This results in fluorescence being obtained with a maximum IQE of 62.5%, which corresponds to a direct emission of the singlet states (25%), plus delayed emission from the triplet states, that is, an additional 37.5% production of singlet excitons. This fluorescence mechanism corresponds to a process of annihilation of two triplet states (TTA – triplet-triplet annihilation), and is often referred to as P-type DF (delayed fluorescence).

The first TTA-OLEDs were produced at low temperature using thin films of MEH PPV (poly[2-methoxy-5-(2′poly[2-methoxy-5-(2′-ethyl-hexyloxy)-1,4 phenylenevinylene]) and with polyfluorene, verifying the validity of the mechanism [SIN 03].

A few years later, Kondalov *et al.* [KON 09] produced a red light emitting TTA-OLED using rubrene (5,6,11,12-tetraphenylnaphthacene) as a source of triplet states. They clearly demonstrated the validity of the TTA mechanism and obtained an EQE above 11%, very close to the maximum quantum yield of 12.5%, taking into account the extraction yield of 20% cited by the authors. However, this mechanism implies a high accumulation of triplet states, which in turn can lead to a roll-off effect and, consequently, a degradation of the yields.

More recently, other modes of using the TTA effect have been explored, transforming fluorophores into TTA materials using a sensitizer or an exciplex[30].

In the case of a fluorescent sensitizer, the approach consists of transferring energy from the triplet state of the fluorophore to the triplet state of the emitter, transforming two triplet states ($^{3}T_{1}^{*}$) into an excited state of higher energy (singlet or triplet state) via the TTA mechanism; this state then emits. In this way, it is possible to obtain two emissions, one from the sensitizer, the other from the emitting material. Using Alq_3 as the sensitizer and an anthracene derivative (ADN or 9,10-bis(2′-naphthyl)anthracene) as the TTA emitter, Bardeen, Lee and coworkers [CHE 18a] obtained two emissions of green and blue light: the authors attributed the blue emission to phosphorescent luminescence resulting from the TTA process, whereas the green emission is due to fluorescence (Figure 4.26a).

30 An exciplex is a complex formed by the interaction of an excited molecular entity with a ground state counterpart of a different structure.

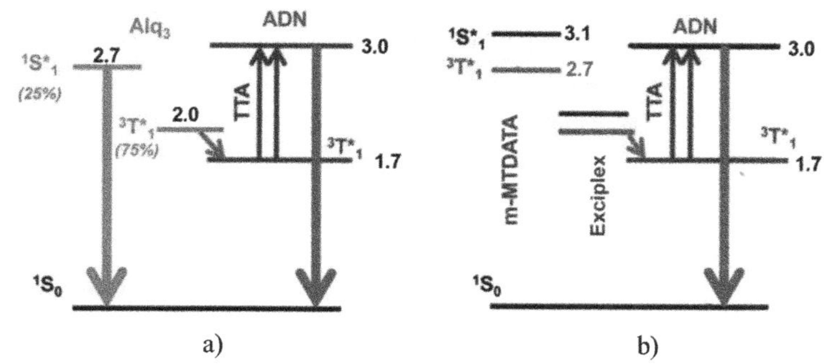

Figure 4.26. *Energy diagrams of two delayed emission mechanisms involving TTA*

COMMENT ON FIGURE 4.26.– *a) Alq_3 sensitizer associated with the ADN emitter. Two light emissions are obtained: one centered at 522 nm (green) from Alq_3, the other, delayed, at 456 nm, probably resulting from a triplet state of order greater than 1 and generated by the TTA process. b) TTA mechanism involving the exciplex of m-MTDATA and ADN generated at the interface of the m-MTDATA/ADN heterojunction. a) Adapted from [CHE 18a]. b) Adapted from [LIN 17a].*

Another means of obtaining blue emissions involves the use of an exciplex state followed by a TTA conversion: this method has been demonstrated by the previous two groups [LIN 17a]. Using an exciplex as a sensitizer presents an advantage in that the voltage of the OLED cell can be significantly reduced. With the exciplex formed at the interface of a heterojunction between the m-MTDATA (4,4′,4″-tris(N-3-methyphenyl-N-phenyl-amino) triphenylamine) donor and the ADN acceptor, a blue light emission centered at 432 nm, corresponding to an energy of the order of 3 eV, is obtained, and the voltage required to trigger electroluminescence is only 2.2 V. This result stems from the fact that the triplet state of the exciplex, with an energy of the order of 2 eV, transfers its energy to the triplet state of the ADN ($^3T^*_1$; 1.7 eV), which then generates a triplet state of much higher energy (3 eV) via the TTA process; this triplet state then deactivates, emitting blue light (Figure 4.26b).

A further delayed fluorescence mechanism involves the use of compounds for which the triplet state $^3T_1^*$ has a slightly lower energy level than the singlet state $^1S_1^*$, and, by thermal activation, can transfer its energy to the singlet state according to a reverse ISC process (RISC). This gives a theoretical maximum IQE of 100% fluorescence, made up of 25% direct emission and 75% delayed emission. This fluorescence process is sometimes referred to as E-type delayed fluorescence (E-type DF) or thermally assisted delayed fluorescence (TADF)[31] (Figure 4.27).

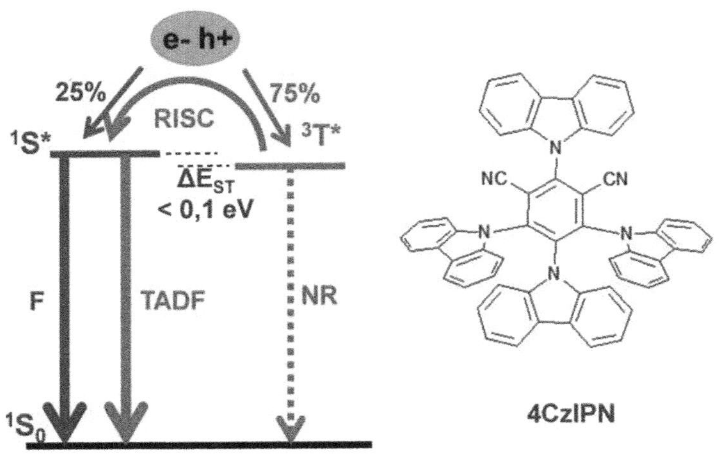

Figure 4.27. *Energy diagram of a TADF fluorescence emission. The composite 4CzIPN has four electron-donating carbazolyl groups and two electron-accepting nitrile (CN) groups. The 3T* excitons are completely reconverted to $^1S^*$ with no phosphorescence emissions or non-radiative deactivation (NR). Adapted from [UOY 12]*

For this to be possible, the RISC process corresponding to the conversion of the triplet state to the singlet state must be very fast, implying that the energy difference between the triplet and singlet states ΔST must be small (under 0.2 eV). Unfortunately, this usually leads to low intensities of radiative emissivity, counteracting the desired result. Adachi and coworkers [UOY 12] showed that this negative effect can be suppressed by using particular molecules of a bipolar type, characterized by a weak orbital overlap of the HOMO and the LUMO. The fluorescent emission in this case is intense and takes advantage of the collection of triplet states, reconverted into fluorescent singlet states.

Using a molecule such as 1,2,3,5-tetrakis(carbazol-9-yl)-4,6-dicyano-benzene (4CzIPN) with four carbazolyl groups and two nitrile groups attached to a phenyl ring, acting respectively as donors and acceptors. Adachi and coworkers [UOY 12] obtained an OLED that emits green light (507 nm), operating exclusively using the TADF principle, with a high EQE quantum yield of about 19%, close to the theoretical maximum of 20%. Just as we saw in the case of the TTA mechanism, the

---

31 The letters E and P of E-type DF and P-type DF designate the eosin and pyrene molecules with which these phenomena have been observed.

emission features two time constants: one, which is extremely short, corresponds to the direct fluorescence emission, while the other longer period corresponds to the RISC process in which triplet states are converted into singlet states [UOY 12].

New D-A compounds have been synthesized in recent years, achieving even higher emission efficiencies in the blue, green and red. A whole family of bipolar compounds with an acceptor group containing a boron atom and a donor group containing phenyl rings were found to be excellent fluorophores meeting the criteria for a TADF mechanism[32]. The bipolar TDBA-DI compound thus includes a first symmetrical, TDBA acceptor functional group structured by oxygen-boron bridging bonds. It is linked to a second donor group DI made up of indolocarbazole functions. The whole unit provides favorable conditions for TADF fluorescence emissions (Figure 4.28).

Compounds of this type are characterized by an assembly in which the acceptor (A) and donor (D) fragments are located in two quasi-orthogonal planes, which has the effect of confining the HOMO and LUMO to A and D, respectively, and leads to a very low ΔEST of 0.073 eV; this favors the RISC process between the triplet and singlet states and prevents non-radiative deactivation of the triplet state (Figure 4.27).

Depending on the nature of the host material, EQE of between 21% and 38% may be obtained in terms of blue light. Energy yields are also high, in excess of 50 lm/W for a luminance of 1 cd/m^2, and remain above 30 lm/W for luminances of 1000 cd/m^2. According to the authors, these high EQE values may result from improved extraction efficiency and from the horizontal orientation of the dipole formed by the emitting molecule.

TADF fluorescent OLEDs thus constitute a new family of LEDs with the potential to outperform phosphorescent OLEDs and TTA OLEDs. These "third-generation" OLEDs do not require the use of expensive heavy metal-based phosphors using materials such as platinum, iridium or ruthenium; furthermore, all of the singlet and triplet excitons resulting from the combination of injected charges can be transformed into radiative emission. The recent development of boron-based bipolar compounds is promising, but the stability of boron derivatives must be improved in order for the technology to become fully effective and applicable.

32 The boron atom is isoelectronic to a positive carbon (carbocation) due to a vacant p-orbital, which giving it a pronounced accepting character. Many D-A compounds have been synthesized using boron as an acceptor element; one typical example is that of a dimesityl borane group $[C_6H_2(CH_3)_3]_2B$ linked to a donor group of the tetraphenyl ethylene type $(C_6H_5)_3(C=C)–C_6H_4–$ [HUO 19].

It is interesting to note that hyperfluorescent OLEDs, based on TADF technology, have been commercially available since 2016; produced by the Wise Chip company, in Japan, these components are used in passive matrix display equipment (PMOLED).

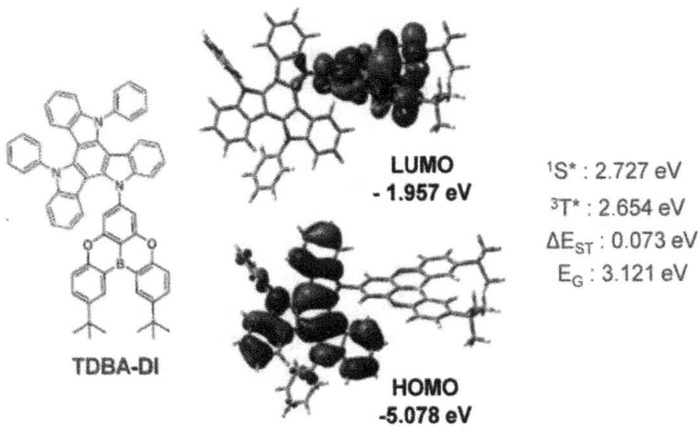

Figure 4.28. *HOMO and LUMO molecular orbitals of TDBA-DI. TDBA (blue) and DI (red) represent the "acceptor" and "donor" part of the molecule, respectively. Note the respective confinements of the HOMO and LUMO to the DI and TDBA. Adapted from [AHN 19]*

4.2.3. *QDot light-emitting diodes*

As we saw in Chapter 2 (QDots or QDs), semiconductor nanomaterials have excellent photoluminescence properties in solution, both in terms of quantum yields – close to one – and in the quality of the emitted colors. However, one major problem with respect to display applications is to determine whether these photoluminescence properties (observed in solution) are preserved if the QDs are used in solid film form, that is, a form suitable for use in electroluminescent devices.

In television displays, QDot diodes can be used in a similar way to OLEDs, where pixels made up of red, green and blue organic electroluminescent compounds (RGB primary color system) are directly excited by an electric current supplied by thin-film transistors. Compared to conventional LCD displays, this presents the advantage of simplified design, faster response times, lower energy consumption and, notably, better color contrast and a wider field of vision [DAI 17].

The Jin and Peng groups described the creation of a red-light emitting electroluminescent pixel produced entirely from colloidal solutions [DAI 14]. The

LED is of classic structure and is characterized by different hole- and electron-conducting layers, located on either side of the emitting QD layer (Figure 4.29).

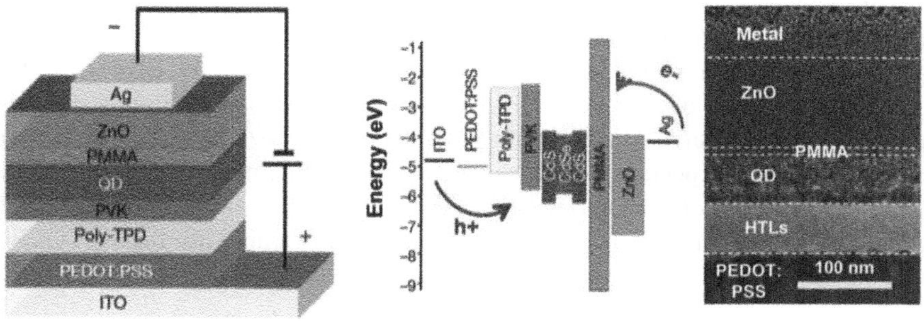

a) b) c)

Figure 4.29. *Multilayer QLED device*

COMMENT ON FIGURE 4.29.– *a) Structural diagram of a QLED showing the different layers inserted between a positive ITO (indium tin oxide) transparent electrode (anode), from which holes (h^+) are injected, and a negative silver electrode (cathode) for electron (e^-) injection. The conductive PEDOT:PSS layer provides good electrical contact between the ITO electrode and the HTL layers; the thin PMMA (insulating) layer reduces electron flow to levels compatible with the hole mobility. b) Energy diagram showing the relative arrangement of the VB and CB of the different semiconductor materials. c) Transmission electron microscopy image of a cross-section of the QLED. Adapted from [DAI 14].*

The p-type HTLs are made of poly(N,N'-bis(4-butylphenyl)-N,N'-bis(phenyl)-benzidine) (poly-TPD, 30 nm) and poly(9-vinylcarbazole) (PVK, 5 nm)[33]. The emissive layer is 40 nm thick, and is based on a core-shell structure QD, CdSe@ZnS. The ETL (electron transport layer) is zinc oxide (ZnO, 150 nm), separated from the emissive layer by a thin layer of poly(methacrylate) (PMMA, 6 nm). According to the authors, this thin layer of PMMA is the key to the proper functioning of the diode, contributing to balancing electron and hole injections, and

33 Hole mobility in PVK and poly-TPD HTLs is 2.5×10^{-6} cm^2/Vs and 10^{-4} cm^2/Vs, respectively, while electron mobility in ZnO is around 1.8×10^{-3} cm^2/Vs, that is, more than 10–100 times higher than the hole mobility in the HTL. The choice of low thicknesses of PVK and PMMA is intended to optimize emission yields by equalizing the intensities of the electric currents due to electrons and holes, respectively, and to prevent the accumulation of one category of load carriers in the emissive layer.

thus avoiding an accumulation of electrons in the emissive layer which would be detrimental to the correct operation of the diode.

This device exhibits excellent performance, emitting an extremely pure red color in the form of a narrow and perfectly symmetrical emission peak, centered at 640 nm and with a width at half-height of only 28 nm. The EQE is 20.5%, comparable to that of the best OLEDs, and decreases only slightly for high intensities (low roll-off). The emission voltage is relatively low (1.7 V at 100 mA cm^{-2}), and the life expectancy of the device is estimated at more than 100000 hours (i.e., over 10 years) of continuous operation with a luminance of 100 candela m^{-2}. According to the authors, this performance is comparable to that of OLEDs manufactured using vacuum vaporization techniques [DAI 14].

Similar results have been obtained for QLEDs emitting green and blue colors. Several teams are now working on producing QLEDs by inkjet printing or transfer printing techniques, which are evidently much cheaper than traditional vacuum evaporation methods and are well suited for producing flexible and portable displays connected to different types of sensors [CHO 18a]. Industrial-scale production of QLED TV displays is a recent development, but for the time being, vacuum evaporation techniques are still being used for their manufacture. One important aim is now the development of colloidal QD-based displays, requiring new printing techniques using QD solutions; these are expected to be more cost-efficient.

4.2.3.1. *Transparent and flexible QLEDs*

The latest generation of portable optoelectronic devices involve increasingly flexible systems that can conform to a variety of non-planar shapes while retaining their intrinsic properties. The development and pursuit of "augmented reality", in which data-based virtual images are superimposed onto a clear real-world background, has led to a new focus on direct coupling of transparent display screens with substrates that may themselves be transparent or opaque. Figure 4.30 shows an example of the operation of a transparent QLED (the butterfly image) affixed to a curved glass panel.

Extremely thin QD screens can now be produced, characterized by very high brightnesses (up to 200000 cd/m^2), high-purity colors (with emission peak widths of around 30 nm at half-height) and low emission voltages of emission (lower than 2 V), characteristics which make them ideally suited for this type of application.

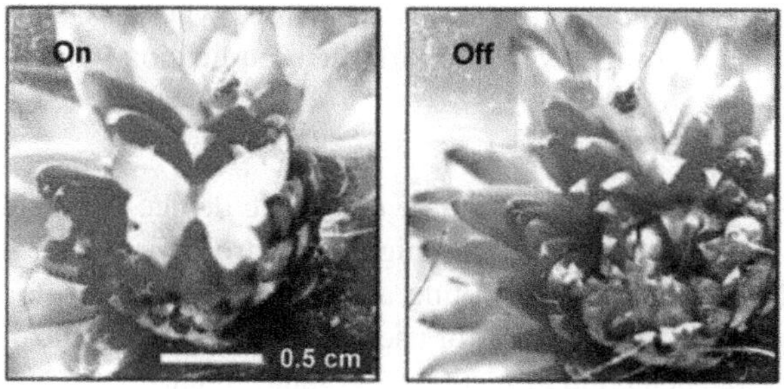

a) b)

Figure 4.30. *Operation of a transparent QLED overlaid on colored glass. a) Butterfly image corresponding to the "on" state of the QLED. b) The same system with the QLED switched off. Adapted from [CHO 18b]*

Various solutions have been put forward for producing transparent and flexible QLEDs. The major challenge lies in combining high brightness and high transparency. Choi *et al.* [CHO 18b] recently produced transparent QLEDs containing core-shell CdSe@ZnS QDs, constructed in a similar way to that shown in Figure 4.29. Once again, ZnO is used for the ETL layer, covered by a thin insulating layer of alumina (2 nm); as before, this reduces electrode injection from the cathode and makes it compatible with the hole injection rate of the anode. The HTL layer is made of TFB[34], a commercial hole-transporting polymer, comparable in properties to PDT, and with lower conduction than ZnO [CHO 18b]. The two ITO electrodes are deposited by evaporation and cause the light emission to emerge from both sides of the QLED; however, slightly different luminances are emitted by the upper (30000 cd/m^2) and lower (43000 cd/m^2) sides, giving a total of 73000 cd/m^2 for an applied voltage of 9 V. These values far exceed those of OLED (organic LED) or PLED (phosphorescent LED) based systems. The overall thickness of the QLED, after encapsulation with a layer of transparent parylene on both sides, is very low and does not exceed 5 μm; the system can thus withstand stress tests with no negative effects on its emission characteristics. The transparency of the whole is also very good; the transmission of light in the green (554 nm) is above 90%, and transmission is in excess of 80% for the whole visible spectrum.

Other QLEDs can also be produced with red and blue emissions, using the same type of core/shell QD with a CdSe core of controlled size, covered either by a CdS/ZnS bilayer for red emissions or a single layer of ZnS for blue emissions. The mixture of the three types of QD leads to a transparent diode emitting white light.

34 TFB is a copolymer derived from fluorene and a diphenyl amine of the formula poly[(9,9-dioctylfluorenyl-2,7-diyl)-co-(4,4′-(N-(4-sec-butylphenyl)diphenylamine)].

The details concerning the preparation of these three types of QDs are described by Choi *et al.* in the supporting information for their publication [CHO 18b].

4.2.3.2. *QLED displays in wearable electronics*

The current high demand for portable electronic devices with integrated display screens raises a need for ultra-thin, highly electroluminescent screens, able to operate under low voltages and emitting a range of very pure colors. QLEDs meet these criteria, making them suitable for use in patch-type biomedical photoelectronic devices for real-time detection of plethysmographic signals[35].

A wearable photoelectronic device designed to display blood pressure variations in real time (a plethysmograph-type device) has recently been put forward by Korean teams working in association with Samsung. The system consists of a QLED (emitting in the red) coupled to a PbS-based photodetector (PhD) (Figure 4.31).

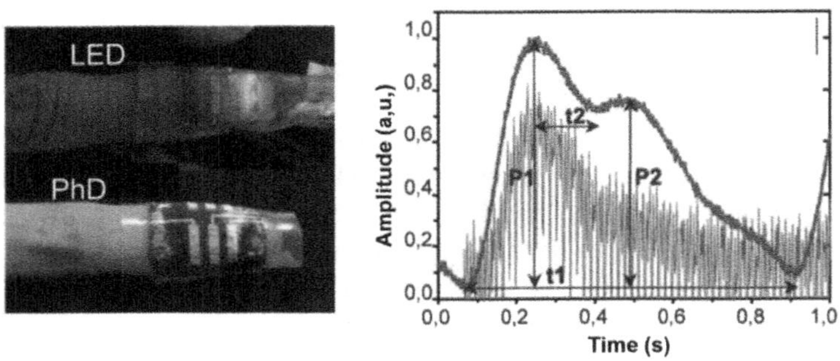

a) b)

Figure 4.31. *Portable plethysmograph device for blood pressure measurement*

COMMENT ON FIGURE 4.31.– *a) The QLED and the photodetector (PhD) are placed close to one another on the distal phalanx of the patient's index finger. b) Blood pressure variation recorded by the PhD. The red curve is obtained using a flexible QLED with graphene monolayer electrodes. The black dashed line corresponds to a rigid QLED with ITO electrodes. Adapted from [KIM 17].*

Both components are flexible, and can be deformed without losing their properties. Attached to the index finger, the light emitted by the QLED is absorbed by the skin and transmitted directly to the PhD which, under weak polarization, produces a current depending on the intensity of the intercepted light.

35 Plethysmography refers to a set of methods used in medicine to measure volume variations, mainly in cardiology for the measurement of heart rhythms.

The QLED is wrapped around the tip of the distal phalanx of the index finger, with the PhD placed near the QLED. The red light is absorbed by the skin and received by the PhD. Depending on blood pressure, a greater or lesser intensity of light is transmitted to the PhD, which then reproduces the blood pressure variations.

4.3. Conclusions on energy conversion

Photovoltaics and electroluminescence are two types of systems used for energy conversion. The first corresponds to a transformation of light into electricity, while the second, the inverse of the first, corresponds to the transformation of electrical energy into light. Significant ongoing developments in the field of photovoltaics aim to reduce the carbon footprint of human activities; however, an old technology based on the use of monocrystalline silicon still remains dominant in the area. Constant improvements to this technique have resulted in cost, reliability and photoconversion yield parameters which currently offer the best solution for commercial use. Other emerging technologies (OPVs, DSSCs and perovskite-based technologies), which use nanotechnology, are not yet able to compete with silicon. For the moment, these approaches are limited to niche markets with specific requirements (powering connected objects, flexible photovoltaic panels or indoor power supplies). Despite the rapid progress made in organic and perovskite technologies in recent years, essentially in terms of yields, problems relating to stability and longevity still remain unsolved. The efficient transformation of electricity into light is a further important area of research aimed at reducing our carbon footprint. LEDs, based on inorganic, organic or QDot materials, have revolutionized the field of display and lighting, often through the use of nanomaterials, nanometric thin films and other forms of nanotechnology. This fundamental shift has resulted in the creation of new, energy-efficient products, and the widespread adoption of these new lighting solutions is likely to lead to significant savings in terms of energy usage.

5

Electro- and Photocatalysis

Catalysis is a vast field, relating to the application of chemical processes to lower the activation energies of reactions and, consequently, to increase transformation rates. This has major implications in terms of energy consumption. The impact of nanotechnologies on various forms of catalysis will be highlighted in this chapter, with a particular focus on the contribution of nanomaterials, described in Chapters 1 and 2 of this book.

In this context, two major themes of active research in connection with more general environmental problems will be addressed. The first relates to the production of hydrogen by catalytic decomposition of water (water splitting), while the second concerns the use of hydrogen as a fuel for electricity production via the use of H_2/O_2 fuel cells. Catalytic processes intended to add value to certain natural and abundant compounds, such as the reduction of CO_2 to produce low-carbon fuels (CO, CH_4 or CH_3OH) or the electrochemical activation of small molecules such as nitrogen, which will not be discussed here.

Water splitting is a highly demanding reaction in terms of energy usage[1]; it can be carried out by heating, by photochemistry or by electrolysis, and is of great interest for use alongside renewable energy sources (RESs), such as solar or wind generators, or if it can profit from the recovery of latent heat sources, for example, from nuclear reactors. Unfortunately, these transformations – intended to provide

For a color version of all the figures in this chapter, see: www.iste.co.uk/lacaze/nanotechnology.zip.

1 The free energy of water splitting ($H_2O \rightarrow H_2 + ½\ O_2$) under standard pressure and temperature conditions is 237 kJ, corresponding to an exchange of $2e^-$ at the thermodynamic potential of 1.23 V.

“green” alternatives – are currently more costly than current industrial methods based on the use of fossil fuels. Minimizing this cost is crucial, and, to this end, the search for appropriate, specific catalysts for each type of transformation is of the utmost importance. The results of this research will condition the development and marketing of a process with huge potential environmental implications. Nanotechnologies have a significant role to play in many approaches.

5.1. Water splitting

Around 95% of the H_2 currently used for industrial applications, out of a global production of around 60 million tons (Mt), is obtained by steam methane reforming (SMR) of natural gas (CH_4). The remaining 5% is obtained by various non-carbon based production methods, including around 4% from electrolysis and 1% from photochemical, photothermal and, to a lesser extent, biomass fermentation transformations[2].

Water splitting and hydrogen production by electrolysis, corresponding to the general Power to Gas (P2G) transformation mode, are considered as a future pathway for “green” hydrogen production. At present, water splitting is the only alternative to SMR for obtaining large quantities of hydrogen using electrolyzers of variable sizes, with powers ranging from 0.3 to 1 MW[3]. It is also the only way to produce large quantities of H_2 in a gaseous state, locally and on demand; this can be stored under pressure, and avoids the need for long-distance transport solutions.

Unlike the SMR process, which simultaneously produces significant amounts of CO_2 and is subject to impurities[4], the electrolytic process results in extremely pure hydrogen which is suitable for use in fuel cells. The fact that it can be produced locally on site also makes it suitable for use in fueling stations for fuel cell powered

2 Hydrogen is mostly used for synthesizing of ammonia (accounting for 55% of production), 90% of which is then transformed into agricultural fertilizers. A further 25% is used for hydrocarbon refining, metallurgy and plate glass industries, and the electronics industry; 10% is used for methanol production, while the remaining 10% is used in various industrial applications (source: Hydrogen Europe).

3 An electrolyzer operating with a current density of $1 A/cm^2$ under a voltage of around 2 V only provides a power of 2 W/cm^2. To reach 1 MW of power, the catalyst must have an equivalent surface of 50 m^2, corresponding to a production of 5.2 g of H_2 per second, or 18.7 kg H_2 per hour.

4 It is estimated that the CO_2 produced by the SMR process ($CH_4 + 2H_2O \rightarrow CO_2 + 4\ H_2$) contributes 0.3% to total CO_2 emissions, which corresponds to the fact that for every 8 g of H_2 produced, 44 g of CO_2 is also released.

electric vehicles. A recent study carried out in South Carolina for BMW [FEL 15][5] shows that over a 15-year period, for a station equipped with electrolyzers operating in an alkaline environment and with the capacity to deliver 25 kg of H_2 per hour, the cost prices of H_2 produced by electrolysis are comparable to those for SMR. The cost of electrolytic H_2 is closely linked to the cost of electricity and, as far as electricity produced from RES is concerned, it seems likely to prove competitive in comparison with other production techniques [MAG 19].

5.2. Electrolysis techniques

Two main types of electrolyzers (monopolar or bipolar) are currently used to produce hydrogen under industrial operating conditions using an alkaline or acidic liquid electrolyte (Figure 5.1).

The oldest and most successful technique is performed in a concentrated potassium alkaline medium (between 20 and 30 wt% KOH in water). The electrolysis cell features cathode and anode compartments, separated by a membrane which is porous to OH^- ions and water molecules. One of the major advantages of this technique is that the electrodes are made from non-precious metals. The cathode, where reduction takes place with the release of H_2 ($2H_2O + 2\ e^- \rightarrow H_2 + 2OH^-$), is generally made from a Ni/C composite, and the anode, where oxidation takes place with the release of O_2 ($2OH^- \rightarrow ½O_2 + H_2O + 2e^-$) generally consists of an alloy based on Ni, Co and Fe.

Acid electrolysis is a more recent development, dating from the 1960s when General Electric proposed the first acid electrolyzer with a polymer membrane acting as a solid electrolyte. This technology is generally referred to as PEM electrolysis, for proton exchange membrane or polymer electrolyte membrane. The PEM is generally made from Nafion, known for its high chemical inertia and its high conductivity with respect to protons (0.1 S/cm), which allows electrolysis to be carried out in acidic media with high current densities above 2 A/cm^2, whereas the current densities in alkaline media are much lower (<0.4 A/cm^2). However, the use

5 The study was conducted in 2014 for a BMW station in South Carolina, equipping a fleet of 350 vehicles running on H_2 fuel cells [FEL 15]. The authors report that using electrolyzers operating in basic or acidic media, with a production of 20–25 kg H_2/hour, the average cost H_2 was between $5.10 and $7.50 per kg, comparable to that of liquefied H_2, transported from an SMR production unit, with an overall cost price of between $5 and $8 per kg. This cost estimate is based on an average kWh price of $0.07, and the electricity consumed accounts for three-quarters of the total cost of H_2.

of a highly acidic electrolyte, in which protons are reduced at the cathode ($2H^+ + 2e^- \rightarrow H_2$) and water is oxidized at the anode ($H_2O \rightarrow ½O_2 + 2H^+ + 2e^-$), means that non-corrodible materials with high catalytic power are required. This is the reason why current PEM electrolyzers use electrodes based on precious metals such as platinum, ruthenium or iridium. This aspect of PEM electrolysis makes it more expensive than alkaline-based electrolysis: the identification of non-corrodible materials outside of the Pt group, with catalytic properties comparable to those of Pt, is a major focus of research.

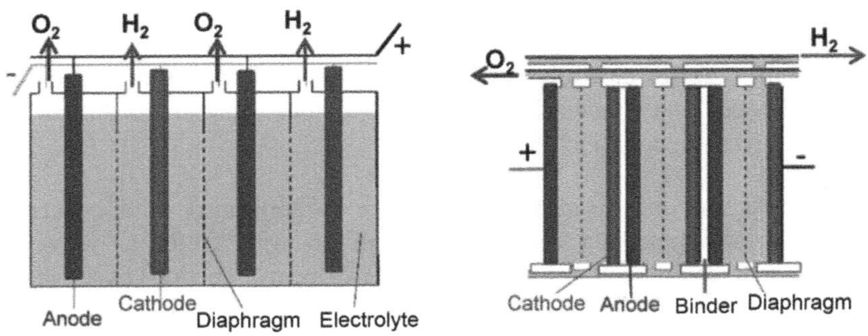

a) b)

Figure 5.1. *Simplified diagrams of electrolyzers*

COMMENT ON FIGURE 5.1.– *a) Monopolar electrolyzer. Each electrode behaves specifically as a cathode (common negative terminal) or as an anode (common positive terminal). b) Bipolar type electrolyzer. The electrode is made up of two distinct materials linked together by an electronically conductive binder. Each side works as a cathode or anode, resulting from the potential gradient established at the cell extremities. Adapted from [KRI 20].*

There are advantages and disadvantages to both processes. These have been analyzed in great detail in two review articles [CAR 13, SCH 16]. One advantage of the alkaline electrolysis technique compared to PEM is the relatively low cost of the system, due largely to the use of non-precious metals and non-fluorinated separation membranes; the technology is known to be reliable over time (guaranteed operation over periods of 80000 hours) and can reach high production levels (with powers of the order of 1 MW for a single electrolyzer). However, the H_2 produced in this way is of lower quality than that obtained by PEM electrolysis, an aspect which may limit its use in fuel cells.

Finally, there is a family of all-solid electrolyzers operating at high temperatures (between 500°C and 1000°C) that were first proposed in the 1980s, which used ceramic electrolytes that conduct O^{2-} or H^+ ions and are known as solid oxide electrolyzer cells (SOEC). The major advantage of high-operating temperatures lies in the conversion rate, which results in a voltage very close to that of the thermodynamic value of water splitting (1.23 V); furthermore, the cells have low ohmic resistance and no catalyst is needed. Solid electrolytes may be proton conductors or O^{2-} ion conductors. The advantage of a proton conductor is that very pure, dry hydrogen is obtained; conversely, an O^{2-} ion conductor produces hydrogen mixed with residual water vapor [LAG 12]. Furthermore, the higher ionic mobility of protons compared to O^{2-} ions allows the former system to operate at temperatures below 500°C; O^{2-} ion electrolytes require higher operating temperatures of between 800°C and 1000°C [MED 19]. However, significant progress has been made in the field of oxide use in recent decades. Electrolysis can be carried out with current densities of the order of 0.5–0.8 A/cm^2 and at operating temperatures between 650°C and 800°C. This is also one of the few electrolysis processes that can simultaneously reduce water vapor and CO_2; in the latter case, a mixture of CO and H_2 (syngas) is produced. Nevertheless, the longevity of these cells is still insufficient to justify large-scale application, as no tests have yet proven sufficient reliability beyond 10000 hours [WAN 19b].

In conclusion, the energy balance for H_2 and O_2 production by water splitting in a liquid medium largely depends on the electrochemical characteristics of the two electrode reactions: the hydrogen evolution reaction (HER) and the oxygen evolution reaction (OER), leading to the release of H_2 and O_2 gas, respectively. The power required for water splitting is equal to the product of the intensity of the electrolysis current I and the applied voltage U, which may be considered as the sum of three components: (1) the Nernst voltage U_N corresponding to the difference between the equilibrium potentials of the two electrode reactions ($U_N = E°(O_2/H_2O) - E°(H^+/H_2)$, i.e., 1.23 V at 25°C); (2) the voltage due to the solution resistance ($U_S = RI$); and (3) the overpotential η resulting from the reduction and oxidation kinetics of the reactions $H^+ + e^- \rightarrow ½H_2$ and $H_2O - e^- \rightarrow ½O_2 + H^+$. While the ohmic drop U_S resulting from the resistance of the solution can be significantly decreased by using solutions with a high concentration of electrolytes, the overpotential η resulting from the kinetics of each of the transformations can only be minimized by using efficient catalysts specific to each of the electrode reactions (Figure 5.2).

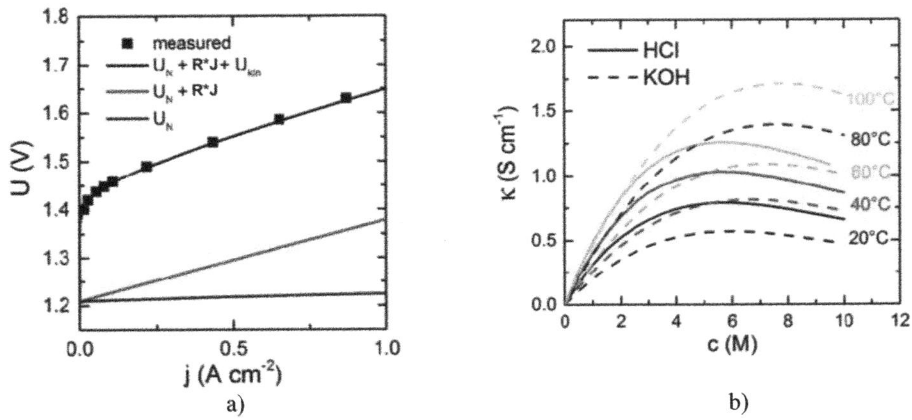

Figure 5.2. *Electrochemical characteristics of an electrolyzer for water splitting*

COMMENT ON FIGURE 5.2.– *a) Voltage at the electrodes as a function of current density corresponding to the sum of the Nernst voltage (U_N, 1.23 V), the ohmic drop due to the solution ($U_S = R^*J$) and the overpotentials resulting from the kinetics of the reduction and oxidation reactions (η). b) Conductivities of HCl and KOH electrolyte solutions between 20°C and 100°C. Note that for an electrolyte concentration of 6 M and a temperature of 60°C, the HCl and KOH solutions are characterized by an average resistance of 1 Ω cm, which, for a current density of 1 A/cm^2, results in an ohmic overpotential of 1 V cm. Adapted from [SCH 16].*

Similarly, the electrochemical reactions that recombine H_2 and O_2 into H_2O, which form the basis for fuel cell operation and correspond to the hydrogen oxidation reaction (HOR) and the oxygen reduction reaction (ORR), are also determinant from an energetic perspective in terms of fuel cell operations. The presence of overpotentials at the electrodes results in a decrease in the electromotive force of the cell, and thus reduces its energy yield.

The use of efficient and cheap catalysts is essential, and the identification and development of these catalysts is a main focus of research into H_2 production and fuel cell operations. This materials-oriented research involves structural problems on a nanometric scale, and many innovations in the area result from the introduction of nanomaterials.

5.3. HER and OER processes in water splitting

Each of the reactions involves an electron exchange process that is characterized by an electric current, an electrode potential and a rate constant k, which is a function of the overpotential applied to the electrode. Electrolysis on an industrial scale requires high current densities (between 1 and 2 A/cm^2), which imply the application of high overpotentials to the electrodes, compared to the standard thermodynamic potential E_0 of the reaction. These overpotentials decrease as the rate constant k_0 measured at the potential E_0 increases (Box 5.1). At present, the highest rate constants k_0 are obtained for precious metals such as platinum and ruthenium/iridium alloys, which thus constitute the most efficient electrochemical catalysts for reducing the required overpotential amplitude, which is a source of energy loss. Unfortunately, these materials are very expensive; even when used in a very divided form, generally mixed with carbon micro-particles (activated carbons) in order to minimize the quantity of precious metal while maintaining their catalytic effect, the cost of the materials remains a significant drawback. For this reason, the replacement of Pt-family metals with electrochemical catalysts which are cheaper, but offer similar performances in terms of HER and OER reactions, is a major focus of research.

A further difference between electrolysis in basic or acidic media concerns corrosion of the electrode metals. Nickel-cobalt based materials, which are good catalysts in alkaline media, cannot be used in acidic media due to corrosion. A variety of other, abundant and cost-effective materials are currently being tested in the hope of matching or exceeding the catalytic properties of platinum group metals (PGMs) for water splitting. The reactions, like the catalysts, are very different in acidic and alkaline media; for this reason, the HER and OER processes will be examined separately for each medium.

5.3.1. *HER in an acidic medium*

The commercial catalysts which are currently available consist of composites containing Pt NPs (nanoparticles) and activated carbon, with between 20% and 40% wt Pt. The main objective of research in this area is to identify new electrode materials with little to no precious metal; one option is to deposit minute quantities of PGMs in the form of isolated atoms or monolayers onto a support with low catalytic power.

Many examples of different materials and their performances can be found in the literature. For the purposes of our presentation, we have limited our discussion to a few particularly promising and well-documented examples.

5.3.1.1. *Main Ni- and Mo-based catalysts*

The use of transition metals, mainly Mo and Ni, in the form of nitrides, carbides, borides and phosphides, has been extensively studied and described by Zeng and Li [ZEN 15].

Among these, mixed nickel and molybdenum nitride electrode materials, synthesized by the Azdic group as carbon-substrate Ni-Mo nitride nanosheets (NiMoNx/C), were found to be very efficient and significantly better than Mo nitride (MoN/C) alone; the authors believe that this is the result of a synergistic effect between Ni and Mo. The material is stable in acidic media, and the overpotential is significantly improved compared to simple nitride (decrease of 150 mV at –5 mA/cm^2); however, it remains higher than that observed with a Pt/C catalyst (around 150 mV at –5 mA/cm^2) [CHE 12].

Comparable results have been obtained using nanostructured molybdenum carbides (Mo_2C), tested in the form of NPs anchored on CNTs (carbon nanotubes) [CHE 13b], or as porous nanowires [LIA 14]. In the latter case, an electrode consisting of an assembly of porous Mo_2C nanowires has been shown to be very stable in an 0.5 M H_2SO_4 medium, with no significant reduction of the electrolysis current over a period of more than 25 hours with an overpotential of 130 mV applied to the electrode and constant production of a 15 mA/cm^2 current. The very clear improvement in performance compared to compact Mo_2C (negligible current at high overpotentials) is attributed to the high porosity of the material, giving a very large specific surface area that considerably increases the number of active sites of the catalyst which are accessible to protons.

In the case of metal borides and phosphides, particularly molybdenum borides [VRU 12] and nickel phosphides (Ni_2P) [POP 13], the latter has been shown to perform best in an acidic medium, highlighting the predominant role of phosphorus. The electrode material, produced using a thermal process, consists of spherical crystalline Ni_2P NPs (average size 21 nm) deposited on a titanium substrate. The reduction current-potential curves obtained in 0.5 M sulfuric acid medium show respective overpotentials of 130 and 180 mV at 20 mA/cm^2 and 100 mA/cm^2, that is, a voltage difference of around 100 mV compared to Pt/C. The exchange current is evaluated at 3.3×10^{-5} A/cm^2, nearly two orders of magnitude below that of a standard Pt/C catalyst (2.7×10^{-3} A/cm^2). The Tafel slope of 46 mV/dec reflects a reduction mechanism different to that observed on Pt/C, for which the slope of 30 mV implies a decisive step of H atom dimerization. According to Popczun *et al.* [POP 13] and Callejas *et al.* [CAL 16], these characteristics are equivalent to, if not better than, those observed using a range of molybdenum-based catalysts, such as Mo_2C in the form of NPs deposited on CNTs, MoB, MoS_2 on a graphene substrate, or Ni-Mo-N sheets.

The rate constant k of an electrochemical oxidation or reduction reaction is given by the Butler-Volmer relation:

$$k = k_0 \exp\text{-}(\alpha\eta F/RT),$$

where k_0 (in s^{-1}) represents the rate constant of the transformation at the standard thermodynamic potential E_0, α is the transfer coefficient of the reaction ($0 < \alpha < 1$), $\eta = E - E_0$ is the overpotential at the electrode as a function of the applied potential E (volts), F is the Faraday constant (96500 C), R is the ideal gas constant, T is the absolute temperature and n is the number of exchanged electrons. This rate constant may also be characterized by the exchange current density j_0:

$$j_0 = nFk_0\, C_0,$$

where C_0 is the concentration of the redox species in M/l.

The potential/current density curves are shown below.

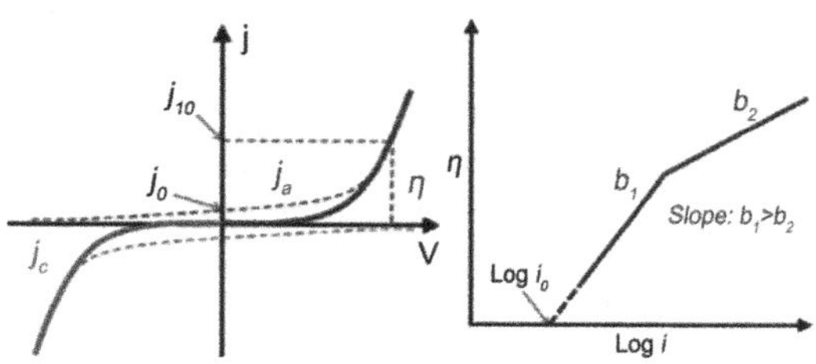

a) b)

a) anodic (dashed, blue) and cathodic j_c/η (dashed, red) curves and the resulting j/η curve (solid line). J_{10} is a current density reference of 10 mA/cm^2 to which corresponds an overpotential value.

b) Tafel lines $\eta = f(\log j)$.

Plotting Tafel lines and determining their slopes, *b* is a means of analyzing reaction mechanisms. For example, in the case of electrochemical proton reduction, a slope of 120 mV per decade (dec) corresponds to an initial, slow step of H^+ adsorption (Volmer) according to the reaction $H^+ + e^- \rightarrow H^\bullet$, followed by a rapid recombination $2H^\bullet \rightarrow H_2$ (Tafel). Conversely, a slope of 30 mV/dec corresponds to a fast H^+ adsorption step followed by slow Tafel recombination. An intermediate value of 40 mV/dec corresponds to a slow step in which $H^\bullet$ is desorbed by H_3O^+ according to the reaction $H^\bullet + H_3O^+ + e^- \rightarrow H_2 + H_2O$ (Heyrovsky) [SUE 17].

Box 5.1. *Intensity/potential curves ($j = f(\eta)$) and Tafel lines*

Chalcogenides (MoS_2, WS_2) are some of the most studied materials; MoS_2, widely used in industry as a catalyst for hydrocarbon desulfurization catalyst, is one example of a two-dimensional (2D) material for which the proton reduction mechanism has been analyzed in detail. This research has led to an improved understanding of the notion of active site of a catalyst, resulting, more recently, in the development of ultimate catalysis by single atoms on a 2D support (single atom catalyst – SAC).

5.3.1.2. *MoS_2 and its active sites*

MoS_2 NPs, described earlier (see Chapter 2, "2D Materials"), generally take the form of flat polygonal nanoplatelets, 3–4 nm thick, with three alternating S-Mo-S atomic layers, with a high density of unsaturated sulfur atoms on the edges of the NP.

One particular feature of this nanomaterial comes from the fact that only part, rather than the entire, surface of the MoS_2 exerts a catalytic effect. These active sites are located around the edges of the particle (Figure 5.3). Theoretical calculations have shown that the active sites of MoS_2 NPs correspond to unsaturated sulfur atoms, located precisely on the periphery of the NPs, a finding which was demonstrated experimentally by Chorkendorff and coworkers [JAR 07].

Measuring the perimeter and area of MoS_2 NPs by scanning tunneling microscopy (STM), the authors showed that the current density corresponding to the proton reduction reaction was proportional to the total perimeter of the MoS_2 NPs and not to their area. This confirms that only a portion of the catalyst surface is active, and that the active hydrogen adsorption sites correspond to the sulfur atoms located on the edges of the NP[6].

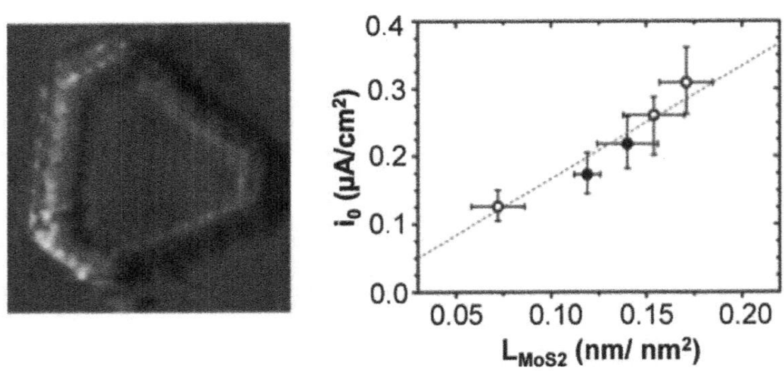

a) b)

Figure 5.3. *Proton reduction on a gold electrode modified with MoS_2 NPs*

---

6 The MoS_2 NPs used by Chorkendorff and coworkers [JAR 07] were deposited on a single-crystal (111) gold surface by thermal vacuum evaporation of molybdenum. This was followed by sulfidation using H_2S in the vapor state and annealing at 400–500°C, resulting in polygonal NPs with an average side length of around 2 nm.

COMMENT ON FIGURE 5.3.– *a) High-resolution STM image of a MoS_2 NP adsorbed on a monocrystalline gold surface (perimeter ~8.6 nm). b) Linear variation of the exchange current density i_0 as a function of the total length ($LMoS_2$) of the NP edges contained in a geometric area of 1 nm^2. The current density is measured for an overpotential of –150 mV versus NHE (normal hydrogen electrode) applied to the electrode. Adapted from [JAR 07].*

The Tafel lines[7] measured for low overpotentials show slopes between 55 and 60 mV per decade of current. The exchange current i_0 calculated using the number of active sites per cm^2 is estimated at 7.9×10^{-6} A/cm^2, far from that of platinum, calculated in the same way to be around 4.5×10^{-4} A/cm^2 for a crystalline face (111). Nevertheless, this current is greater than that obtained for most transition metals [JAR 07], and its TOF (Turnover Frequency, see Box 5.2), calculated from the exchange current (i.e., for a zero overpotential) taking into account the number of active sites per cm^2, is equal to 0.02 s^{-1}, compared to 0.9 s^{-1} for Pt(111)

Alongside the catalytic effect resulting from the chemical nature of the elements, there is an additional effect, resulting from the surface structure, which comes into play in gas release and can influence the process rate. Lu *et al.* [LU 14] showed that the proton reduction rate increased considerably when the electrode surface was made of a film of MoS_2 nanoplatelets, arranged perpendicular to a titanium substrate (Figure 5.4b)[8].

7 The current-potential curves obtained for surfaces of this type show high overpotentials in excess of 150 mV for low current densities (0.3 mA/cm^2), which, *a priori*, might indicate the absence of any significant catalytic effect. The reported values were measured for a monolayer of MoS_2 NPs deposited on gold (inactive at a potential of 150 mV), while in practice, the electrodes include much larger catalyst masses and therefore much higher current intensities.

8 Both MoS_2 films (thickness 200 nm) are prepared thermally. The nano-structured film is obtained on a titanium substrate by reacting $(NH_4)_6Mo_7O_{24}$ and $CS(NH_2)_2$ in an autoclave at 180°C, while the compact film is obtained at the same temperature by reacting $CS(NH_2)_2$ on a molybdenum substrate. The MoS_2 nanoplatelets have a thickness of 4 nm and a size of 200 nm. The gas bubbles formed on the MoS_2-NS electrode are sized between 50 and 60 µm, while on MoS_2-Planar they reach sizes 10 times larger (between 500 and 600 µm) before detaching from the surface. This corresponds to the respective adhesion forces of the bubbles (10.8 µNewton on MoS_2-NS and 124.8 µN on MoS_2-Planar), corresponding to respective contact angles (CA) of 158° and 135°; the wettability of MoS_2-NS is slightly higher than that of MoS_2-Planar. By analogy with a super-hydrophobic surface, the authors state that the MoS_2-NS surface is "super-aerophobic".

This nanostructuring (MoS_2-NS) effectively reduces the adhesion forces between the H_2 bubbles and the surface, thus facilitating gas release in the form of tiny bubbles. In contrast, a compact, flat MoS_2 electrode (MoS_2-Planar) (Figure 5.4a) has a higher adhesion force and the bubbles produced are larger. This results in a large part of the electrode surface being masked, making it less accessible to protons, causing an additional ohmic drop and an increase in the overpotential (Figure 5.4d).

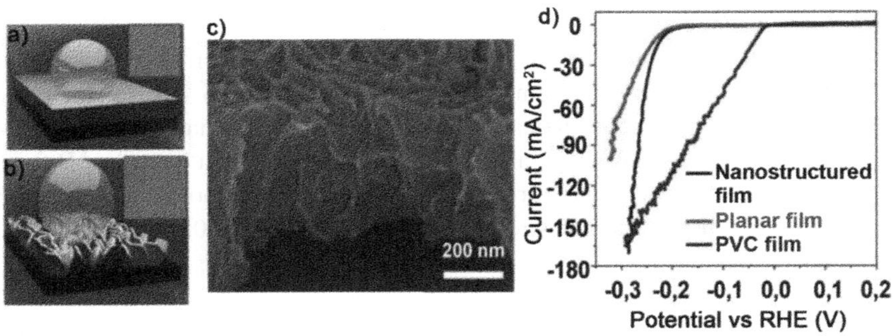

Figure 5.4. *Proton reduction on a MoS_2 electrode*

COMMENT ON FIGURE 5.4.– *a) and b) Formation of an H_2 bubble on a planar MoS_2 surface and on a nanostructured surface respectively. c) SEM image of the nanostructured surface. (d) Comparison of H^+ (0.5 M H_2SO_4) discharge curves on a titanium electrode covered by a planar MoS_2 film, nanostructured film and on a Pt/C composite electrode (20 wt% Pt). Adapted from [LU 14].*

The i/V curves show a higher proton reduction rate on MoS_2-NS than on MoS_2-Plan (Figure 5.4d). Furthermore, despite a higher overpotential than on Pt/C at low current values (–0.25 V for –10 mA/cm^2)[9], the overpotential is equal to that of Pt/C (around –0.27 V) when the current density reaches –150 mA/cm^2.

More recently, He *et al.* [HE 17] showed that the catalytic activity of MoS_2 could be significantly enhanced by combining it with black phosphor (BP). The material, synthesized thermally, consists of BP nanosheets (~10 μm) "decorated" with MoS_2 nanoflakes of much smaller size (100 nm) on the surface.

9 The reference current density of 10 mA/cm^2 (very low compared to the 1 and 2 A/cm^2 produced with industrial electrolyzers) was chosen based on the average performance provided by photovoltaic systems that might be used in conjunction with electrolyzers for the purposes of hydrogen production.

The overpotential observed for a current of –10 mA/cm^2 is 85 mV, with no observable degradation of characteristics after 10000 polarization cycles. The exchange current i_0 (6.6×10^{-4} A/cm^2) is 20 times higher than that of undoped MoS_2. The authors attribute this improvement to the electron donor effect of BP to MoS_2, resulting from the fact that BP has a higher Fermi level than MoS_2.

TOF (Turnover Frequency) s^{-1}

The TOF is a value used to characterize the efficiency of a catalyst with respect to a particular chemical reaction. It is defined in s^{-1}, and is expressed as the ratio between the number of molecules produced per second and the number of active sites of the catalyst.

In the case of an electrochemical reaction carried out with a determined current density, and assuming a faradic yield of 100%, the number of molecules produced per second is given by the relation

$$j(A/cm^2)/[(1.602\times10^{-19}\ C)\times(\text{number of e-exchanged})]$$

Taking N_C to be the number of active atoms (or molecules) contained in the mass of the catalyst, the respective TOF of the H_2 and O_2 formation reactions, with exchanges of two and four electrons, respectively, are given by the relations

$$H_2 : TOF(s^{-1}) = j(A/cm^2)/[(1.602\times10^{-19}\,C)\ \times 2\times N_C]$$

$$O_2 : TOF(s^{-1}) = j(A/cm^2)/[(1.602\times10^{-19}\,C)\ \times 4\times N_C]$$

The TOF increases with the overpotential in a monotonic way and must be defined as a function of the applied potential. The variation of TOF relative to the H_2 formation reaction in 0.5 M H_2SO_4 medium and on a 9.4 nm thick MoS_2 film is shown below. Adapted from [SHI 15].

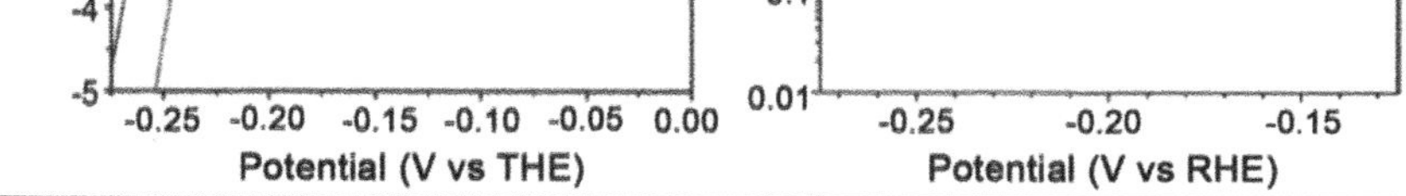

Box 5.2. *Definition of TOF*

Similarly, Han *et al.* [HAN 18] showed that MoS_2 NPs anchored perpendicularly on hydrogenated graphene sheets acquire significant catalytic power for proton reduction, comparable to that observed with BP-MoS_2.

Tungsten sulfide, WS_2, which is structurally comparable to MoS2, has also been shown to be a good catalyst for proton reduction in a highly divided form (nanoflakes with dimensions between 5 and 20 nm). Cheng *et al.* [CHE 14e] showed that an electrode material comprising highly conductive carbon powder, Nafion and ~0.35 mg/cm^2 of WS_2 presents an overpotential, compared to a 20% Pt/C electrode, of the order of 120–130 mV for a current density of -10 mA/cm^2. For bulk WS_2, the overpotentials for the same current density are above of 600 mV.

It is thus apparent that the abundant and inexpensive molybdenum sulfide-based electrode materials currently used in the petroleum industry for hydrodesulfurization are still effective electrode materials when used in a highly divided form, or when mixed with other elements to modify their electronic properties. We now have a better understanding of the proton reduction mechanisms at work on this type of material, and its combination with other elements such as phosphorous has resulted in considerable improvements in performance, which now almost matches that of industrial Pt/C catalysts. However, the TOFs measured for an overpotential of 200 mV are still lower than those of Pt (around 50 s^{-1}) and do not exceed 20 s^{-1}.

5.3.1.3. *Porous, conductive and metal-free 2D materials*

The use of materials with hierarchical porosity is a further possible solution for catalyzing H_2 formation. The combination of highly conductive nitrogen-doped graphene (N-GR) sheets with a non-conductive porous 2D compound such as (g-C_3N_4)[10] via van der Waals force results in a 3D electrode hybrid material with good catalytic activity with respect to the HER reaction (Figure 5.5). Duan *et al.* [DUA 15] showed that a material of this type, denoted PCN@N-GR, can be prepared using a relatively simple wet protocol which consists of filtering a nanosheet suspension of both PCN (g-C_3N_4) and GR, followed by nitrogen doping

---

10 g-C_3N_4 is a graphitic material constituted by a monolayer assembly of heptazine (C_6N_7) nuclei, tri-coordinated by trivalent nitrogen atoms. This assembly has the effect of creating pores of regular dimensions in the g-C_3N_4 monolayers delimited by an environment with three or six nitrogen atoms, respectively (Figure 5.3a). Unlike graphene, they are weakly conductive but highly luminescent, and, moreover, exhibit conductivity variations under laser irradiation, which have led to numerous applications, including in photoactive memory devices. More details on the synthesis methods and properties of these materials can be found in a recent review article by Inagaki *et al.* [INA 19].

of the graphene, leading to a thick and flexible (30 μm thick), large-area (3.4 cm diameter) composite film.

The PCN@N-GR film resulting from the assembly of the PCN and N-GR layers is characterized by a double porosity: nanoporosity from the PCN layers, comprising pores of nanometric dimensions (15–20 nm) created during the synthesis of the C_3N_4 material, and micro-porosity, estimated at a few μm, resulting from the spacing between PCN and N-GR layers[11].

The electrocatalytic properties of PCN@N-GR are comparable or better than those observed with previous catalysts. The overpotential required to produce a current density of −10 mA/cm^2 is under 100 mV (75 mV). However, this electrode efficiency is only optimized after a "forming" period comprising a large number of voltage cycles. According to the authors, this has the effect of increasing the wettability of the film throughout its thickness, resulting in improved material-electrical charge exchange, and thus an increase in the proton reduction rate.

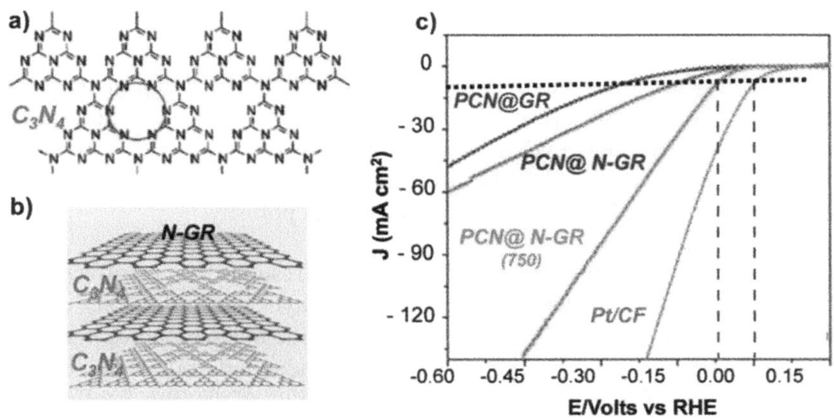

Figure 5.5. *Electrochemical characteristics of the PCN@N-GR composite material*

11 The PCN@N-GR hybrid material described by Qiao's group [DUA 15] is obtained by a multi-step process. Microporous PCN sheets are obtained by mixing microparticles of SiO_2 with cyanamide (N≡C-NH$_2$) in the molten state; after prolonged heating at 550°C, cooling and the removal of SiO_2 using NaOH, microporous bulk PCN is obtained. The material is then treated using ultrasonic stirring to release PCN nanosheets. N-GR is obtained using a classic protocol from a suspension of graphene oxide, which is reduced to graphene (GR) using hydrazine in ammonia medium. The PCN and GR nanosheet suspension is filtered to produce a thick film of PCN@GR. The GR is doped with nitrogen by subjecting the film to a solvothermal process (150°C) in the presence of NH_3 [DUA 15].

COMMENT ON FIGURE 5.5.– *a) Structure of a g-C_3N_4 film. The film consists of a sequence of heptazine nuclei linked together by nitrogen atoms, which creates pores inside the film, the size of which can be adjusted by lithographic synthesis in the presence of SiO_2 spheres (120 nm in diameter); in this way, the pores which are formed have dimensions calibrated on those of the SiO_2 NPs. b) Structure of the PCN@N-GR composite. c) Proton reduction curves (H_2SO_4 0.5 M) for different electrode materials. Note that the PCN@N-GR (750) material (750 = the number of "pre-forming" voltage scan cycles) is the most similar to Pt/CF (CF: carbon fiber) in terms of its reduction characteristics. Adapted from [DUA 15].*

Between the initial PCN@N-GR film and the "cyclized" film noted PCN@NGR (750), obtained after 750 scan cycles, the overpotential at –10 mA/cm^2 decreases by about 150 mV, from 225 to 75 mV, and the exchange current i_0 increases from –0.36 to –0.42 mA/cm^2. Furthermore, the slope of the Tafel lines changes from 88.6 to 49.1 mV per decade. This shows that a mechanism change takes place during the forming step, toward a Heyrovsky-type mechanism, which exhibits slower kinetics than those observed for Pt/C, for which the Tafel slope is of the order of 30 mV/dec (see Box 5.1).

The stability of the system, evaluated based on the behavior of the current-potential curves as a function of the number of discharge cycles, indicates good reproducibility after 5000 cycles. This shows that the electrode is robust, and confirms the interest of these 3D structures obtained by assembling graphitic, conductive, metal-free 2D materials with hierarchical porosity.

According to the authors, these properties are the result of several effects specific to this structure: (1) almost total exposure of the catalytic sites to protons, due to the hierarchical porosity (natural PCN nanopores alongside the micropores intentionally created during the assembly); (2) good electrical conductivity of the film due to the presence of graphene nanosheets; and (3) high flexibility of the film, conferring high robustness and likely contributing to the preservation of integrity over long periods of electrolysis. However, the nature and the number of catalytic sites in the material remain unidentified, making it difficult to evaluate the TOF of the catalyst.

5.3.1.4. *HER and single atom catalysts*

Catalysts are known to be more efficient the smaller their size. Unfortunately, their surface energy also increases as their size decreases, resulting in a pronounced tendency to aggregate, which causes size disparities and thus significant variations in their catalytic properties; this sometimes results in the formation of undesirable by-products.

One way to prevent aggregation is to use a support that interacts with the metal catalyst and allows the atoms to be adsorbed as isolated entities, resulting in an ultimate atomic state catalyst. This is a very recent form of heterogeneous catalysis, based on the catalytic properties of isolated metals (SACs). The most frequent substrates used for the atomic catalysts are oxide surfaces, metals or 2D inorganic or organic materials [YAN 13b, CHE 18c, JIA 18] (Figure 5.6).

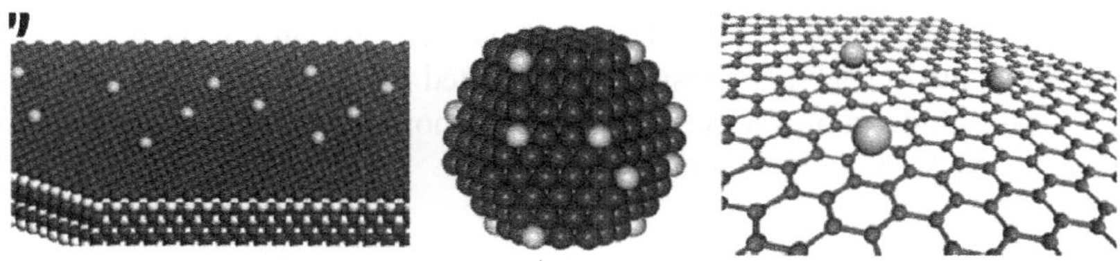

a) b) c)

Figure 5.6. *Different types of SACs (single atom catalysts). Diagrams a), b) and c) show (in yellow) the metal atoms (SAC) deposited, respectively, on an oxide surface, on a metal NP and on a two-dimensional organic material such as graphene. Adapted from [YAN 13b]*

A first justification for this new form of catalysis comes from the need to reduce the concentrations of precious metals used in commercial platinum, iridium or ruthenium-based catalysts. A second, more research-oriented reason is that it provides the means of exploring the catalytic behavior of a certain number of transition metals (Ni, Fe, Co, etc.), known to have a certain catalytic power, which can be analyzed more closely when the metals are used in the form of SAC deposited on various substrates.

A surprising result of this type of catalysis was observed by Deng *et al.* [DEN 15b] using an MoS_2 electrode doped with a very low concentration of platinum atoms. The catalytic effect of 2D MoS_2 had previously been attributed to the presence of unsaturated sulfur atoms on the periphery of the MoS_2 nanoplatelets; researchers found that the sulfur atoms located in the central part of the MoS_2 platelets, which are usually inert, take on high catalytic power when in proximity to platinum atoms. This entirely new phenomenon has been interpreted as a result of the modification of the electronic properties of the sulfur atoms, induced by the presence of Pt atoms. The adsorption energies of the hydrogen atoms on these sulfur atoms fell to almost zero[12], leading to high HER rates. Electrodes made of this new

12 The adsorption energies of a hydrogen atom (ΔG_H) on the sulfur atom closest to a platinum atom are close to 0 eV, while the adsorption energies on the outer sulfur atoms are higher, of the order of 0.1 eV [YAN 13b].

material, in which the proportion by weight of platinum is only around 1.7%, have a proton reduction rate comparable to that of conventional Pt/C catalysts, which contain a much higher proportion of Pt (40% by weight)[13]. In practical terms, the stability of the catalyst is very good and, moreover, is unaffected by the presence of "poisons", such as alcohols, which have a negative effect on precious-metal only electrodes [DEN 15b].

More recently, Zhang *et al.* [ZHA 17b] obtained very promising results using a catalyst noted PtSA@NT-FN consisting of isolated Pt atoms adsorbed on a network of cobalt phosphide (CoP) NTs, themselves anchored perpendicularly to a porous nickel foam (Ni-Foam, NF) support (Figure 5.7a).

The process is innovative in terms of the way Pt atoms are deposited on the NT, produced in isolated form and without the formation of aggregates. Unusually, cyclic voltammetry is used for anodic dissolution of a Pt electrode into Pt^{2+} ions which migrate and adsorb, very progressively, onto the CoP NT. The authors observed a stabilization of the catalytic effect of the "platinized" CoP NT electrode after a minimal number of polarization cycles (5000), which, in their view, corresponds to an optimal deposition of the Pt^{2+} ions on the CoP NTs[14].

The electrochemical characteristics of this electrode material are remarkable, both in neutral and in acidic or alkaline media, and it outperforms conventional Pt/C electrodes in terms of both overpotential and stability (Figure 5.7). In a neutral buffer medium (1 M sodium phosphate) and after stabilization of the electrode, the overpotential for a current of 10 mA/cm^2 is only 24 mV versus NHE (normal hydrogen electrode) and, for high current densities (j ~ 200 mA/cm^2), it remains below 250 mV, whereas for Pt/C it exceeds 350 mV. The slopes of the Tafel lines are 30 mV/dec, corresponding to the same reduction mechanism observed with Pt/C electrodes (which have Tafel slopes of 31 mV/dec). The mass activity of the catalyst, expressing its catalytic power per 1 g of catalyst, is much higher than that of Pt/C. For an overpotential of 50 mV, the authors cite a figure of 70 A/g of

13 The MoS_2 electrode doped with Pt atoms contains only 0.018 mg of Pt per cm^2, while a Pt/C 40% electrode contains 0.64 mg/cm^2, that is, 35 times more.

14 Pt atoms are deposited in a three-electrode electrochemical cell comprising a reference electrode (calomel), a Pt anode, and a cathode consisting of CoP NTs anchored on a nickel foam. The electrolyte is a 1 M buffer solution at pH 7.2. The polarization cycles applied to the CoP NTs are carried out in such a way that the Pt remains at a positive and the CoP NTs at a negative potential, resulting in the anodic dissolution of the Pt followed by the transport of Pt^{2+} ions and their deposition on the Co NTs, without reduction to Pt^0. This result seems to be confirmed by XPS analyses, and contrasts with the behavior of Pt in traditional catalysts where it is encountered in the 0 oxidation state (Pt^0).

Pt, compared to the 17 A/g obtained for a Pt aggregate catalyst; this equates to a four-fold increase in a catalytic power[15].

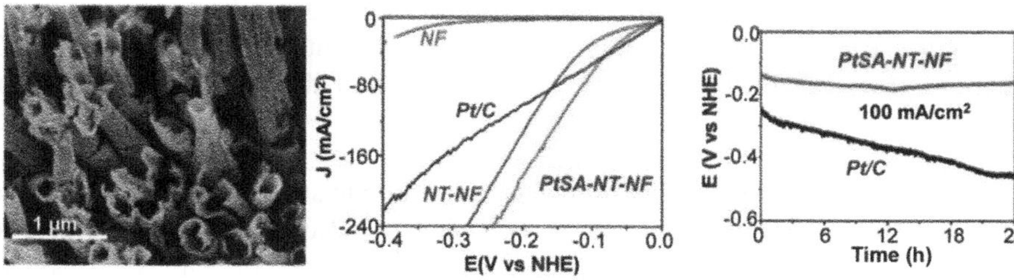

a) b) c)

Figure 5.7. *Electrocatalytic properties of the platinized CoP NT network*

COMMENT ON FIGURE 5.7.– *a) SEM image of CoP nanotubes (NT) anchored on a nickel foam (NF) before platinization. b) Polarization curves in neutral medium (pH 7) of NF, Pt/C, NT-NF and PtSA@NT-NF electrodes. Note that at high current densities, the overpotential observed on a Pt/C electrode is considerably higher than for NT-NF and PtSA@NT-NF electrodes. c) Evolution of the overpotentials of PtSA-NT-NF and Pt/C electrodes over time for a current density of 100 mA/cm², demonstrating the very high stability of the PtSA@NT-NF electrode. Adapted from [ZHA 17b].*

The stability of the catalyst is also superior to that of the Pt/C; for electrolysis at 100 mA/cm^2 over a period of 24 hours, a constant overpotential of the order of 150 mV is maintained. Under the same conditions, the overpotential of the Pt/C electrode increases from 250 mV to more than 400 mV.

In conclusion, the material described above, in which tiny amounts of Pt are dispersed in atomic and oxidized form onto CoP NT walls anchored on a nickel foam, is one of the first SAC catalysts to exhibit catalytic properties superior to those of commercial Pt/C type electrodes. According to the authors, the material should be easy to produce on a larger scale than their centimetric sample, and no binders are required; it thus shows immense promise as a catalyst, and could come to replace traditional Pt/C catalysts in the fairly near future.

---

15 A mass activity of 1 A/g Pt, defined for a given overpotential, is equivalent to a TOF of 1×10^{-3} s^{-1} for the HER. In the present case, the Pt content of PtSA-NT-NF is 0.14 mg/cm^2 compared to 1.62 mg/cm^2 in Pt/C (20%). From the curves in Figure 5.4b, with a measured current of 180 and 100 mA/cm^2, respectively, for PtSA-NT-NF and Pt/C (20%) at an overpotential of 200 mV, we deduce a TOF of the order of 7.8 s^{-1} for PtSA-NT-NF and 0.38 s^{-1} for Pt/C (20%).

The SAC technique has also been tested on certain non-precious transition metals such as iron, cobalt and nickel. In this case, 2D graphitic substrates, such as porous graphene and, more recently, graphdiyne (GDY) (see section 1.6.3) have been used as novel substrates for anchoring catalysts in the metallic state in a regular and isolated manner. Very encouraging results have recently been obtained for HER reactions in an acidic medium using nickel and iron deposited on graphene or graphdiyne.

However, optimal deposition of metals in the atomic state on a 2D substrate can only occur when the substrate features vacancies or structural defects favorable to strong insertion of the metal. In the case of graphene, there are variety of ways to synthesize porous material, either by CVD or by wet process; the pores or vacancies formed in each case are different, resulting in distinct adsorption properties and consequent variations in catalytic power.

This is apparent in the case of nickel on porous graphene, as studied by others [QIU 15; ZHA 18a]. The first solution developed by Chen's group [QIU 15] consists of producing graphene by CVD on a porous nickel substrate (pore size ~300 nm). The nickel is then slowly dissolved in concentrated HCl until the Ni atomic state is reached, while the graphene retains the initial porous structure of the nickel. Nickel atoms are adsorbed, either at substitution sites (replacement of a C atom by nickel), or anchored in the structural defects corresponding to the vacancies. The energy corresponding to the fixation of nickel in a substitution site is very high, a fact which accounts for the high stability of the catalyst during prolonged electrolysis in 1 M H_2SO_4 acid medium. The low adsorption energy of hydrogen atoms on Ni, comparable to that of Pt, accounts for the catalytic effect. The catalyst is thus characterized by an exchange current of 5.3×10^{-2} mA/cm^2, one order of magnitude higher than that of MoS_2 (5.1×10^{-3} mA/cm^2), a Tafel slope of 45 mV/dec and a TOF of 0.8 s^{-1} at 300 mV (much higher than the TOF of MoS_2, which is between 0.01 and 0.3 s^{-1}). The interest of this form of catalyst is further confirmed by its high stability; for an overpotential of 150 mV, a current density of about 9 mA/cm^2 can be maintained at 90% of its initial value for 120 hours.

Similarly remarkable results were obtained by Zhang *et al.* [ZHA 18a], using a porous "defective" graphene (DG), obtained by heating nitrogen-doped graphene at high temperature (1150°C). The insertion of nickel atoms via a series of reactions described in Figure 5.8a, similar to those used by Chen's group, results in a catalyst noted A-Ni@DG (nickel atoms adsorbed on DG).

As before, the stability of the A-Ni@DG catalyst in 0.5 M H_2SO_4 medium is excellent, with no observable variation in overpotential during a 10-hour electrolysis period at constant current (5 mA/cm^2) and for an overpotential of 60–70 mV versus NHE. Note that nickel atoms adsorbed on graphene behave very differently than Ni alloy electrodes in acidic media; the latter degrade and corrode rapidly when subjected to electrolysis under these conditions [QIU 15]. Another factor which contributes to the stability of A-Ni@DG in acidic media is the high interaction energy of Ni with the sp^2 atoms of graphene, close to a carbide bond.

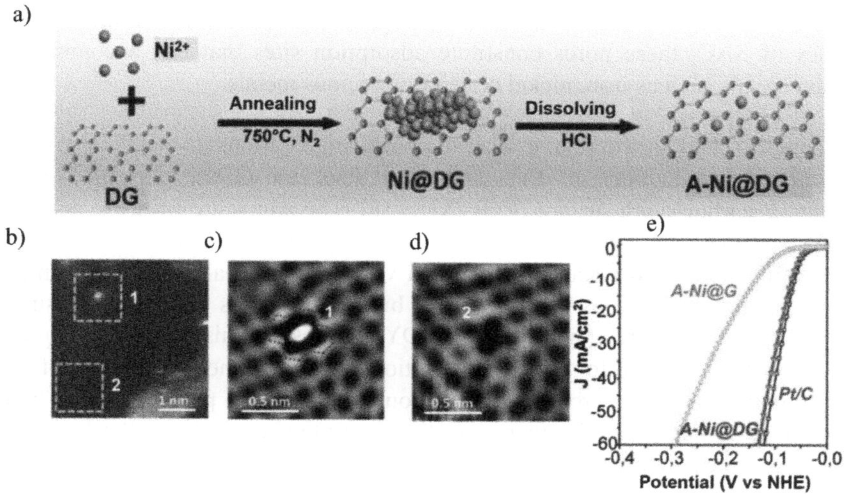

Figure 5.8. *Synthesis and electrocatalytic characteristics of A-Ni@DG*

COMMENT ON FIGURE 5.8.– *a) Catalyst synthesis. The porous graphene DG is impregnated with Ni^{2+} and then annealed at 750°C under nitrogen, resulting in Ni° NPs adsorbed on DG (Ni@DG). Slow dissolution of Ni° NPs in concentrated HCl medium (6 hours) results in the A-Ni@DG structure, in which isolated nickel atoms are adsorbed in the graphene defects. b) TEM localization of a nickel atom (1) and a defect (2). c) and d) High-resolution TEM images of the Ni atom (1) and the structural defect (2). e) Reduction curves in 0.5 M H_2SO_4 medium obtained on defective (A-Ni@DG) and non-defective (A-Ni@G) graphene, compared to a Pt/C electrode. The masses of Ni and Pt catalyst are identical and equal to 1.24 wt%. Adapted from [ZHA 18a].*

This strong interaction of Ni with the surrounding carbons also results in low hydrogen atom adsorption energy, and in a rapid hydrogen evolution rate; the Tafel slope of 31 mV/dec is characteristic of a determining kinetic step corresponding to the dimerization of the hydrogen atoms into H_2 (see Box 5.1). Note that for equal mass charges of Ni and Pt catalysts (1.24 wt%), the proton reduction curves in 0.5 M H_2SO_4 medium are almost identical, and the overpotential is less than 150 mV for a current density of 60 mA/cm^2. The calculated TOF of 5.7 s^{-1} for an overpotential of

100 mV and of the order of 40 s^{-1} at 200 mV is particularly high, exceeding that of many catalysts, including molybdenum-based materials [ZHA 18a][16].

The notion of defects favoring the adsorption of a meta-catalyst also applies in the case of GDY; compared to defective or porous graphene, GDY presents the additional advantage of regular pore distribution across its surface. As in previous examples of SAC, these pores constitute adsorption sites that can accommodate transition metals such as iron, nickel or certain precious metals.

Yuliang Li and coworkers group attached single atoms to GDY by direct electrochemical reduction of Ni^{2+} and Fe^{3+} ions on GDY, deposited on a microporous carbon cloth surface (CC) [XUE 18]. The atoms are firmly attached to the GDY, with no aggregation of the atoms or degradation of the GDY network (Figure 5.9). Theoretical calculations show a very precise localization of the metal atoms inside each pore, as well as high binding energies with the substrate, guaranteeing the high strength of the M@GDY composite. This anchoring of metal atoms (M) is attributed to orbital interactions between the *sp*-orbitals of the diethynyl bonds and the *d*-orbitals of transition metals; unlike previous cases, bond destruction does not occur. Furthermore, the adsorption energies of the hydrogen atoms on the metal atom are close to 0, strongly favoring a rapid H_2 formation process [XUE 18][17].

Remarkably, the atoms attached in this way are perfectly resistant to prolonged electrolysis reactions in acidic media and to high current intensities. The voltammetric curves in Figure 5.9b show that GDY "decorated" with iron atoms exhibits electrocatalytic properties close to those of a Pt/C electrode containing 20 wt% Pt. The overpotential with respect to the redox potential of the H^+/H_2 couple is of the order of 90 mV for a current density of 20 mA/cm^2; in comparison, that of the "non-metallized" GDY electrode is very high (580 mV). The overpotential compared to the Pt electrode remains essentially constant at 80 mV up to fairly high current densities of 60 mA/cm^2; this is not the case for the Ni@GDY electrode, for which the overpotential with respect to the Pt/C electrode increases with current

16 The low catalyst content of SACs may constitute a drawback in that high catalyst concentrations, such as those seen in the case of Pt mixed with carbon particles, are difficult to reach. Pt concentrations of 20% or more by weight are needed in order for commercial electrolyzers to operate with current densities of around 2 A/cm^2 in acidic media.

17 The electrode is a composite inserted in a commercial carbon cloth (CC) substrate, made of overlapping carbon-based fibers, onto which the GDY is deposited using a classic method developed earlier by the same group [LI 10a], resulting in the CC/GDY composite. The Ni and Fe metals were deposited directly by electrochemical reduction of Ni^{2+} or Fe^{2+} ions on the composite. The CC/M@GDY electrode (M = Ni or Fe) presents a double hierarchical porosity, combining the microporosity of the CC with the nanoporosity of the GDY.

intensity, rising from 150 to 280 mV when the current density increases from 20 to 50 mA/cm^2.

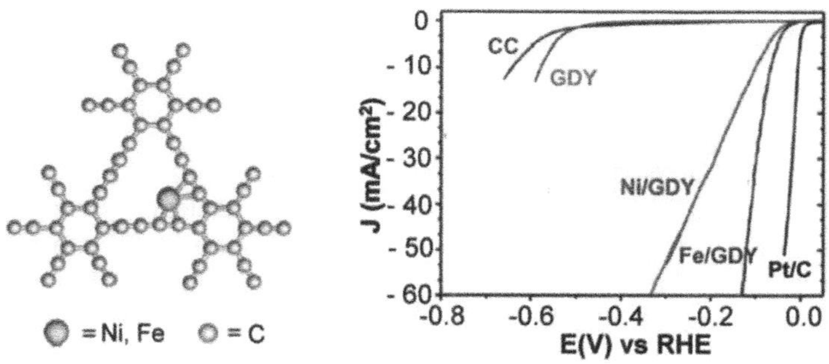

a) b)

Figure 5.9. *Electrocatalytic properties of a graphdiyne electrode modified with Ni or Fe atoms*

COMMENT ON FIGURE 5.9.– *a) Preferential position of Ni or Fe atoms on a GDY mesh resulting from interactions with sp orbitals of two –C≡C– clusters. b) Current/voltage curves of H^+ ion reduction on Pt/C, Fe@GDY, Ni@GDY, GDY, and CC (carbon cloth) electrodes in 0.5 M H_2SO_4 medium. Adapted from [XUE 18].*

In terms of stability, the authors show that an Fe@GDY type electrode can perform a very high number (5000) of reduction cycles between 0 and 90 mA/cm^2 without any loss of efficiency; similarly, no degradation was observed when the material was subjected to a constant reduction current at 30 mA/cm^2 for over 60 hours. The Tafel lines are characterized by a slope of 38 mV/decade, very close to those of Pt/C (34 mV/decade), confirming rapid hydrogen evolution kinetics. The authors also observed high mass activity of the catalyst: 80 A/mg Fe for an overpotential of 200 mV (giving a TOF of 20 s^{-1}), compared to 16.6 A/mg Ni for nickel and only 2.3 A/mg Pt in the case of Pt/C.

These results show Fe@GDY to be one of the most promising precious metal-free catalysts for use in acid electrolysis hydrogen production, due to its easy accessibility, robustness and good performance in terms of HER. The low overpotential of 60 mV at 10 mA/cm^2 means that it outperforms many systems, such as those based on various forms of MoS_2, and other SAC-type systems made with nickel or cobalt atoms deposited on graphene or nitrogen-doped graphene [FEI 15]. However, its TOF of 20 s^{-1} for an overpotential of 200 mV is lower than that of the previous SA-Ni@DG catalyst (50 s^{-1}).

In conclusion, these innovative approaches are still in their infancy, and their reliability remains to be tested under operating conditions close to those of commercial electrolyzers, with high current densities and continuous operation over long periods of at least 1 week.

5.3.2. *HER in alkaline media*

The reaction leading to the formation of H_2 in an alkaline medium involves a water reduction mechanism, different from the proton reduction that takes place in acid media. In energy terms, the reaction in alkaline media is more demanding due to the need for an initial water-splitting step, in which the H-O-H bond must be broken before $H^•$ can be adsorbed; this step replaces the direct reduction of protons adsorbed on the electrode. Unlike reduction reaction in acidic media, where protons are the only species adsorbed on the electrode surface, the presence of OH^- ions and their formation during the water reduction process interfere with the adsorption of H_2O and $H^•$, contributing to a reduction of the rate of the Volmer reaction (see Box 5.1)[18]. Although the overall reaction scheme is similar to that encountered in acidic media, different catalysts must be used.

As we indicated at the start of our presentation, most commercial electrolyzers used for hydrogen production operate in alkaline media, and the majority of electrodes feature Pt-based catalysts. To date, these catalysts have proved most efficient catalysts, although their stability is lower than in acidic electrolyte media [MAH 17].

Once again, the identification of efficient catalysts which do not rely on the use of precious metals is a key aim. Various solutions have been put forward, although none have yet progressed beyond the laboratory stage; these are described in detail in recent papers by Mahmood *et al.* [MAH 18] and Colli *et al.* [COL 19], who provide a more realistic, critical analysis.

The catalytic effect of Pt in a basic medium was poorly understood for many years, but the situation has now changed. Sheng *et al.* [SHE 13b] have shown that the catalytic effect is mainly due to an optimal adsorption energy of the $H^•$ hydrogen atoms on the metal; furthermore, water molecules play a decisive role in the global kinetics of H_2 evolution in an alkaline medium (see Box 5.3).

18 As in the case of HER in an acidic medium, there are three steps, identified by Volmer, Heyrovsky and Tafel. The overall rate of the reaction is determined by the slowest step, which, depending on the case, leads to different slopes of the Tafel lines (V/logi). If the first water reduction step (Volmer) is kinetically limiting, that is, $H_2O + e^- \rightarrow H^•_{ads} + OH^-$, the slope is 120 mV/dec; it is 40 mV if the Heyrovsky reaction, that is, $H_2O + H^•_{ads} + e^- \rightarrow H_2 + OH^-$, is determinant, or 30 mV/dec if the terminal Tafel reaction ($H^•_{ads} + H^•_{ads} \rightarrow H_2$) is slowest.

Surprisingly, the activity of platinum is considerably enhanced when its surface is decorated with nickel hydroxide clusters (with a surface coverage of about 40%). This effect, discovered by Subbaraman *et al.* [SUB 11], is believed to result from a double action of the hydroxide on both the adsorption of H_2O and that of the OH^- resulting from the dissociation of H_2O.

The Pt neighboring the hydroxide plays a role in the adsorption of $H^{\bullet}_{ads}$ atoms that recombine to form H_2. The adsorption energy of OH^- on the hydroxide must not be too strong, or it would prevent the adsorption of H_2O and its dissociation. By comparing different transition metal hydroxides, the authors show that the catalytic effect decreases in the order Ni > Co > Fe > Mn, and that this catalytic effect is inversely correlated with the binding energy of $OH\text{-}M^{2+\delta}$, where $2 + \delta$ is the oxidation state of the metal ($0 \leq \delta \leq 1$) and OH^- is the hydroxyl group resulting from the dissociation of water.

Exchange current i_0 and $H^{\bullet}$ adsorption energies

Sheng *et al.* [SHE 13b] showed that the change in the exchange current (i_0) of the HER reaction in acidic or basic media, correlated as a function of the adsorption energy ΔE_H of $H^{\bullet}$ on a metal in the atomic state, can be represented by two line segments forming a "virtual" triangle, comparable to a Δ (volcano plot). Pt is at the top, with an H° adsorption energy of –0.45 eV, while tungsten (W) is at the bottom left of the Δ with a very strong adsorption energy of –0.8 eV, favorable to hydride formation and thus unfavorable to the H_2 evolution reaction. Similarly, gold and silver are located toward the bottom right of the Δ branch, with positive adsorption energies of 0.15 and 0.25 eV, respectively, which in this case constitutes an unfavorable situation for $H^{\bullet}$ adsorption. The fact that platinum is located in an intermediate position between these two extremes confirms that the rate of the HER reaction is optimal for adsorption energies of $H^{\bullet}$ on the metal that are neither too strong ($E < 0$) nor too weak ($E>0$) (Sabatier principle).

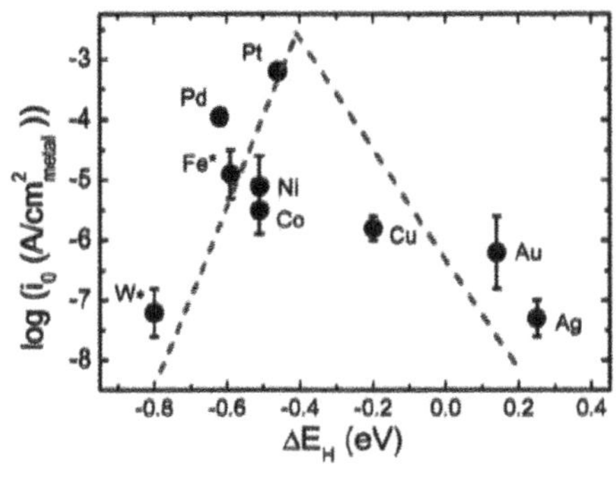

Adapted from Sheng *et al.* [SHE 13b]

Box 5.3. *Exchange current i_0 as a function of adsorption energies ΔE_H*

They observed that in 0.1 M KOH medium, the overpotential on Pt (111) for a current of 10 mA/cm^2 is relatively high, of the order of 300 mV, compared to 170 mV for $Ni(OH)_2$ clusters[19].

The addition of a very low concentration of Li^+ ions also contributes to improving the activity of the Pt electrode, lowering the overpotential to 100 mV; this effect is attributed to the complexation of water molecules by Li^+ and, presumably, to stronger adsorption on $Ni(OH)_2$.

Ruthenium, which is much cheaper than Pt[20], also offers very good catalytic properties for the electrochemical reduction of water in acidic or basic media. Deposited on various 2D graphitic materials, such as C_3N_4 [ZHE 16] and C_2N [MAH 17], it results in significant catalytic effects in alkaline media, superior to those observed with a 20% Pt/C electrode.

Mahmood *et al.* [MAH 17] recently created an innovative electrode material by assembling ruthenium nanocrystals on a porous graphitic 2D lamellar compound, made up of a regular assembly of nitrogen and carbon atoms, with the empirical formula C_2N (Figure 5.10a). The resulting composite material (Ru@C_2N) is conceptually close to an SAC catalyst, with the difference that the ruthenium is used in crystallite form with a very small number of atoms (sized between 1 and 1.5 nm), rather than as isolated atoms. As before, a good balance between the adsorption energies of H_2O and OH^- on ruthenium, along with a strong ability to adsorb hydrogen atoms (comparable to Pt), mean that this composite material is an excellent catalyst for water reduction in both alkaline and acidic media. The intensity-potential curves obtained for this electrode material reveal remarkable electrochemical behavior, superior to that of a Pt/C electrode and to that of other M metals (Pd, Co, Ni and Pt), inserted into C_2N in the same way as ruthenium (Figure 5.10a).

The overpotential at a current density of 10 mA/cm^2 is only 17 mV in 1 M KOH medium, 3.7 mV lower than the value obtained for Pt/C (20.7 mV). The TOF is around 2 s^{-1} for an overpotential of 50 mV, compared to 1 s^{-1} for Pt/C under the same conditions. Furthermore, no degradation of the polarization curves is observed following 10000 polarization cycles between 0.2 and –0.10 V, in either acidic

19 For comparison, note that the overpotential required to reduce protons in an acidic medium (0.1M $HClO_4$) for a current density of 10 mA/cm^2 and with the same Pt(111) electrode is only 0.06 V.

20 As an indication, in April 2020, one troy ounce (31.1 g) of Pt cost around $750 USD, compared to $250 USD for the same weight of Ru. Iridium, palladium and rhodium are even more expensive, with respective prices of $1500, $2250 and $7000–$10000 per ounce during the same period.

(0.5 M $HClO_4$) or alkaline (1 M KOH) media. As the authors indicate, compared with all known non-Pt/C catalysts, Ru@C_2N appears to be one of the best H_2O reduction catalysts in alkaline media; it therefore has the potential to be used as a replacement for Pt/C catalysts in commercial electrolyzers. The only possible hindrance concerns the greater scarcity of Ru compared to Pt.

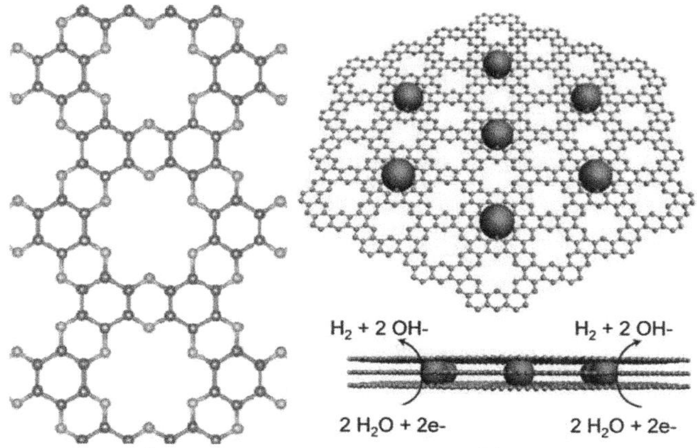

Figure 5.10. *Structures of C_2N and the M@C_2N composite*

COMMENT ON FIGURE 5.10.– *a) Porous structure of C_2N showing the presence of cavities in the compound (size 0.87 nm) surrounded by six nitrogen atoms (blue dots), metal crystallite adsorption sites. b) Structure of the M@C_2N composite (where M is a Pd, Co, Ni, Pt crystallite of size 3–4 nm). The Ru@C_2N composite is obtained by means of a simple condensation reaction between hexaaminobenzene and hexacetocyclohexane, followed by the electrochemical reduction of $RuCl_3$, filtering, then annealing at 900°C. Adapted from [MAH 17].*

The results cited above provide a clearer understanding of the water reduction mechanisms at work in alkaline media. They also fueled important research in terms of the identification of non-precious metal catalysts with the ability to promote bi-functional and optimal adsorption of water and hydrogenated species.

The transition metals nickel, iron, cobalt, molybdenum, copper and tungsten, associated with other non-metallic elements such as boron, carbon, nitrogen, phosphorus, sulfur and selenium, are generally used in porous alloy or composite form. An extensive list of these materials can be found in recent publications [ZEN 15], [MAH 18] and [COL 19]. Furthermore, McCrory *et al.* [MCC 15] provide a very useful comparative analysis of the different catalysts used for HER and OER processes in both acid and alkaline media. This study shows that a very large number of catalysts can reduce water with overpotentials lower than 100 mV and current densities of 10 mA/cm^2, while those used for water oxidation require much higher overpotentials, ranging between 300 and 500 mV, at the same current densities, as we shall see later.

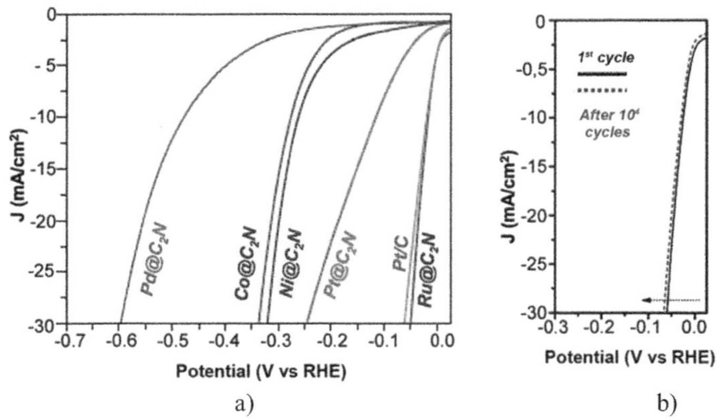

Figure 5.11. *Electrocatalytic properties of different M@C_2N cathode materials. a) Comparative reduction curves of H_2O in alkaline medium (1 M KOH) using the M@C_2N (M: Pd, Co, Ni, Pt or Ru crystallite) and Pt/C composite structures. b) Stability of Ru@C_2N evaluated after 10000 polarization cycles between 0.2 and –0.1 V. Adapted from [MAH 17]*

Girault and coworkers [COL 19] compared the performances of several catalysts in operating conditions close to those of commercial electrolyzers. They show that a monopolar electrolyzer with a sponge nickel cathode (Raney Ni type) and a stainless steel anode can be operated continuously in 6 M KOH medium and at 75°C over a period of 1 month at a current density of 300 mA/cm^2 with no appreciable variation in voltage. The overpotential is lowest with spongy Raney Ni catalysts, to which other transition metals (cobalt) are added to increase the catalytic power of the alloy via a synergistic effect.

The best results, in terms of energy yields, obtained with this type of catalyst correspond to overpotentials of the order of 100–120 mV for current densities of 100 mA/cm^2; these values are comparable to those observed in acidic media with non-precious metal catalysts. However, the specific porosity of Raney nickel appears to be a determining factor, and the contribution of a co-catalyst associated with nickel is not necessarily beneficial, in the case where the latter modifies the structure of the nickel and can cause a decrease in its initial porosity; this is evident in the case of Ni-Raney-Co alloys [HER 13].

5.3.3. *Conclusions on HER reactions*

The examples described above demonstrate the continued relevance of research into new catalysts for the H_2 evolution reaction, in both acid and alkaline media. The development of catalysts based on the use of metals in isolated atom state fixed onto 2D conducting substrates is certainly promising and deserves further attention. However, obtaining dispersions of atoms at concentrations not exceeding 1%–2% by weight of the catalyst may prove to be an obstacle to obtaining very high current densities (between 1 and 2 A/cm^2 for commercial electrolyzers); to the best of our knowledge, no catalysts of this type can currently withstand current densities of 1–2 A/cm^2. The scarcity of long-term standardized tests is also regrettable. Catalyst performances (in terms of overpotential as a function of a given current density) are generally often evaluated over periods of less than 24 hours and for current densities very different from those of commercial electrolyzers, and tell us little about their real reliability for continuous operation over periods of a week or more. The concentrations of electrolytes in acid and alkaline media also vary widely between research teams (although these often fall within the range of 5–6 M KOH for an alkaline electrolyte and around 1 M H_2SO_4 for an acid electrolyte). A standardization of the concentrations of acid and alkaline electrolytes or a correction of the overpotentials to take account of different ohmic drops would make it easier to evaluate overpotentials based on catalysts alone. In short, despite great innovations in this area, much research is still required before the standard Pt/C type cathodes used in commercial catalysts can be replaced by precious metal-free electrodes.

5.3.4. *Catalysts for oxygen evolution reaction*

The oxidation of water in an acidic or basic medium, leading to the release of oxygen, is a slower process than the H_2 evolution reaction. Compared to the proton reduction reaction at equal current intensity, this reaction requires much higher overpotentials and thus increases the energy cost of the overall water decomposition reaction. Different catalysts are used depending on the medium (alkaline or acidic). Many of the metal oxides used in alkaline media, the most common option used in commercial electrolyzers, cannot be used with acidic electrolytes due to their rapid degradation; iridium and ruthenium oxides are among the few materials which are reliable in these media. The overpotentials required to achieve electrolysis densities greater than 10 mA/cm^2 are high, ranging from 250 to 300 mV [SON 20]. As in the case of the HER process, the search for non-precious metal catalysts is an important objective, and has attracted the attention of many research teams.

5.3.4.1. *OER in an alkaline medium*

The O_2 formation reaction in an alkaline medium involves a four-electron balance consisting of four successive single-electron steps, each corresponding to the formation of an intermediate species (noted *) adsorbed on a particular catalytic site (also noted *):

$$OH^- + * \rightarrow *OH + e^-$$

$$*OH + OH^- \rightarrow *O + H_2O + e^-$$

$$*O + OH^- \rightarrow *OOH + e^-$$

$$*OOH + OH^- \rightarrow O_2 + H_2O + e^-$$

Or, for the whole process:

$$4OH^- \rightarrow O_2 + 2H_2O + 4e^-$$

One of the most widespread mechanisms, the AEM (adsorbate evolution mechanism), involves the evolution of successive adsorbates corresponding to the intermediate species MOH (or *OH), MO (or *O) and MOOH (or *OOH); the decomposition of the latter releases oxygen (Figure 5.12a).

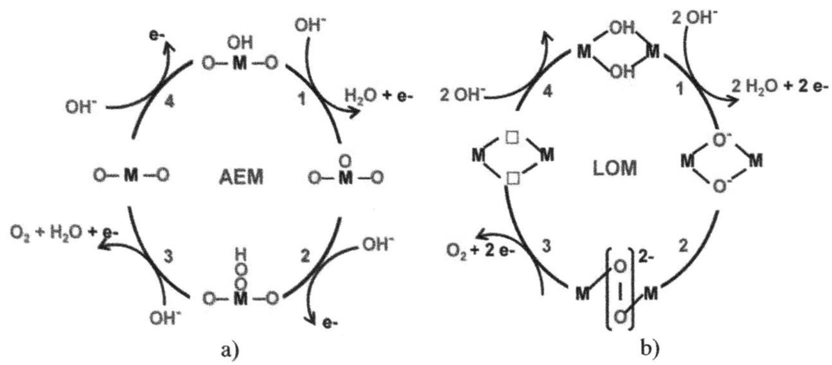

Figure 5.12. *AEM and LOM mechanisms of water oxidation. a) AEM mechanism. b) LOM mechanism. Oxygenated species from the catalyst are shown in red, those originating from the solution are shown in blue. Adapted from [SON 20] and [GRI 16]*

In this mechanism, the forced transition from MOH to MOOH means that the overpotential cannot theoretically be less than 0.37 V, corresponding to the theoretical energy difference between MOH and MOOH [MON 16, SON 20]; this also explains why the overpotential required for the OER is much higher than that needed for proton reduction.

One solution to this problem is to stabilize MOH and MOOH separately on different unsaturation sites, thus avoiding the direct link between MOOH and MOH. The corresponding reaction scheme, LOM (lattice oxygen-mediated mechanism), differs from AEM in that it involves a peroxy group –OO–, originating from the oxygen atoms in the lattice rather than those in the solution. Moreover, it can be stabilized by interactions with two metal unsaturation sites. Its oxidation releases O_2 while the two catalytic sites are regenerated, before adsorbing new OH groups (Figure 5.12b). Note that although the –OO– group involved in the oxidation process comes from unsaturated sites in the crystal lattice, it is renewed using the OH^- in the solution.

The validity of these mechanisms is supported by numerous theoretical calculations, aimed at defining useful characteristics to guide the choice of metals with the potential to constitute highly efficient catalysts. ABO_3-structured perovskites[21], considered to be catalysts for water oxidation in alkaline media, could react according to a LOM-type mechanism.

Grimaud *et al.* [GRI 17] recently used ^{18}O isotope labeling followed by *in situ* mass spectrometric analysis to demonstrate that the oxygen released does, in fact, largely originate from the crystal lattice of the oxide, thus confirming a LOM-type mechanism is at work in the case of perovskites. The mechanisms involved in O_2 evolution are far from being fully understood, with multiple effects which may combine to result in increased catalytic power. In the case of perovskites, [CHE 15b] noted that progressive modification of the structure of a perovskite such as $LaCoO_3$ by inserting strontium, up to the point where $La_{0.2}Sr_{0.8}CoO_3$ was obtained, resulted in an increase in the electrolysis current due to improved conductivity of the catalyst. This effect was attributed to a reduction in the band gap of the material, resulting from a progressive alignment of the O–Co–O atoms and an increase in the oxidation state of Co beyond Co^{3+}. The authors explain that these effects cause a significant overlap between the 3d band of the metal and the p-band of the oxygens, contributing to a narrowing of the band gap [CHE 15b, GRI 16].

5.3.4.1.1. Two-dimensional transition metal catalysts

Song and Hu [SON 14] designed new catalysts for the OER reaction in a KOH medium using thin layers of Co and Mn hydroxides. Their material, synthesized from a variety of cobalt hydroxide (β-$Co(OH)_2$), was obtained by replacing a

21 ABO_3 structured perovskites correspond to a formulation in which A is generally a rare earth metal (La, Y, etc.) or alkaline-earth metal (Ba, Sr, etc.) and B is a transition metal (Ni, Co, Fe, etc.). Depending on the degree of oxidation of A (2+ or 3+), the oxidation state of B can vary between 4+ and 3+, making it possible to modulate its adsorption properties with regard to intermediate species produced during water oxidation.

fraction of the Co^{2+} ions with Mn^{3+} ions, resulting in a 2D structure noted CoMnLDH (CoMn-layered double hydroxides). This structure consists of $CoMn(OH)_2$ nanosheets (of irregular size, between 50 and 100 nm, but with an average thickness of 3–4 nm), separated by an anionic layer that compensates for the Mn^{3+} ions (Figure 5.13a)[22].

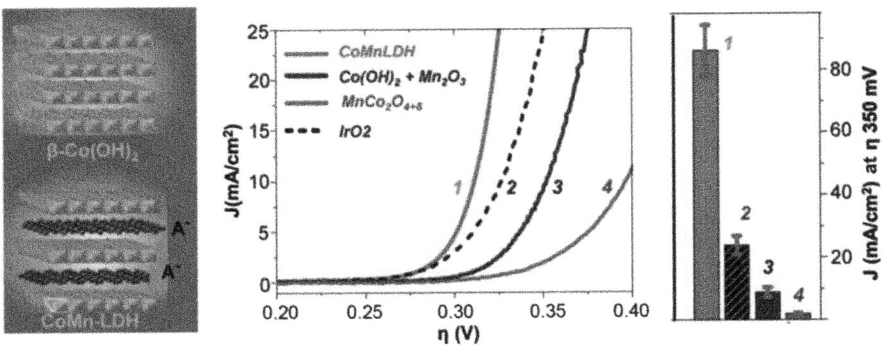

a) b) c)

Figure 5.13. *Electrocatalytic characteristics of Co and Mn-based electrode materials*

COMMENT ON FIGURE 5.13.– *a) Schematic representations of the lamellar structures of β-$Co(OH)_2$ and CoMnLDH. The CoMnLDH layers are formed by MO_6 octahedra (M: Co or Mn); A^- is the anionic layer which compensates for the excess charges introduced into β-$Co(OH)_2$ by Mn^{3+}. b) Oxidation curves of the CoMnLDH, $Co(OH)_2 + Mn_2O_3$, $MnCo_2O_{4+\delta}$ and IrO_2 catalysts in 1 M KOH medium. c) Current densities obtained with the catalysts in (b) for an overpotential of 350 mV. The catalysts are deposited on a porous carbon fiber paper (CFP). The Co loading in the CoMnLDH electrode is 0.222 mg/cm². Adapted from [SON 14].*

The strongest catalytic power is obtained for an atomic concentration ratio of 2 between Co and Mn, and following an anodic conditioning process which corresponds to a kind of "forming". Figure 5.13c shows that for an overpotential of 350 mV, the current density of 85 mA/cm² obtained using the CoMnLDH catalyst

22 CoMnLDH is obtained simply, at room temperature, by adding hydrogen peroxide H_2O_2 to a mixture of two salts, $Co(NO_3)_2$ and $Mn(NO_3)_2$, in solution, resulting in the oxidation of Mn^{2+} to Mn^{3+}. The addition of NaOH to the mixture results in CoMnLDH in the form of a yellowish precipitate. The proportions of Co^{2+} and Mn^{3+} in CoMnLDH correspond to the initial proportions of the two salts in solution. Calcination of CoMnLDH at 450°C results in a non-stoichiometric spinel $MnCo_2O_{4+\delta}$ ($\delta = 0.5$), while at 90°C the same preparation method results in a mixture $Co(OH)_2 + Mn_2O_3$.

far exceeds those observed with the other catalysts. Notably, it is three times higher than that obtained with IrO_2.

The catalyst loading (CL) in the electrode material can also have an important role on its catalytic activity. Morales-Guio *et al.* [MOR 16] also showed that the mass activity of thin films of oxy(hydroxides) of different transition metals was dependent on the CL, expressed in μg/cm^2 of the catalyst. The authors observed that the mass activity (MA) measured in A/(g of oxide) decreases when the CL increases, reflecting a loss of efficiency of a fraction of the catalyst. The experiment, carried out using metal oxides of increasing activity, NiOx, CoOx and CoFeOx, confirms this tendency, which amounts to considering that optimal catalyst activity is attained within a limited range of CL (Figure 5.14b).

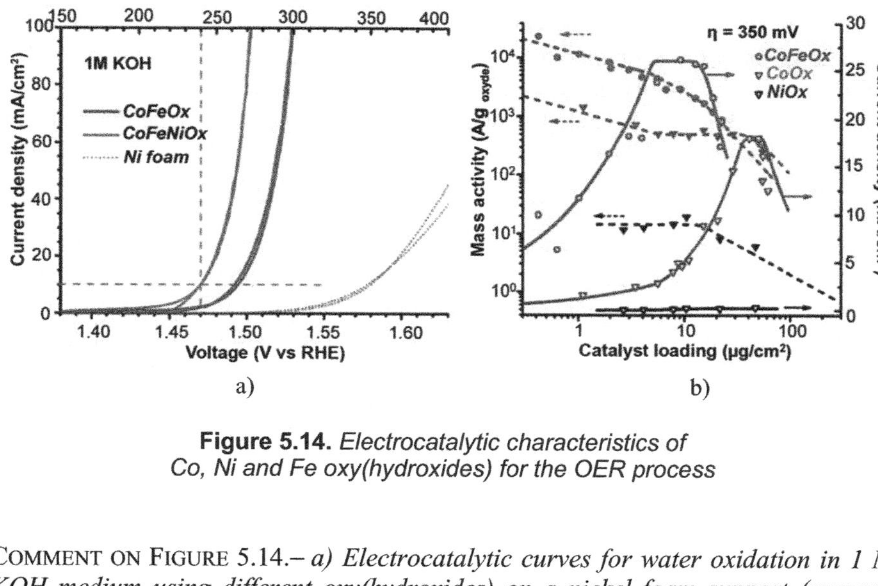

Figure 5.14. *Electrocatalytic characteristics of Co, Ni and Fe oxy(hydroxides) for the OER process*

COMMENT ON FIGURE 5.14.– *a) Electrocatalytic curves for water oxidation in 1 M KOH medium using different oxy(hydroxides) on a nickel foam support (average CL: 5 μg/cm^2, except FeNiOx for which CL = 3 μg/cm^2). b) Relation between the mass activity and the mass charge of the catalysts for an overpotential of 350 mV. The dashed and solid curves represent the variations of MA and current density, respectively, as a function of CL. Adapted from [MOR 16].*

The CoFeOx catalyst, seen to be the most efficient, can sustain current densities of 10 mA/cm^2 in 1 M KOH medium with low overpotentials of 320–330 mV, corresponding to an optimum loading between 5.5 and 9 μg/cm^2 (Figure 5.13b). This catalyst presents a high TOF of 3–4 s^{-1} at an overpotential of 350 mV, well above those of oxides such as NiOx, FeOx and CoOx, which are characterized by low catalytic capacities and exhibit TOFs in the range of $1–2\times10^{-3}\ s^{-1}$.

5.3.4.1.2. MOF or graphene-structured catalysts

Precise control of the arrangement of metals such as Ni, Co or Fe inside a catalyst is often the key to obtaining strong catalytic effects. MOFs (metal-organic frameworks, see Chapter 2), in which the metals are perfectly positioned within a lattice, provide a means of achieving this level of control.

Zhao *et al.* [ZHA 16b] obtained an extremely efficient catalyst for the OER process in a basic medium using an MOF with Ni and Co nodes. The hydrothermal approach to MOF synthesis is widely accepted, and, in this case, consists of a condensation reaction between benzene dicarboxylic acid (BDA) and nickel and cobalt ions in equal concentrations. After sonication and various conditioning operations, the MOF is obtained in the form of ultra-fine nanosheets (NiCo-ultrathin MOF nanosheets) of irregular size varying between 200 and 500 nm but of a constant thickness of around 3.1 nm, corresponding to four superimposed and coordinated monolayers. A suspension of these nanosheets in a solvent is then deposited onto a solid conductive substrate, such as glassy carbon (Figure 5.15a).

A detailed study of the structure of the NiCo-UMOFN shows that the Ni and Co atoms are regularly arranged in the lattice and are close together, inducing electronic coupling between the two metals through the 2p orbitals of the oxygens, with a transfer of electronic charges from nickel to cobalt. The small size of the nanosheets means that numerous unsaturation sites, that is, catalytic sites, are present on their surface. The polarization curves obtained for MOFs containing both Ni and Co, Ni alone and Co alone clearly show a synergistic effect between the two metals; as a result, the NiCo-UMOFN catalyst has a significantly lower overpotential compared to that of a "mono-metallic" Ni-UMOFN or Co-UMOFN (Figure 5.15b).

In the case where the NiCo-UMOFN catalyst is adsorbed on a highly conductive porous Cu substrate, the overpotential measured for a current of 10 mA is still better than that obtained using a glassy carbon substrate. In this case, it falls below 200 mV (189 mV vs. E° H_2O/O_2), outperforming commercial RuO_2 electrodes and the vast majority of the non-precious metal catalysts discovered previously.

The nanostructuring of the MOF also contributes to increased efficiency, probably due to improved contact between OH^- and the catalytic sites when compared to the same MOF in the bulk state (bulk NiCo-MOF), which presents a higher overpotential than NiCo-UMOFN (Figure 5.15b).

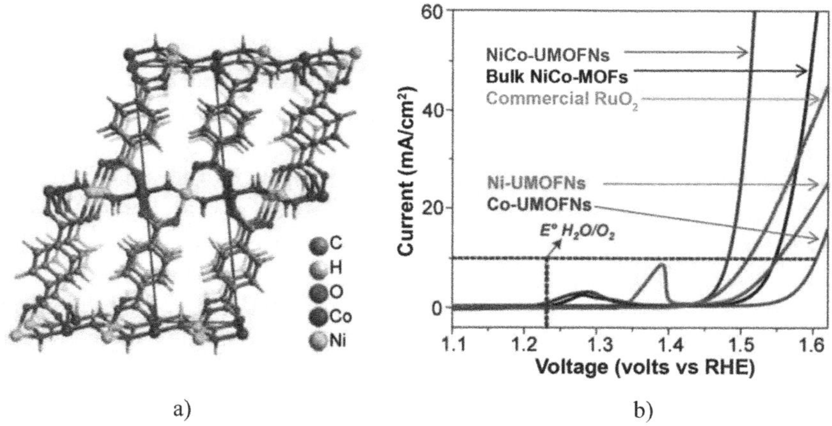

Figure 5.15. *Electrocatalytic characteristics of MOFs with Ni and Co nodes. a) Structure of a NiCo-UMOFN nanosheet formed by a condensation reaction between benzene dicarboxylic acid and Ni^{2+} and Co^{2+} ions. b) Oxidation curves of different MOFs for an OER process in 1 M KOH medium and on a glassy carbon substrate. Adapted from [ZHA 16b]*

The faradic efficiency of O_2 formation reaction is close to 100% (99.3%) and the TOF of the reaction for an overpotential of 0.3 V is 0.86 s^{-1}; for Ni-UMOFN and Co-UMOFN, the TOFs are only 0.18 and 0.09 s^{-1}, respectively. In terms of reliability, the authors observe that the current remains substantially constant over a long period of more than 1 week. Maintaining the overpotential at 0.25 V, the electrolysis current of around 10 mA/cm^2 decreases by only 2.6% after 200 hours, whereas with a RuO_2 electrode under the same conditions, the current drops to 60% of its initial value after only 10 hours of operation.

Recent works have also confirmed the interest of MOFs for the OER process in alkaline media. One novel approach involves the creation of unsaturation sites at the metal level by the action of a plasma and the elimination of some ligands coordinated to the metal on the surface, conferring catalytic properties on the metal atoms. Tao *et al.* [TAO 17] have shown that an MOF (ZIF-67)[23] consisting of Co nodes and imidazole ligands acquires significant catalytic power after irradiation by a plasma produced locally by the dielectric barrier discharge (DBD) technique. The

23 MOF ZIF-67 is classically synthesized by a condensation reaction between Co^{2+} ions and 2-methyl imidazole ($C_4H_6N_2$), resulting in a powder made up of dodecahedral particles with an average size of around 500 nm.

loss of ligands in the new product has the effect of increasing the positive character of cobalt, improving its catalytic properties with respect to the oxidation reaction.

The plasma-irradiated MOF presents much better electrocatalytic properties than the same MOF without plasma treatment. For a current density of 10 mA/cm^2, the overpotential is 410 mV (giving a potential of 1.64 V vs. NHE) in a potassium borate medium (0.5 M KBi, pH ~9), while the overpotential of the untreated MOF is 570 mV. The plasma-treated MOF is characterized by a TOF of 0.46 s^{-1} for an overpotential of 300 mV, more than 10 times that of the untreated MOF (0.043 s^{-1}). These values confirm the correlation between improved catalytic properties and the increased degree of cobalt unsaturation resulting from the loss of imidazole ligands. As before, the MOF is very robust, resulting in a high stability of the discharge current of 10 mA/cm^2 over a period of more than 24 hours with no material degradation or increase in the overpotential.

These examples show that MOF chemistry, which is relatively easy to implement and scalable, may offer the means of producing robust and efficient catalysts to replace the commercial catalysts currently used for the water oxidation reaction in alkaline media.

Other research into catalysts has focused on the use of nanoporous graphene as a substrate for metals. Compared to MOF, graphene is extremely conductive, and this may constitute an advantage in terms of improving electron exchange between catalytic sites.

Following this approach, a catalyst of the SAC type, in which Ni, Co or Fe atoms are adsorbed, has recently been proposed by a group of researchers from 10 different laboratories [FEI 18]. Using a particular mode of graphene transformation, nanopores, defined by cyclic structures of the C_4N_4 type, were created in the graphitic network of graphene; following adsorption of the metal M, this results in a hybrid material known as a metal n-doped holey graphene framework (M-NHGF) (Figure 5.16) [24].

24 The M-NHGF catalyst is obtained by means of a multi-step process. The first stage is to produce a 3D hydrogel of graphene, obtained using hydrothermal techniques (180°C) by reacting the graphene oxide (GO) suspended in water with the metallic precursor M and H_2O_2; the hydrogen peroxide, H_2O_2, is responsible for the formation of the pores in the graphene. The hydrogel is then lyophilized and annealed (900°C) in the presence of NH_3, resulting in M-NHGF. As the authors indicate, this type of preparation from GO can be scaled up to produce large quantities of the catalyst. Further details of the production technique can be found in [FEI 18].

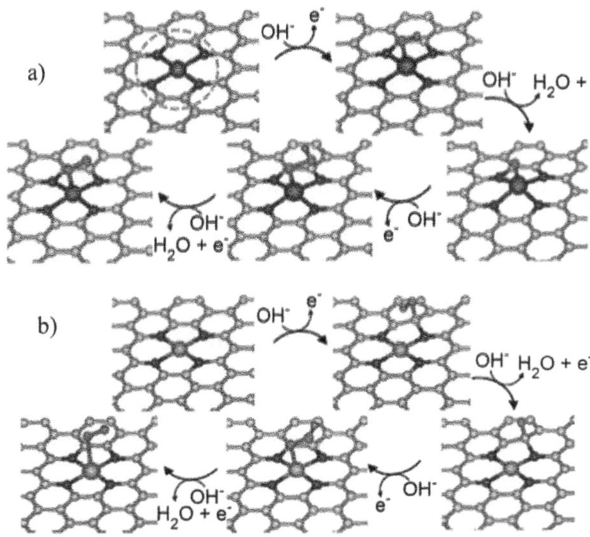

Figure 5.16. *Presumed mechanisms of OH^- oxidation on cobalt and nickel adsorbed in porous nitrogen-doped graphene (NHGF) nanosheets*

COMMENT ON FIGURE 5.16.– *a) and b) show the four oxidation steps leading to the intermediate species *OH, *O, *OOH and *OO finally resulting in O_2. a) Oxidation of OH^- on the single Co site in the Co-NHGF. The dashed light blue circle highlights the C_4N_4 arrangement in which the metal is inserted. b) Sequential adsorption of OH^- on two C and Ni sites in Ni-NHGF. The adsorption of *OH occurs first on the carbon closest to the Ni, and is followed, as before by the formation of *O on the same carbon, then *OOH and *OO adsorbed on Ni. Purple, green, red, dark gray and yellow dots represent Co, Ni, O, C and H, respectively. Adapted from [FEI 18].*

The nitrogen atoms arranged on the periphery of the nanopores are able to adsorb metal atoms in a robust manner, resulting in catalytic sites. Ni was seen to be the most efficient of the three metals covered by the study. The global mechanism of the reaction, established using a model of the energies involved in the various oxidation steps, features two catalytic sites in the case of Ni; one of these is a C in the vicinity of Ni, which, in energy terms (according to theoretical calculations), would be more favorable than the case where the metal (Co) is considered as the only catalytic site.

In terms of performance, a current density of 10 mA/cm^2 was obtained for an overpotential of 331 mV with respect to the equilibrium potential (1.23 V), which the authors consider to be comparable to other known catalysts based on different transition metals such as nickel and cobalt, evaluated using standardized criteria [MCC 13]. With respect to other MOFs, its performance is comparable to that of Ni-ZIF-67, but lower than that of the bimetallic MOF NiCo-UMOFN, for which the same current density of 10 mA/cm^2 is obtained with an overpotential of less than 200 mV. Its TOF, calculated for an overpotential of 300mV, is equal to 0.72 s^{-1}, which is also lower than the corresponding value for NiCo-UMOFN (0.86 s^{-1}).

In conclusion, these (non-exhaustive) examples show the new solutions available for developing precious metal-free catalysts. Bimetallic MOFs in the form of nanosheets or NPs, treated on the surface to create a large number of unsaturation sites, offer an alternative to the traditional metal oxides used in alkaline media for the OER process. They are easier to produce than porous graphenes, without the need for high temperatures, and the diversity of potential structures offers an additional advantage, enabling exploration of a wide variety of materials with as-yet unknown catalytic properties.

5.3.4.2. *OER in an acidic medium*

Iridium (IrO_2) and ruthenium (RuO_2) oxides have long been considered to be the best catalysts for water oxidation in acid and alkaline media. As we have seen, PEM electrolyzers present many advantages over alkaline electrolyzers. They are much more compact, can deliver much higher current intensities than alkaline electrolyzers and are better adapted to intermittent operation, making them more suitable for use with renewable energy systems (solar and wind power) [BAR 05]. The only obstacle to their development is a financial one, linked to the exclusive use of iridium-based catalysts, which are currently the only known means of ensuring long lifetimes and high current densities at relatively low overpotentials.

There are thus two main areas for research in this field: one prospective, with the aim of replacing precious metal-based catalysts with less costly and more readily available materials, and one rather more pragmatic, consisting of minimizing the quantities of precious metals inserted into a catalyst without reducing its performance.

5.3.4.2.1. Developing non-precious metal catalysts

Most of the Ni- and Co-based oxides used in alkaline media become unusable in acidic media due to their rapid dissolution. Much research has focused on manganese oxides and mixed derivatives of cobalt and other metals, stabilized by insertion into crystalline structures which are resistant to corrosion in acidic media. A threshold of 10 mA/cm^2 is widely accepted as a minimum performance target for this type of OER catalyst, compatible with photo-electrochemical renewable energy systems.

Manganese-based catalysts

Manganese oxides, which are stable in an acidic medium, have attracted attention in the context of water oxidation in an acidic medium. Unfortunately, at high overpotentials, the oxide dissolves into MnO_4^- resulting in its rapid destruction. Nevertheless, Huynh *et al.* [HUY 14] and Chan [CHA 18] have shown that a Mn oxide (δ-MnO_2) obtained by an electrodeposition process, initially very weakly catalytic, can be activated by electrochemical reduction leading to an intermediate species (α-Mn_3O_4); during the O_2 evolution reaction, this species transforms into a much more active Mox oxide, with increased activity seemingly resulting from the presence of Mn^{3+} ions in the crystal lattice. The slopes of the Tafel lines evolve from 650 to 90 mV/dec at pH 2.5 (Na phosphate buffer solution) and an intensity of 1 mA/cm^2 is obtained for an overpotential of 600 mV, 100 times higher than the intensity of the initial oxide. However, despite this clear improvement, the low values of the electrolysis currents and the impossibility of applying high voltages constitute a serious handicap for commercial applications.

More recently, Li *et al.* [LI 19a] have improved these characteristics using another crystalline variety of Mn oxide (Υ-MnO_2). With this oxide as a catalyst, a constant current of 10 mA/cm^2 can be produced continuously over very long periods of time (over 8000 hours) with no observed deterioration of the system. This crystalline variety of MnO_2 is simple to prepare and is already available commercially, due to its use in manufacturing of Leclanché batteries (Zn/MnO_2). In practical terms, this shows that the product is both abundant and inexpensive. Its stability in acidic media is very good provided that the voltage is maintained within a narrow window between 1.65 and 1.75 V versus RHE; above 1.75 V, MnO_2 is also oxidized and dissolves, resulting in the soluble species MnO_4^- [LI 19a].

In conclusion, the use of manganese oxides in an acidic environment is limited by their incompatibility with high overpotentials, a fact which necessarily limits their operation to low current densities. Nevertheless, in the case of Υ-MnO_2 oxide, the 10 mA/cm^2 threshold is reached without difficulty and can be maintained over very long periods by applying an overpotential of the order of 420 mV, comparable to that observed with IrO_2 or RuO_2 electrodes for the same current densities.

Cobalt-based catalysts and stabilization techniques

Cobalt oxides, which are very good catalysts for water oxidation in basic media, also become more difficult to use in acidic media because of their propensity to corrode, finally resulting in dissolution. Corrosion can be slowed or prevented by incorporating cobalt into a robust crystalline structure, or by using it in nanostructured form.

Cobalt ferrocyanates $Co_2[Fe(CN)_6]$, similar to materials derived from Prussian Blue, have attracted attention as potential catalysts for water oxidation in acidic or neutral media. Several teams have identified catalysts which perform well in this context, although performances are still well below those attained in a basic medium: their stability remains low, and current densities rarely exceed 10 mA/cm^2.

A recent improvement involves doping the cobalt ferrocyanate with iron, that is, transforming $Co_2[Fe(CN)_6]$ into $Co_{2-x}Fe_x[Fe(CN)_6]$, which leads to a catalytic power which is twice that of $Co_2[Fe(CN)_6]$ at a neutral pH. While this result is interesting, the material is still relatively unstable in an acidic environment (pH 1.5): an initial current of 10 mA/cm^2 obtained with an overpotential of 600 mV decreases regularly over time and drops to 70% of the original value after one hour of operation [HAN 20].

The use of robust host structures for cobalt, such as a polyoxometalate (POM), has produced better results. Blasco-Ahicart *et al.* [BLA 18] synthesized a barium salt of cobalt 290olyoxometalate (Ba[Co-POM]) which is insoluble and very stable in 1 M H_2SO_4. Mixed with a carbon paste (CP) in a proportion of 40 wt%, it can generate an oxidation current of 10 mA/cm^2 with an overpotential of only 360 mV, 100 mV lower than that observed using an IrO_2 electrode under the same conditions. These values are obtained from potentio-dynamic curves. Stability tests, carried out over a continuous period of 24 hours and at low overpotentials (250 mV), correspond to current densities which are relatively low, but still much higher than those obtained with other cobalt- or iridium-based catalysts. Thus, for an overpotential of 250 mV, IrO_2 mixed with the same carbon paste produces a current of 0.03 mA/cm^2, compared to 0.3 mA/cm^2 with Ba[Co-POM)]/CP, a value 10 times higher. Unfortunately, no indication of stability has been given for higher overpotentials, and this information is essential for better evaluation of this new catalyst.

Paradoxically, extreme division of a cobalt oxide is another way to increase its resistance to corrosion in highly acidic media. Co_3O_4 oxides in the form of 200–250 nm thick films[25] and made up of Co_3O_4 grains of about 25 nm in size exhibit greater stability than the bulk oxide in a highly acidic medium (0.5 M H_2SO_4), and dissolve at a much slower rate [MON 17]. At constant potential (1.8 V vs. RHE, i.e., 570 mV surge), a film ~300 nm thick can withstand a water oxidation reaction for over 12 hours at 10 mA/cm^2 with a mass loss of around 70 µg, representing a dissolution rate of around 100 ng/min. For comparison, the dissolution rate of a RuO_2 electrode in acidic media for comparable current densities is much higher, around 540 ng/min (Figure 5.17).

25 The electrode consists of an FTO (fluorine-doped tin oxide) substrate onto which cobalt is deposited by spin coating in a colloidal state. A thermal treatment in air at 400°C results in a nanostructured Co_3O_4 film.

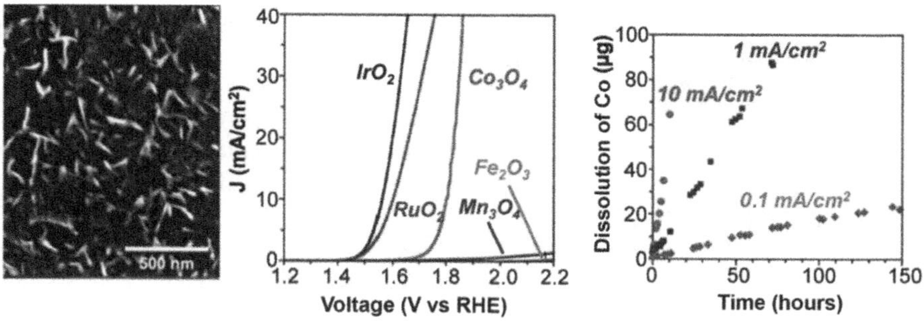

a) b) c)

Figure 5.17. *Electrocatalytic properties of a nanostructured cobalt oxide film*

COMMENT ON FIGURE 5.17.– *a) SEM image of Co_3O_4 nanocrystals. b) Water oxidation curves in 0.5 M H_2SO_4 medium for different electrodes containing Co_3O_4, IrO_2, RuO_2, Fe_2O_3 and Mn_3O_4 catalysts, respectively. c) Dissolved weights of Co for different O_2 formation rates in the case of the Co_3O_4 catalyst. Adapted from [MON 17].*

These examples show the problems inherent in the use of precious metal-free catalysts for water oxidation in acidic media. Cobalt-based catalysts appear to be the most promising option. However, they are subject to corrosion processes, which shorten their useful lifetime. Inserting these materials into robust crystalline structures and nanostructuring in the form of nano-crystallites are two methods that can be used to increase their stability. Nevertheless, their use appears to be limited to low current densities and, as a result, they cannot currently be considered as a true alternative to commercial precious metal catalysts; this justifies ongoing research work in the area.

5.3.4.2.2. Optimizing precious metal catalysts

The majority of electrolyzers operating in an acidic environment using PEMs feature iridium oxide deposited on a TiO_2 support, with an Ir content of the order of 1.5–3 mg/cm^2, as the anode [BÖH 20]. A key research objective is to reduce the amount of iridium used in the catalyst while maintaining or improving performance.

Many of the various catalysts proposed are based on the idea of replacing the titanium oxide substrate, on which IrO_2 is deposited, by a solid, structured, extremely porous and highly conductive substrate, favoring mass transport between the electrolyte and the catalyst; the substrate must also be resistant to acid corrosion. Catalyst efficiency is expected to increase as the contact surface between the IrO_2

and the substrate increases, and depends on the technique used to incorporate the metal oxide on the substrate.

Lee *et al.* [LEE 12] were among the first to define parameters for accurate comparison of different catalysts. They also showed that replacing the IrO_2/C and RuO_2/C composites with small (6–7 nm) crystalline (rutile) NPs resulted in increased stability.

Subsequent research focused on current commercial IrO_2/TiO_2 catalysts, and alternative formulations were proposed, both in terms of replacing TiO_2 (an insulator) with a more conductive substrate and in terms of decreasing the Ir content in the catalyst.

Very recently, Böhm *et al.* [BÖH 20] developed a multi-step method for preparing an IrO_2 catalyst deposited inside and outside porous antimony-doped tin oxide (ATO) microparticles (µPs). A unique feature of this type of conductive catalyst is its high porosity, giving a void volume fraction of around 83%; this gives it a very low density, and, paradoxically, although the proportion of IrO_2 relative to the total mass of the catalyst is high[26], the amount of Ir in the catalyst is very low. Thus, for an Ir content of 25% by weight in the catalyst, the Ir density is 0.08 gIr/cm^3, compared to 1.5 gIr/cm^3, that is, almost 20 times more, for a conventional TiO_2/IrO_2 catalyst.

The authors' method consists of "lining" the walls of ATO µPs with a thin layer of IrO_2, with a thickness between 12 and 16 nm, and consisting of NPs of size 1–2 nm (Figure 5.18).

The authors compared the electrocatalytic characteristics of the 25-IrO_2/ATO catalyst (the most efficient form), with 25 wt% Ir and cavities of around 180 nm in diameter, to those of a commercial 75-IrO_2/TiO_2 catalyst containing 75% Ir (Elyst Ir75). 25-IrO_2/ATO clearly out-performed 75-IrO_2/TiO_2 in terms of both overpotential and stability under the same conditions. For a current density fixed at 10 mA/cm^2, the overpotential for the first catalyst remains essentially constant with

---

26 ATO µPs are obtained by combining the colloidal lithography technique with sol-gel chemistry. A first step results in µ-Ps consisting of PMMA (Polymethyl methacrylate) microspheres coated with ATO. After removal of the PMMA microspheres (120 or 280 nm in diameter), an inverse opal type structure is obtained corresponding to the ATO µPs. The ATO µPs are then surface modified using a colloidal solution of IrOx. Treatment in a reducing medium composed of a ternary mixture of water/ter-butyl alcohol/methanol leads to the deposition of a homogeneous layer of Ir on the ATO µPs cavity walls. Air calcination of the Ir/ATO iridium µPs at 175°C results in the IrO_2/ATO catalyst. The proportions of Ir/ATO by weight can be modulated between 7% and 43%.

an average value of 420 mV over a period of 2 hours; for the second catalyst, the overpotential increases steadily from 445 to 485 mV. These results also indicate significantly higher mass activities than those of Elyst Ir75. For an applied voltage of 1.53 V (i.e., an overpotential of 300 mV), the mass activity of the 25-IrO_2/ATO catalyst is 65 A/g_{Ir}, compared to 8 A/g_{Ir} for Elyst Ir75.

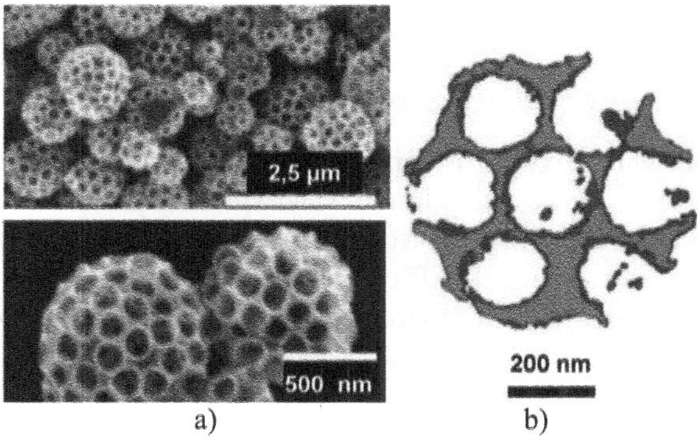

Figure 5.18. *Microparticles (µPs) used as IrO_2 substrates. a) SEM images of antimony-doped SnO_2 µPs with different magnifications. b) Cross-section of a µP coated with a thin layer of IrO_2 (purple) with a thickness between 12 and 16 nm, highlighting the nanopores (size ~200 nm) and the distribution of IrO_2 over the entire inner and outer surface of the µPs. Adapted from [BÖH 20]*

Hufnagel *et al.* [HUF 19] proposed another composite, made up of extremely thin layers of IrO_2. Beginning with carbon NPs extracted from soot and deposited on a titanium substrate, alternating layers of TiO_2 and niobium oxide (NbOx) were then deposited using ALD (atomic layer deposition)[27], resulting in mixed Nb and Ti oxide µPs (NTO). A final ALD of a very thin layer of IrO_2 (thickness <1 nm) results in an IrO_2/NTO catalyst with an Ir content of 0.153 mg/cm^2, compared to values of between 1.5 and 3 mg/cm^2 for commercial IrO_2/TiO_2 catalysts [BÖH 20]. In 0.5 M H_2SO_4 medium, a current of 100 mA/cm^2 was obtained for a voltage of 1.64 V (i.e., an overpotential of 410 mV), corresponding to a high mass activity of 654 A/g_{Ir}, much higher than that of most known systems. This current of 100 mA/cm^2 remains almost constant over a period of 12 hours. However, a slight degradation of the catalyst is observed during electrolysis; according to the authors, this is not due to the dissolution of IrO_2, but probably to an adhesion defect.

27 The ALD technique allows very precise control of the thickness of deposited layers, which follow the geometry of the substrate completely.

These examples demonstrate the ability of IrO_2-based catalysts to resist the high overpotentials required to produce O_2, and thus H_2, in acidic media in PEM electrolyzers, which is far superior to that of other known catalysts. The development of conductive and porous supports is one of the most promising ways of reducing Ir requirements while maintaining catalytic properties equivalent to those of current commercial catalysts. However, progress is still needed with respect to long-term reliability before these new systems can be commercialized.

5.3.4.3. *Conclusions on OER processes*

The OER, in which oxygen is formed by oxidation of water in an alkaline or acidic medium, is the slowest step in the water splitting process and requires much higher overpotentials than HER. Corrosion phenomena, most pronounced in acidic media, represent an additional difficulty and considerably limit the choice of catalysts for electrolysis in these environments. Those materials which perform well in alkaline media are often unusable as catalysts in acidic media as they corrode rapidly. Cobalt-based catalysts appear to be the most promising alternative, although they are still subject to corrosion. Inserting these materials into robust crystalline structures or nanostructuring in the form of nano-crystallites are two methods that can be used to increase their stability. Nevertheless, their use appears to be limited to low current densities and, as a result, they cannot currently be considered a true alternative to commercial precious-metal catalysts; this justifies ongoing research work on precious-metal catalysts, with a view to reducing their loading while increasing efficiency.

5.4. Photoelectrochemical water splitting

Oxygen and hydrogen production by means of water splitting, powered by solar radiation, is seen as one of the main technologies of the future, offering the means of supplying hydrogen fuel in a renewable manner and at very low cost. Fujishima and Honda [FUJ 72] first demonstrated this technique in the 1970s, describing decomposition of water into H_2 and O_2 by the UV irradiation of a titanium oxide electrode (n-type semiconductor), in contact with an electrolyte solution. The PEC cell is divided into two compartments, separated by an ion-permeable membrane. One compartment contains the TiO_2 semiconductor electrode, while the other contains a platinum electrode; the two are electrically connected by an external circuit. Light excitation of the semiconductor (SC) results in the generation of an electron-hole pair (e^-, h^+). After separation, the holes ($^+$) migrate to the surface of the SC, while the electrons are transferred through the external circuit to the Pt electrode. Water, acting as an electron donor, neutralizes the produced holes on the

TiO_2 surface and decomposes to O_2; the electrons injected into the Pt reduce the water, releasing H_2[28].

This process can only occur if the radiation energy hv is higher than the band gap (E_g) of the semiconductor; in the case of TiO_2, for which E_g is equal to 3 eV, this implies radiation with a wavelength of less than 415 nm (UV range). The energy yield in this case is very low (around 2%); a theoretical maximum of 32% can be reached if the SC can absorb over the whole UV-visible spectrum, which would imply light absorption between 200 and 800 nm. For this to be possible, an SC with a band gap of 1.5 eV would be required, compatible with the dissociation energy of water into $H_2 + ½O_2$ (238 kJ/mol), electrically equivalent to a two-electron transfer at a potential of 1.23 V.

Various approaches may be used for photochemical decomposition of water into H_2 and O_2. Some use a single SC, to which a co-catalyst is added; others are based on the formation of heterojunctions between several SCs, giving rise to different types of mechanisms, depending on the arrangement of the energy levels of the conduction band (CB) and valence band (VB) of the SCs. These mechanisms, along with the different materials used for photochemical decomposition of water, have been described recently [HE 18, JIA 18, KAL 18, XU 19a]. The energy characteristics of these mechanisms differ depending on the number of steps involved in the water splitting process. In the first case, where the goal is to produce H_2, water may be decomposed using semi-conducting particles on which a co-catalyst is adsorbed, intended to facilitate the reduction of H_2O by electrons. Systems of this type are known as heterogeneous photocatalysts (Figure 5.19a). For systems with two or more SCs, multiple combinations are involved. In some cases, the mechanisms at work do not require an external voltage source, while in "assisted" systems, water splitting is dependent on an external voltage input. Many of these systems are inspired by the photosynthesis mechanism in plants, in which two photosystems, PSI and PSII, operate according to a Z-scheme.

A redox mediator may be included in the case where the two SCs are separated by a solution, or excluded when the two SCs are in direct contact, possibly with an interfacial metal contact between the two SCs (Figure 5.19b).

28 This operating scheme is very similar to that of a dye-free Grätzel solar cell, where water is used as a redox mediator, being oxidized on the TiO_2 surface and reduced on the platinum electrode. The presence of a catalyst on the TiO_2 or on the platinum increases the rate of reaction. Note, too, that H_2 is produced on the metal electrode, in this case platinum; once again, scarcity is an issue, and, as in the case of water reduction, the search for non-precious metal catalysts is key.

After Fujishima and Honda's initial discovery, research quickly came to focus on the implementation of SCs able to perform photocatalytic water decomposition using visible light. The objective is to develop a low-cost PEC system which can compete with a system combining a photovoltaic cell and an electrolyzer, can operate over periods of several years and can provide an energy yield of 10% for transforming solar energy into fuel, particularly in the field of solar-to-hydrogen (STH).

Many different SC materials and strategies have been tested in the aim of obtaining SCs that are stable in liquid media and are characterized by an E_g which is compatible with a broad absorption band of visible light. Large-gap SCs can be doped in order to function in visible light, using elements which create impurity levels corresponding to new electronic states. Photoexcitation of these electronic states, within the band gap, by visible light can then result in a photocurrent.

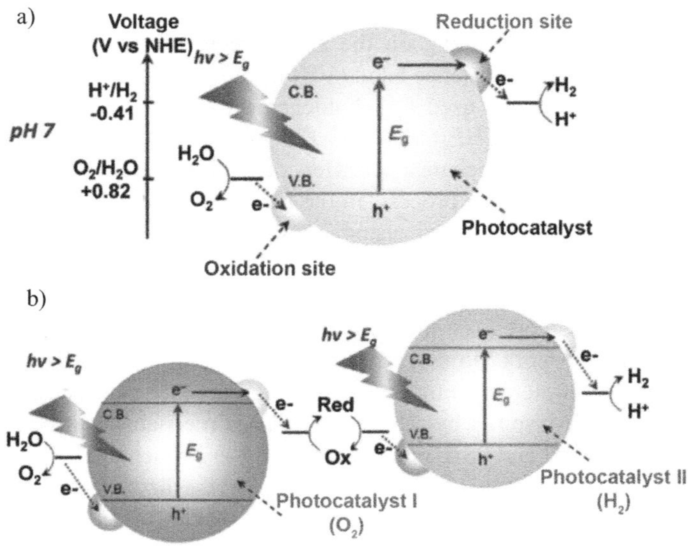

Figure 5.19. *One- and two-step photocatalysis mechanisms*

COMMENT ON FIGURE 5.19.– *a) Operation of a heterogeneous photocatalyst consisting of an SC and co-catalysts for water oxidation (O_2) and reduction (H_2). (b) Two-step mechanism corresponding to a Z-type scheme, with a redox shuttle (Red/Ox) in the liquid medium transferring photoelectrons from the conduction band (CB) of photocatalyst I to the valence band (VB) of photocatalyst II. Adapted from [MAE 10].*

Many photocatalysts use TiO_2 as the SC, doped with various elements or used in conjunction with sensitizing dyes which, after excitation, can transfer photoelectrons into the conduction band of the TiO_2, in a process similar to that described in connection with Grätzel photovoltaic cells (see Chapter 4).

A variety of materials can be used in this type of mechanism; detailed descriptions may be found in Abe's review paper [ABE 10]. In what follows, we shall simply describe some examples of photocatalysts, corresponding to systems with one or more SCs which carry out water splitting in one or more steps. These examples have been chosen to illustrate a number of problems encountered during their implementation, along with possible solutions; they also involve the use of nanomaterials described in Chapters 1 and 2.

5.4.1. *Heterogeneous photocatalysts*

Heterogeneous photocatalysts consist of SC particles (the photocatalysts), locally decorated with metals or metal derivatives (co-catalysts) with the ability to catalyze proton reduction, according to the scheme shown in Figure 5.19a. Several conditions are necessary to convert solar energy into chemical energy with good yields. First, the band gap of the SC must be as narrow as possible in order to absorb a maximum amount of light energy. Second, proton reduction, by electrons, and water oxidation, by holes, can only take place if the lower and upper levels of the conduction and valence energy bands are favorably positioned with respect to the redox potentials of H^+/H_2 and O_2/H_2O. This implies that the respective energies of the lower and upper levels of the CB and VB must, respectively, be higher and lower than the corresponding energies of the redox potentials of H^+/H_2 and O_2/H_2O, or of any other "sacrificial" redox system used as a substitute for O_2/H_2O.

Very few SCs actually correspond to these criteria in full. Moreover, even when the energy conditions are met, there is no guarantee of long-term stability due to possible degradation of the SC by photocorrosion.

Cao *et al.* [CAO 15] obtained exceptional results using Cd sulfide (CdS) and transition metal phosphide co-catalysts (Mo, Ni and Co). They achieved extremely high H_2 production rates using CdS NPs decorated with the metal phosphide co-catalysts CoP, Ni_2P and Cu_3P as co-catalysts [CAO 15] (Figure 5.20).

Using the CoP-CdS(5) photocatalyst[29] in the presence of lactic acid (a sacrificial compound added in solution as an electron donor to prevent O_2 formation), the H_2

29 The catalysts were obtained in powder form by grinding a mixture of CdS and metal phosphide in the solid state. CoP/CdS(5) corresponds to a catalyst containing 0.05 wt% CoP.

formation rate is 254 mmol h^{-1} g^{-1} (giving a TOF of around 0.01s^{-1}) during the first 4 hours, and remains very high over a period of 100 hours. The same CdS SC with Pt as a co-catalyst gives much lower H_2 release rates of only a few tens of mmol h^{-1} g^{-1}(~75 mmol h^{-1} g^{-1} over a 10-hour period) (Figure 5.20b). These very high transformation rates confirm the catalytic role of the transition metal phosphides with respect to the proton reduction reaction [CAL 16] (see section 5.3.1.1). Furthermore, they contribute significantly to increasing light absorption between 500 and 800 nm, complementing the absorption properties of CdS (Figure 5.20c). This may explain the high photocatalytic power of the CoP/CdS (5) catalyst, which has a quantum yield of over 25% for wavelengths above 420 nm. However, the presence of Cd in the catalyst is an ecological drawback and is likely to limit interest in the system.

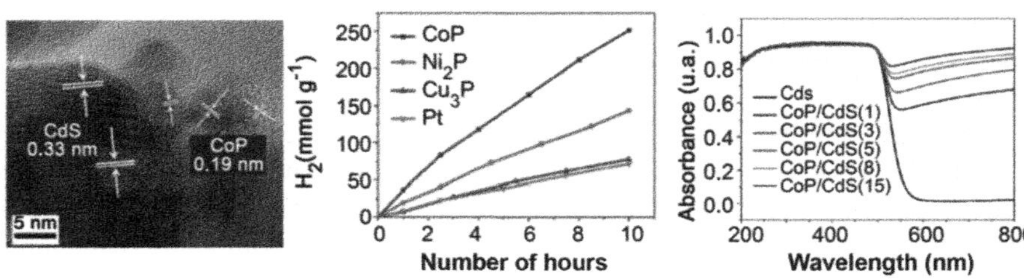

a) b) c)

Figure 5.20. *Heterogeneous MP/CdS photocatalysts*

COMMENT ON FIGURE 5.20.– *a) High-resolution TEM image (HRTEM) of CoP/CdS NPs highlighting the crystalline nature of the NPs. b) H_2 production rates by M/CdS catalysts with M = CoP, Ni_2P, Cu_3P and Pt. c) Comparative absorption curves of CdS and CoP/CdS(x) catalysts, where x is the mass ratio of CoP to photocatalyst (1, 3, 5, 8 and 15: 0.01, 0.03, 0.05, 0.08 and 0.15). Adapted from [CAO 15].*

5.4.2. Photocatalytic systems with two SC heterojunctions

While a whole range of SCs made from multinary oxides exist, systems comprising $BiVO_4$ (BVO) associated with other absorbing SCs have received the most attention, and have achieved energy yields between 7% and 8% [JIA 18a, MAL 18]. Different methods have been used to improve the energy conversion yields of these systems, including nanostructuring, the use of multiple

The CoP co-catalyst takes the form of NPs (size ~10 nm) and is in close contact with CdS, synthesized as hexagonal sheets (sizes ~50 nm).

heterojunctions, doping to modify energy band characteristics, surface modifications and even the adjustment of light absorption characteristics.

5.4.2.1. *Nanostructuring*

BVO, in its monoclinic crystalline form (scheelite), is particularly attractive because of its relatively low E_g value (2.4 eV) and low production cost which could result from wet processing [KUD 99]. However, it also has a high charge recombination rate due to a small mean free path L_D. This short charge diffusion range (~70 nm) is responsible for premature charge recombination, resulting in low conversion efficiency in thick BVO photocatalysts.

Nanostructuring of SCs offers the means of limiting charge recombination while maintaining high light absorption, leading to a clear improvement in yields in the case of WO_3/BVO heterojunctions. Pihosh *et al.* [PIH 15] obtained energy yields of the order of 8% by realizing a photoelectrochemical cell featuring a $WO_3/BiVO_4$+CoPi photoanode connected to a Pt electrode[30] (Figure 5.21).

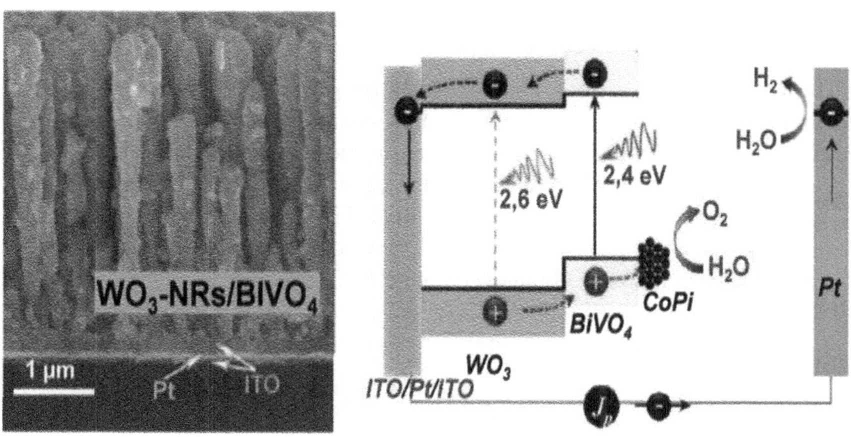

a) b)

Figure 5.21. *$WO_3/BiVO_4$/CoPi photoelectrocatalyst in neutral medium*

COMMENT ON FIGURE 5.21.– *a) Cross-section of ITO/Pt/ITO/WO_3-NR/$BiVO_4$, showing WO_3 nanorods coated with $BiVO_4$. b) Energy diagram of the SCs WO_3 and $BiVO_4$. ITO/Pt/ITO is the (conductive) window through which light arrives with absorption occurring between 300 and 500 nm. Note the cumulative effect of*

30 A voltage of 1.23 V is applied between the two electrodes. In the absence of light, the current intensity is zero and, as soon as irradiation is applied, a current of 6.72 mA/cm^2 is produced (90% of the theoretical value).

generation of (+) and (–) charges in the two SCs (type II heterojunction). The (+) charges are neutralized by H_2O and release O_2 on contact with the CoPi NPs (co-catalyst); the (–) charges are transferred to the Pt electrode and reduce H_2O, producing H_2 on contact with the Pt. The solution has a pH of 7. Adapted from [PIH 15].

The photoanode corresponds to a core/shell structure, using WO_3 in the form of an array of nanorods (NR) of a few microns in length (~2.5 μm) and about 200–300 nm in diameter, covered by an extremely thin layer of BVO (~25 nm). This layer is also modified on the surface by CoPi (Co phosphate) NPs, which have the effect of catalyzing the OER process.

The amounts of H_2 and O_2 obtained for AM1.5G irradiation verify the stoichiometric ratio of 2; for H_2, they are of the order of 100 $\mu mol\ h^{-1}\ cm^{-2}$, that is, a mass catalytic power of 130 $mmol\ h^{-1}\ g^{-1}$ evaluated with respect to the mass of BVO, which is approximately the same order of magnitude obtained with CdS/metal phosphide photocatalysts[31].

5.4.2.2. *Graphene as a support for photocatalysts*

The adsorption of various photocatalysts on graphene (GR) has been a major subject of research in recent years; in this context, GR is used as an all-solid electronic mediator in a Z-type photocatalytic mechanism. Iwase *et al.* [IWA 11] showed that GR oxide (GO) reduced by photolysis (r-GO) and connected to $BiVO_4$ (O_2 generator) and rhodium (Rh)-doped Ru/$SrTiO_3$ (H_2 generator) photocatalysts exhibited this type of mechanism, attenuating electron-hole recombination and thus increasing the production of O_2 and H_2 (Figure 5.22).

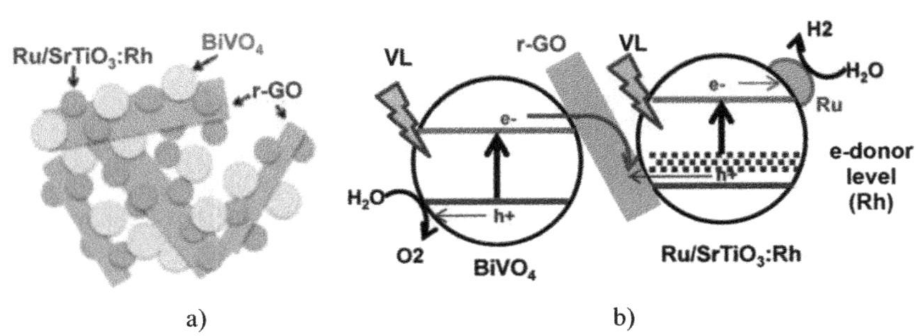

Figure 5.22. *Photocatalyst consisting of two SCs, $BiVO_4$ and $SrTiO_3$:Rh connected by a graphene sheet (r-GO) which acts according to a Z-type mechanism*

31 This calculation is approximate and is based on WO_3 cylinders of 200 nm diameter and 2.5 μm length covered with a homogeneous 25 nm layer of BVO, with a mass density of 6.1 g/cm^3.

COMMENT ON FIGURE 5.22.– *a) Image of the photocatalyst showing $BiVO_4$ and $Ru/SrTiO_3$:Rh NPs adsorbed on the r-GO sheets. b) Energy diagram of the two SCs connected by a conductive r-GO sheet irradiated in visible light (VL). The $SrTiO_3$ SC is doped with Rh, which introduces an electron donor level into the energy band gap from which photoelectrons are produced. The Ru adsorbed on the SC acts as a catalyst for water reduction. Adapted from [IWA 11].*

The authors observed a threefold increase in O_2 and H_2 generation due to the presence of r-GO and to the good chemical inertia of r-GO over a 24-hour period, although H_2 and O_2 production rates remained modest (~0.3 mmol $h^{-1}g^{-1}$).

Other photocatalysts, described in detail by Li *et al.* [LI 16d] in a recent paper, have also been created, generally by combining metal sulfide semiconductors with graphene. Graphene is usually obtained by reduction of GO oxide (r-GO), and the catalytic effect of the SC/r-GO composite is enhanced by the adsorption of precious metal (Pt) NPs on the r-GO surface. The presence of r-GO greatly improves charge separation and leads to an increase in H_2 production rates, reaching between 20 and 50 mmol $h^{-1}g^{-1}$.

Titanium oxide (TiO_2) presents certain advantages in that it is a very stable SC and, unlike CdS, it is non-toxic. Unfortunately, its high gap limits the exploitation of visible light. This material has also been combined with graphene to enhance its photocatalytic properties. Xiang *et al.* [XIA 12] have shown that its catalytic activity for H_2O reduction is enhanced when it is used in the form of NPs, adsorbed onto MoS_2/graphene sheets which act as co-catalysts. For an optimal composition by weight of the different components[32], the mass catalytic activity of the photocatalyst is maximized and is of the order of 2 mmol $h^{-1}g^{-1}$, that is, 39 times that of TiO_2 alone and four times that of TiO_2/MoS_2 without GR. This confirms the synergistic effect between TiO_2, MoS_2 and GR, along with the mediating role played by GR in improving charge separation and increasing the H_2 production rate. Converted into power density, expressed in $A/g_{catalyst}$, this result is equivalent to around 0.1 A/g (giving a TOF of about 0.16 s^{-1}); this is significantly lower than the values usually reached with electrolysis catalysts. Other, more valuable, applications may be found in the field of oxidative wastewater treatment. In this case, the holes generated during the photonic process are used to degrade organic matter in low concentrations, and the relevance of TiO_2 in conjunction with GR in this case is well established [KUS 19].

32 The TiO_2/MoS_2/GR photocatalyst has a mass of 80 mg and, in its optimal composition, consists of 99.5% TiO_2, 0.45% MoS_2 and 0.05% GR.

5.4.3. *Conclusions*

The issue of energy storage in chemical form is an important one, notably in terms of reducing greenhouse gas emissions. Electrochemical or photochemical water splitting techniques for hydrogen production have been a major focus of research in recent years, with the ultimate aim of replacing the fossil fuel-based industrial processes which currently dominate H_2 production. At present, the operational industrial electrolyzers used to produce large quantities of H_2 rely on the use of precious metal electrodes to decrease the overpotentials at these electrodes and thus limit the energy needed to transform water into H_2 and O_2. The current trend in developing PEM electrolyzers raises a further issue: the use of acidic media means that non-corroding anode catalysts are required for use in the O_2 formation reaction, for which the overpotential is the highest. The creation of non-precious alternatives to precious metal catalysts presents a considerable challenge, and despite numerous advances, no new solution has yet been applied on an industrial scale. A great deal of research is currently directed toward the identification and manufacture of electrodes that are a match for current commercial catalysts in terms of efficiency but that also contain increasingly small proportions of precious metal. In this respect, nanostructured electrode materials, in which minute quantities of Pt or Ir are disseminated, provide a promising avenue.

In terms of photocatalysis, a wide variety of techniques have been developed and tested with the aim of attaining energy yields of 10%, the threshold for commercial exploitation. The use of one or more SCs in conjunction with co-catalysts to generate H_2 or O_2 generators has been widely explored. A number of systems, including CdS (which has unfortunate environmental drawbacks), have proved to be extremely efficient, with record H_2 production rates of the order of 250 mmol $h^{-1}g^{-1}$. The advantage of photocatalysis compared to electrolysis is that there is no need for an external power source, but the rate of H_2 production is very low, as we see from the comparative TOF values. Finally, the combination of electrolysis cells with photovoltaic electricity sources is another possible approach to H_2 production and storage. The energy efficiency of this technique is close to that of the photovoltaic mechanism, and can exceed the 10% target, but at a higher cost.

5.5. Fuel cells

The reaction involved in hydrogen fuel cells involves chemical recombination of water from oxygen and hydrogen, following the opposite approach to that used in water splitting. The chemical energy released by recombining H_2 and O_2 into H_2O is transformed into electrical energy. The reaction can be carried out in acid or alkaline media and, as in the case of water decomposition, specific catalysts are required for each medium.

Work is currently underway on five major types of fuel cells, with key aims such as cost reduction and increasing reliability. These types are distinguished by the nature of the electrolyte, which may be alkaline, acid, solid or liquid. The oldest form is the alkaline fuel cell (AFC), which was already being used by NASA in the context of the Apollo mission in the 1970s. This time-honored technology has recently been improved by the use of anion exchange membranes with high ionic conductivity (AEMFCs: anion exchange membrane fuel cells), which can compete with acid-medium proton exchange membrane fuel cells (PEMFCs) in terms of energy efficiency [DEK 18].

The slowest step of the OER process involves the formation of O_2 by oxidation of water or OH^-. Similarly, oxygen reduction is the slowest step in the water recombination process, and is the reason for a large part of the overpotential required for the global recombination reaction. It can be minimized by the use of precious metal catalysts, and is eliminated when the reaction is carried out at very high temperatures (between 700°C and 1000°C). In the latter case, liquid electrolytes are replaced by solid ionic conductors in the form of metal oxides (SOFCs: solid oxide fuel cells) or in the molten state in the case of alkali carbonates (MCFCs: molten carbonate fuel cells) [SUN 10]. However, the high temperatures required raise other difficulties, and usually result in premature degradation of the materials. Research has enabled operating temperatures to be lowered to between 400°C and 600°C, via the use of catalyzers.

5.5.1. *Operating principle of a fuel cell*

The fuel cells which are currently available on the market use H_2 or an alcohol (methanol) as fuel, an aqueous acidic or alkaline medium as the electrolyte and precious metal catalysts as electrodes. An electrochemical cell operating using H_2 and O_2 is divided into cathodic and anodic compartments, separated by a membrane which is semi-permeable to H^+ or OH^- ions.

Hydrogen or methanol is injected into the anode compartment and is oxidized at the anode, releasing H^+ protons, according to the reactions:

$$H_2 \rightarrow 2H^+ + 2e^- \text{ or } CH_3OH + H_2O \rightarrow 6H^+ + CO_2 + 6e^-$$

Simultaneously, oxygen is introduced into the cathodic compartment and is reduced at the cathode by a mechanism similar to that involved in water oxidation. The $4e^-$ reduction of O_2 to H_2O, corresponding to the reaction $4H^+ + O_2 + 4e^- \rightarrow 2H_2O$ and characterized by a thermodynamic equilibrium potential of 1.23 V versus RHE, consists of a succession of steps involving the intermediate species *OOH, *O and *OH; these intermediate species are the same as those appearing in the OER reaction, but the order is reversed. Formed successively according to the reactions $O_2 + 4(H^+ + e^-) \rightarrow {*OOH} + 3\,(H^+ + e^-) \rightarrow {*O} + H_2O + 2(H^+ + e^-) \rightarrow {*OH} + H_2O + (H^+ + e^-) \rightarrow 2H_2O$, calculation of their energies shows that the reduction to H_2O only becomes effective and spontaneous for a potential of the order of 0.8 V, that is, well below 1.23V; this corresponds to a theoretical overpotential of around 0.4 V. In an acidic medium, H^+ ions diffuse to the anodic compartment and react with *OH (Figure 5.23a), while in an alkaline medium, it is the OH^- ions that diffuse to the anodic compartment (Figure 5.23b). A detailed analysis of these mechanisms can be found in a recent article by Kulkarni *et al.* [KUL 18].

The specific overpotentials for each of the reactions, notably that due to the ORR, mean that the energy yield of the reaction is well below 100%. Depending on the chosen electrolyte medium (acid or alkaline) and electrode materials, the overpotential required for O_2 may vary between 300 and 600 mV for current densities of 10 mA/cm^2, whereas H_2 oxidation, a much faster process, occurs with overpotentials of less than 100 mV for the same current density conditions.

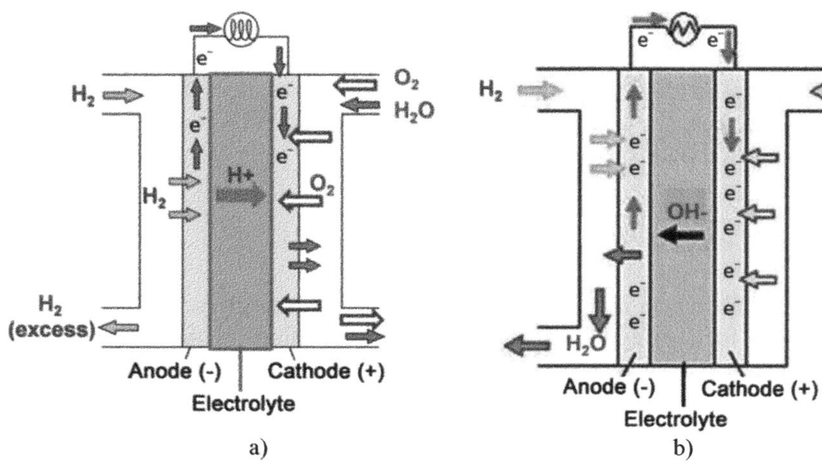

Figure 5.23. *H_2/O_2 fuel cells*

COMMENT ON FIGURE 5.23.– *a) H_2/O_2 cell operating in an acid medium. The electrolyte is generally enclosed in a proton exchange membrane, corresponding to a PEMFC (proton exchange membrane fuel cell). b) H_2/O_2 cell operating in an alkaline medium. The electrolyte is in contact with a membrane which is permeable to OH^- ions. Adapted from a U.S. report on fuel cells [DOE 16].*

The yield ξ of the H_2/O_2 system is thus determined as a function of the overpotentials $\eta_{cathode}$ and η_{anode} and the difference ΔE^0 (1.23 V) between the redox potentials of the H^+/H_2 and O_2/H_2O reactions [GEW 18]. It is given by the relation:

$$\xi = 1 - (\eta_{cathode} + \eta_{anode})/\Delta E^0$$

In practice, yields can drop to values of around 45%, mainly due to the high overvoltages required for O_2 reduction and notably for the ORR process. The problems discussed earlier in the context of electrolyzer design also apply in the case of fuel cells; the key to optimal operation lies in the choice of catalysts for oxygen reduction in an acidic or alkaline medium (AEMFC or PEMFC).

Both technologies have advantages and disadvantages. For some time, AFC technology was overshadowed by developments in the field of acidic and PEM batteries, due to difficulties caused by residual traces of CO_2 in the oxygen: these react with the basic medium, resulting in the formation of solid carbonates which can deactivate the catalysts and clog the pores of the membranes, leading to battery malfunction. This problem has since been solved with the development of anion exchange membranes. New formulations, notably featuring interpenetrated polymer networks (IPN) functionalized by ammonium ($-N^+R_3$), phosphonium ($-P^+R_3$) or sulfonium ($-S^+R_2$) groups, have been used to improve both mechanical properties and ionic conductivity to levels which are now comparable to those of PEMs, of the order of 0.1 $S.cm^{-1}$ [MER 11, DEK 18, OMA 18].

Another significant advantage is the possibility of using non-precious metal catalysts, stable in alkaline media, along with the option to use alternative fuels (alcohols) instead of H_2. Finally, there are many chemical options for producing anion exchange membranes, eliminating the need for expensive fluorinated compounds, such as Nafion, which play an essential role in PEMFCs.

The advantages of PEMFCs mean that the technology is widely used in electric vehicles and has been adopted by many automotive companies. Compared to other hydrogen fuel cells, PEMFCs have the highest energy density and the longest lifetime. They have the potential to operate almost instantaneously (within seconds) and their assembly cost is also lower than that of other devices [WAL 17]. However, the catalysts used for O_2 reduction are precious metal-based, usually including platinum. Fluorinated membranes (Nafion) also account for a significant part of the cost of the cell; the development of new polymers with the same proton conduction properties as Nafion is thus a significant research goal.

The following paragraphs provide an overview of recent work in the design of O_2 reduction catalysts. A wide range of materials have been developed; we have chosen to focus on examples that illustrate different research trends and involve the use of nanomaterials, broadly similar to those used in catalysts for water oxidation in the OER process.

5.5.2. *Choice of O_2 reduction catalysts*

As we saw in the case of electrolyzer technology, reducing the cost of O_2 reduction catalysts is an important goal, and a variety of strategies have been adopted. One consists of reducing the mass of Pt while compensating for the loss of reactivity by increasing the reactive surface[33]. Others draw on the fact that many metal oxides (Co, Ni or Fe-based) are stable in alkaline medium and can thus be used as reduction catalysts in this medium. The latest and most innovative solutions, still at the laboratory stage, involve the use of metal-free graphitic-type nitrogen-doped lamellar structures, and/or SAC structures building on new developments in electrochemical catalysis [LI 19c].

There are two main approaches to reducing the cost of catalysts. In the case of PEMFCs (acidic medium), research is focused on minimizing the useful mass of precious metal while maintaining the characteristics of current commercial catalysts. In the case of AEMFCs (alkaline medium), the aim is to identify alternative catalysts based on non-precious metals (Co, Ni and Fe) to replace the current Pt/C models.

5.5.2.1. *Catalysts with low precious metal content*

An efficient way of reducing useful mass while increasing the electrochemically active surface area (ECSA) of the catalyst involves nanostructuring of the precious metal. A cross-laboratory research group obtained significant improvements in reduction current by replacing the traditional Pt/C catalyst with jagged Pt nanowires (J-PtNW)[34]; these present a particularly high ECSA compared to classic Pt/C catalysts [LI 16a].

---

33 The replacement of Pt by less expensive materials essentially concerns the cathodic catalysts used for O_2 reduction. Much larger quantities of precious metal are required in this case than in anodic catalysts. The H_2 oxidation reaction is very fast and uses minimal quantities of Pt, with very little effect on the overall cost of the cell.

34 Nano-structuring is achieved using NWs made from a NiPt alloy. Selective dissolution of nickel is carried out by applying a series of polarization cycles in an acidic medium, resulting in a new nickel-free NW structure; at the same time, the Pt atoms are rearranged, resulting in an irregular surface texture on the NW. The NWs thus formed have a "jagged" appearance; they are crystalline and characterized by an active surface area far greater than that of commercial Pt/C catalysts. Their diameters decrease from 5 to 2.3 nm as the nickel is dissolved.

The ECSAs of regular Pt nanowires (R-PtNW) (Figure 5.24a) and those with additional nanostructuring (J-PtNW) (Figure 5.24b) are essentially the same, with respective values of 110 m^2/g_{Pt} and 118 m^2/g_{Pt}; the respective value for traditional Pt/C catalysts is only 74 m^2/g_{Pt}.

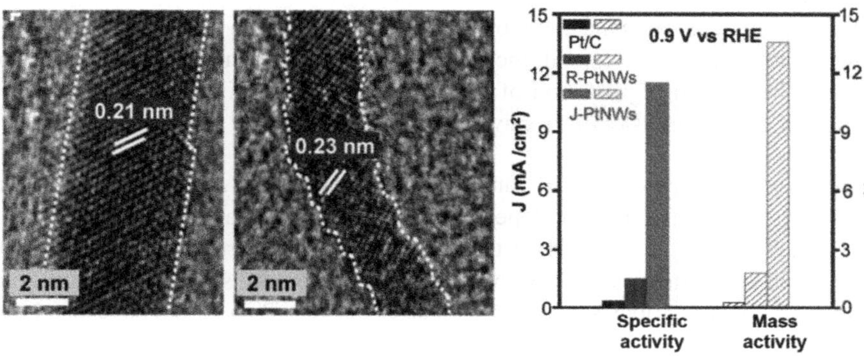

a) b) c)

Figure 5.24. *Crystallographic and electrochemical characteristics of Pt nanowires*

COMMENT ON FIGURE 5.24.– *a) TEM image of a NiPt nanowire. b) TEM image of the Pt nanowire after dissolution of the Ni. Note the irregular structure of the nanowire contour in the latter case (white dotted line) compared to the very regular structure of NiPt nanowire (a); we also note a reduction in diameter. The values of 0.21 nm (a) and 0.23 nm (b) correspond to the distances between (111) crystal planes of NiPt alloy and Pt, respectively. c) Specific and mass activities of the three classes of Pt-based catalysts, highlighting the considerable amplification of reduction currents obtained at 0.9 V versus RHE with the J-PtNW catalyst. Adapted from [LI 16a].*

This difference between the active surfaces results in a significantly higher specific activity of 11.5 mA/cm^2 using the J-PtNW catalyst, compared to 1.59 mA/cm^2 for the R-PtNW catalyst and only 0.35 mA/cm^2 with Pt/C at a voltage of 0.9 V versus RHE. The mass activities obtained with J-PtNW (13.6 A/mg_{Pt}) and R-PtNW (1.76 A/mg_{Pt}) at 0.9 V in 0.5 M $HClO_4$ medium far exceed the target value of 0.44 A/mgPt at 0.9 V recommended by the U.S. Department of Energy in 2017; the mass activity for the Pt/C catalyst is much lower, at 0.26 A/mg_{Pt}, 52 times lower than the value for J-PtNW (Figure 5.24c)[35]. Accelerated reliability tests show high reproducibility of polarization curves after 6000 polarization cycles performed

35 The current densities obtained with the different catalysts correspond to Pt masses of 2.2 $\mu g/cm^2$ for J-PtNW, 2.5 μg for R-PtNW and 7.65 μg for Pt/C.

between 0.6 and 1V versus RHE at the rate of 100 mV/s in 0.5 M $HClO_4$ medium; this indicates good stability for the J-PtNW catalyst. Surprisingly, the mass activity of J-PtNW (13.6 A/mg_{Pt}) is much higher than that of R-PtNW (1.76 A/mg_{Pt}) despite comparable ECSAs. The authors attribute this difference in activity, corroborated by theoretical energy calculations, to an activation effect in the case of J-PtNW, resulting from the reorganization of the Pt crystal lattice, which induces stress between atoms inside the crystal, increasing the adsorption energies of intermediate species. The surface organization of Pt atoms in the form of rhombi could also be responsible for this increase in reactivity[36].

These results exceed those of earlier systems based on Pt_3Ni alloys synthesized in nanocrystal form [CHE 14a], doped, in some cases, with transition metals such as Mo-Pt_3Ni/C [HUA 15], considered to be among the best-performing alloys for O_2 reduction in acidic media [WU 13]. Numerous systems using small amounts of Pt in the form of NPs of various shapes but with sizes between 2 and 5 nm have also been tested and, in many cases, have been found to perform better than traditional Pt/C catalysts [XIA 17b, WAN 19b].

5.5.2.2. *Non-precious metal-based catalysts*

Many solutions have been proposed in this area, generally based on graphitic structures onto which metal NPs or single atoms are adsorbed, using the SAC approach.

Several authors have shown that graphitic compounds doped with N, B, S, P heteroatoms offer potential solutions for the improvement of the ORR. Graphene doped with pairs of elements comprising nitrogen (N) alongside fluorine (F), boron (B) or sulfur (S) or in 3D aerogel form [WU 12b], has been shown to perform better than commercial Pt/C catalysts, as have porous networks of nitrogen-doped carbon nanotubes [MOR 11]. More generally, the most promising alternatives to Pt/C catalysts for O_2 reduction are based on Fe-N-C assemblies [SHE 19], which have been listed and described in multiple journal articles [GEW 18, LI 19c, MA 19, WAN 19c].

Several of these catalysts, similar to SAC-type technologies, exhibit exceptional performance. Liu *et al.* [LIU 18] recently synthesized a catalyst derived from an iron-doped MOF (ZIF-8), which after pyrolysis, results in a new material in which the iron is coordinated exclusively by four nitrogen atoms (Fe-N_4). This arrangement is comparable to porphyrinic molecular catalysts in which one metal is coordinated with the nitrogens of four pyrrolic groups. The structure of the Fe-N_4 unit has been

36 A rhombus, in this context, is a tetrahedral arrangement of four atoms at the vertices of equilateral triangles.

confirmed by multiple spectrometric analyses, and this reactive species is responsible for the very high catalytic activity of the catalyst (Figure 5.25). Disseminated throughout the composite at very high concentrations, it is also suited to large-scale production[37].

According to the authors, an electrode developed with this catalyst outperformed known Fe-N-C catalysts (Figure 5.25a). The half-wave potential ($E_{1/2}$), in both acidic and alkaline media, is very similar to that observed with a commercial Pt/C electrode. In 0.5 M H_2SO_4 medium, the $E_{1/2}$ of Fe2-Z8-C is slightly lower than that of Pt/C (0.805 V compared to 0.822 V vs. RHE for Pt/C); however, in 0.1 M KOH alkaline medium, it is slightly higher (0.871 V compared to 0.847 V vs. RHE for Pt/C). Furthermore, the stability of the catalyst observed over several hours of operation (10000 s) is also better than that of a Pt/C electrode; for an applied voltage of 0.5 V versus RHE, the intensity of the reduction current remains at 98% of its initial value, whereas for the Pt/C electrode it falls to 82%. The maximum power density of 1,140 mW/cm^2 is also, according to the authors, a record value for this type of cell (Figure 5.25b).

Yang *et al.* [YAN 18] obtained similar results using an electrode material named SA-Fe/NG, which is produced in a different way and also features structures with FeN_4 units, in which the N atoms are part of pyrrolic and pyridine rings. As before, the catalyst can operate in acidic or alkaline media; it is characterized by a half-wave potential for O_2 reduction of 0.80 V versus RHE in 0.5 M H_2SO_4, 30 mV lower than that of Pt/C (0.83 V vs. RHE). In 0.1 M KOH, however, it is 30 mV higher than that of Pt/C (0.88 V compared to 0.85 V for Pt/C). The power density of a cell consisting of the SA-Fe/NG cathode, a 20% Pt/C anode and a Nafion type membrane in an acidic medium (0.5 M H_2SO_4) is also high, reaching a maximum of 823 mW/cm^2 for a respective voltage and current density of around 0.35 V and 2.5 A/cm^2. Theoretical calculations, confirmed by spectral analyses, highlight the value of the structure which, according to the authors, is most active when the Fe sites are coordinated to four N atoms linked to pyrrolic rings.

---

37 The catalyst is synthesized from ZIF-8 (Zeolitic Imidazolate Framework, a commercial MOF), which is easy to obtain by condensing Zn oxide (ZnO) with 2-methylimidazole ($C_4H_6N_2$) and iron oxalate at 220°C without solvent. This has the effect of doping the ZIF-8 with iron and results in Fe2- ZIF-8, which corresponds to an atomic proportion of 2% iron. After coating the Fe-2 ZIF-8 particles with PVP (poly-vinylpyrrolidone), carbonization at 1000°C under argon results in the Fe2-Z8-C catalyst, in which the Zn has been removed and the iron is coordinated by four nitrogen atoms and dispersed in a carbonaceous matrix. The porosity of the final compound is very high (1,265 m^2/g) and improves the catalytic efficiency. A catalyst with slightly different properties, denoted Z8-Fe2-C, is obtained using the same method but with FeOx (iron oxalate) in the place of ZnO.

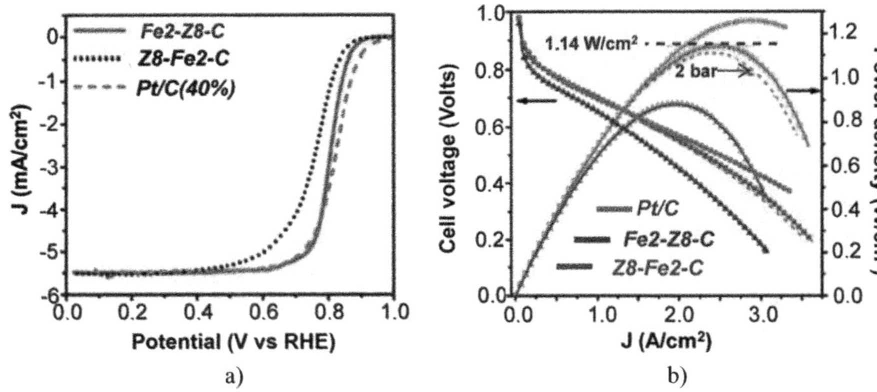

Figure 5.25. *Electrochemical characteristics of Fe-N$_4$ catalysts and corresponding PEMFC cells*

COMMENT ON FIGURE 5.25.– *a) Polarization curves of O_2 reduction (at saturation in 0.5 M H_2SO_4 medium) obtained using a rotating electrode (1600 rpm) with Fe2-Z8-C (2.8 mg/cm^2), Z8-Fe2-C (2.7 mg/cm^2) and Pt/C (0.33 mg/cm^2) catalysts (polarization rate 10 mV/s, P = 1 bar). b) Power density curves of cells made with Fe2-Z8-C, Z8-Fe2-C and Pt/C cathodes, respectively, and a Pt/C anode. The flow rates of H_2 and O_2 are 0.3 and 0.4 L/min, respectively. T = 80°C, P = 2.5 bar (2 bar, dashed curve), 0.5 M H_2SO_4 electrolyte, Nafion® type membrane (NRE 211). Adapted from [LIU 18].*

Precious metal-free materials of this type, containing N atoms, are currently the subject of extensive research. They present a further advantage in that they are generally capable of catalyzing O_2 reduction in acid or alkaline media; some metal-free forms can also be multifunctional, as Hu and Dai [HU 17] showed in a recent article. New pathways for investigation, based on innovative materials such as GDY (see Chapter 1, section 1.1.7.2.1), also show promise in the search for cheaper, efficient alternatives to Pt.

5.5.3. *Conclusions on electrocatalysis and photocatalysis*

Nanomaterials and nanotechnologies have played a significant role in new developments in electrocatalysis and photocatalysis over the last 20 years. The production of hydrogen by water splitting (by electrolysis or photoelectrolysis) and the development of hydrogen fuel cells are two key areas of research, aimed at offering an alternative to fossil fuels and eliminating emissions of CO_2 into the atmosphere.

The Pt content of current water splitting catalysts is problematic in terms of cost; nanostructuring is one way of minimizing the use of costly precious metals while maintaining efficiency. Important results have been obtained in this area; new catalysts are just as efficient as commercial Pt/C catalysts, but use much less Pt. In the same way, non-precious metal-based catalysts, based on nanostructures in which the metal is coordinated with four nitrogen atoms, also offer a promising solution for use in hydrogen fuel cells. However, while the findings made by research teams in this area are impressive, the emergence of an industrial energy sector based on electrochemical hydrogen production and its conversion into electricity is still some way off. This research thus has less of an immediate effect on our day-to-day energy consumption (or production) than the energy storage and conversion technologies presented in the previous two chapters.

Conclusion

Twenty years have passed since the U.S. National Nanotechnology Initiative was launched. This initiative, considered as a national development priority in the United States and promising considerable scientific progress in response to a range of societal challenges, triggered a worldwide upsurge in interest. Many countries aimed at exploiting the entirely new properties of nanomaterials, discovered in the 1980s and 1990s, which became observable thanks to developments in near-field microscopy and electron microscopy.

Research in the nanotechnology sector has continued to grow in importance in fields such as electronics, optoelectronics, chemistry, biology, medicine and energy. In this final area, the focus of our analysis, nanotechnologies are likely to play an important role in the transition to more sustainable forms of energy and in environmental preservation. Nanotechnologies have also been key to several important scientific breakthroughs in terms of energy usage, in terms of both storage and conversion.

The first part of this book provided a description of nanomaterials and their properties. We showed that new or relatively new carbon chemical structures, some discovered well before the 2000s (fullerenes, carbon nanotubes, nanodiamonds, carbon dots, etc.), are strategic materials for use in solving energy problems; one major advantage of these materials is the ease of surface modification using functionalization techniques, allowing their properties to be modified. Graphene, first isolated in the 2000s, is considered to be the leading nanomaterial of the 21st century, thanks to its extraordinary conductive properties and mechanical strength. This is, in all likelihood, the material which has been met with the most enthusiasm, due to the wide range of applications and products to which it lends itself; it can also contribute to various electrochemical and photochemical catalysis processes. Graphidyne is an even newer member of the family of two-dimensional (2D) materials, and has already demonstrated its potential; it can be produced simply and on a large scale, and, used in conjunction with metals in the atomic state, can give rise to a new and innovative type of catalysis (single atom catalysis). Like graphene, this paves the way for new solutions in terms of energy storage or conversion reactions.

Inorganic nanomaterials have developed in parallel to these carbon-based systems. Atomic clusters made up of a small number of atoms, metallic or semiconductor NPs of various sizes and shapes, 2D inorganic materials and luminescent quantum dots, which are now used commercially in television screens, are some of the main members of this family. Recent discoveries include new families of hybrid materials of the MOF type, characterized, on a nanometric scale, by 3D organometallic chains: this gives them a nanoporous structure that can be modulated *ad infinitum* to create new alternatives to zeolites for gas absorption and separation. They may also be efficient catalysts for water splitting.

In the second part of the book, we described the main modes of energy storage and conversion, showing how the involvement of nanomaterials and nanotechnologies has contributed to the improvement and development of various energy-related systems. The problem of energy storage, which has yet to be completely solved, conditions the development of intermittent renewable energies as well as the development of electric vehicles; this major issue is still the subject of fundamental research, notably with respect to rechargeable batteries and supercapacitors. Lithium-ion batteries, used in the majority of electric vehicles and portable electronic objects, are one area where a mastery of the interfacial structures of composite materials, developed at the nanometric scale, has resulted in major improvements in performance.

Energy conversion is another important area, covering photovoltaics, lighting devices, electrocatalysis and electro-photocatalysis processes, the latter applied essentially in the context of "green" hydrogen production. Once again, nanostructuring of the materials that make up electroluminescent and photovoltaic devices has been at the root of much of the progress made in recent years; new lighting systems are more efficient and consume less energy than fluorescent tubes or incandescent bulbs. In photovoltaics, real alternatives to the all-silicon approach have emerged, based on the use of polymeric organic nanomaterials and dyes. These offer performances well in excess of the 10% energy efficiency threshold and are well suited for portable and flexible electronic devices, as well as for efficient indoor operation. Perovskites, a still more recent development, look set to compete with silicon and should become more cost effective than the previous materials once certain stability issues have been resolved.

The final challenge relates to the development of a CO_2 emission-free energy chain, based on water splitting (electrolysis) and hydrogen production by means of electrocatalysis and photocatalysis. Promising advances have been made in an academic context, but at present are not sufficient for the development of an industrial energy sector based on electrochemical hydrogen production alone. There is still a long way to go before we can meet humanity's needs for heat, light or transportation simply by performing energy conversion operations on a pool of water. For now, the greatest visible effects of nanomaterials and nanotechnologies in energy consumption and production are found in the field of lighting and, to a lesser extent, photovoltaics. However, one thing is certain: the next 20 years will bring a whole host of new discoveries, profoundly altering the current state of affairs (2020) as described in this book.

References

[ABA 20] ABAKUMOV A.M., FEDOTOV S.S., ANTIPOV E.V. *et al.*, "Solid state chemistry for developing better metal-ion batteries", *Nat. Commun.*, vol. 11, no. 4976, 2020.

[ABE 10] ABE R., "Recent progress on photocatalytic and photoelectrochemical water splitting under visible light irradiation", *J. Photochem. Photobiol. C: Photo-chem. Rev.*, vol. 11, pp. 179–209, 2010.

[ABE 19] ABE J.O., POPOOLA A.P.I., AJENIFUJA E. *et al.*, "Hydrogen energy, economy and storage: Review and recommendation", *Int. J. Hydrogen Energy*, vol. 44, pp. 15072–15086, 2019.

[ABR 15] ABRAHAM K.M., "Prospects and limits of energy storage in batteries", *J. Phys. Chem. Lett.*, vol. 6, pp. 830–844, 2015.

[ACQ 17] ACQUAH S.F.A., PENKOVA A.V., MARKELOV D.A. *et al.*, "Review – The beautiful molecule: 30 years of C_{60} and its derivatives", *ECS J. Solid State Sci. Technol.*, vol. 6, pp. M3155–M3162, 2017.

[ADA 09] ADACHI C., NAKANOTANI H., "Organic semiconductor laser materials", *Material Matters (Aldrich)*, vol. 4, pp. 74–79, 2009.

[ADE 15] ADELHELM P., HARTMANN P., BENDER C.L. *et al.*, "From lithium to sodium: Cell chemistry of room temperature sodium-air and sodium-sulfur batteries", *Beilstein J. Nanotechnol.*, vol. 6, pp. 1016–1055, 2015.

[AHN 19] AHN D.H., KIM S.W., LEE H. *et al.*, "Highly efficient blue thermally activated delayed fluorescence emitters based on symmetrical and rigid oxygen-bridged boron acceptors", *Nat. Photonics*, vol. 13, pp. 540–546, 2019.

[ALI 04] ALI S.S., HARDT J.I., QUICK K.L. *et al.*, "A biologically effective fullerene (C_{60}) derivative with superoxide dismutase mimetic properties", *Free Radical Biol. Med.*, vol. 37, pp. 1191–1202, 2004.

[ALL 09] ALLENDORF M.D., BAUER C.A., BHAKTA R.K. *et al.*, "Luminescent metal-organic frameworks", *Chem. Soc. Rev.*, vol. 38, pp. 1330–1352, 2009.

[ALO 14] ALOTTO P., GARNIERI M., MORO F., "Redox flow batteries for the storage of renewable energy: A review", *Renewable Sustainable Energy Rev.*, vol. 29, pp. 325–335, 2014.

[ANI 18] ANICHINI C., CZEPA W., PAKULSKI D. *et al.*, "Chemical sensing with 2D materials", *Chem. Soc. Rev.*, vol. 47, pp. 4860–4908, 2018.

[ARD 90] ARDIZZONE S., FREGONARA G., TRASATTI S., "'Inner' and 'Outer' active surface of RuO_2, electrodes", *Electrochim. Acta*, vol. 35, pp. 263–267, 1990.

[ARN 06] ARNOLD M.S., GREEN A.A., HULVAT J.F. *et al.*, "Sorting carbon nanotubes by electronic structure using density differentiation", *Nat. Nanotechnol.*, vol. 1, pp. 60–66, 2006.

[AUG 14] AUGUSTYN V., SIMON P., DUNN B., "Pseudocapacitive oxide materials for high-rate electrochemical energy storage", *Energy Environ. Sci.*, vol. 7, pp. 1597–1614, 2014.

[BAC 98] BACH U., LUPO D., COMTE P. *et al.*, "Solid-state dye-sensitized mesoporous TiO_2 solar cells with high photon-to-electron conversion efficiencies", *Nature*, vol. 395, pp. 483–485, 1998.

[BAC 02] BACHILO S.M., STRANO M.S., KITTRELL C. *et al.*, "Structure-assigned optical spectra of Single-Walled Carbon Nanotubes", *Science*, vol. 298, pp. 2361–2368, 2002.

[BAC 16] BACHMAN J.C., MUY S., GRIMAUD A. *et al.*, "Inorganic solid-state electrolytes for lithium batteries: Mechanisms and properties governing ion conduction", *Chem. Rev.*, vol. 116, pp. 140–162, 2016.

[BAE 10] BAE S., KIM H., LEE Y. *et al.*, "Roll-to-roll production of 30-inch graphene films for transparent electrodes", *Nat. Nanotechnol.*, vol. 5, pp. 574–578, 2010.

[BAE 13] BAE W.K., PADILHA L.A., PARK Y.-S. *et al.*, "Controlled alloying of the core-shell interface in CdSe/CdS quantum dots for suppression of Auger recombination", *ACS Nano*, vol. 7, pp. 3411–3419, 2013.

[BAK 07] BAKRY R., VALLANT R.M., NAJAM-UL-HAQ M. *et al.*, "Medicinal applications of fullerenes", *Int. J. Nanomed.*, vol. 2, pp. 639–649, 2007.

[BAL 98a] BALCH A.L., OLMSTEAD M.M., "Reactions of transition metal complexes with fullerenes (C_{60}, C_{70}, etc.) and related materials", *Chem. Rev.*, vol. 98, pp. 2123–2165, 1998.

[BAL 98b] BALDO M.A., O'BRIEN D.F., YOU Y. *et al.*, "Highly efficient phosphorescent emission from organic electroluminescent devices", *Nature*, vol. 395, pp. 151–154, 1998.

[BAL 99] BALDO M.A., LAMANSKY S., BURROWS P.E. *et al.*, "Very high-efficiency green organic light-emitting devices based on electro-phosphorescence", *Appl. Phys. Lett.*, vol. 75, pp. 4–6, 1999.

[BAN 17] BANG E., CHOI Y., CHO J. *et al.*, "Large-scale synthesis of highly luminescent InP@ZnS quantum dots using elemental phosphorus precursor", *Chem. Mater.*, vol. 29, pp. 4236–4243, 2017.

[BAO 18] Bao L., Peng P., Lu X., "Bonding inside and outside fullerene cages", *Acc. Chem. Res.*, vol. 51, pp. 810–815, 2018.

[BAR 01] Bard A.J., Faulkner L.R., *Electrochemical Methods, Fundamentals and Applications*, John Wiley & Sons, Hoboken, 2001.

[BAR 05] Barbir F., "PEM electrolysis for production of hydrogen from renewable energy sources", *Sol. Energy*, vol. 78, pp. 661–669, 2005.

[BAT 00] Bates J.B., Dudney N.J., Neudecker B. *et al.*, "Thin-film lithium and lithium-ion batteries", *Solid State Ionics*, vol. 135, pp. 33–45, 2000.

[BAU 87] Baughman R.H., Eckhardt H., "Structure-property predictions for new planar forms of carbon: Layered phases containing sp^2 and *sp* atoms", *J. Chem. Phys.*, vol. 87, pp. 6687–6699, 1987.

[BER 17] Bera M., Maji P.K., "Graphene-based polymer nanocomposites: Materials for future revolution", *MOJ Poly. Sci.*, vol. 1, pp. 94–97, 2017.

[BET 93] Bethune D.S., Klang C.H., De Vries M.S. *et al.*, "Cobalt-catalysed growth of carbon nanotubes with single-atomic-layer walls", *Nature*, vol. 363, pp. 605–607, 1993.

[BLA 18] Blasco-Ahicart M., Soriano-López J., Carbó J.J. *et al.*, "Polyoxometalate electrocatalysts based on earth abundant metals for efficient water oxidation in acidic media", *Nat. Chem.*, vol. 10, pp. 24–30, 2018.

[BÖH 20] Böhm D., Beetz M., Schuster M. *et al.*, "Efficient OER catalyst with low Ir volume density obtained by homogeneous deposition of iridium oxide nanoparticles on macroporous antimony-doped tin oxide support", *Adv. Funct. Mater.*, vol. 30, art. 1906670, 2020.

[BOL 08a] Bolotin K.I., Sikes K.J., Jiang Z. *et al.*, "Ultrahigh electron mobility in suspended graphene", *Solid State Communications*, vol. 146, pp. 351–355, 2008.

[BOL 08b] Bolotin K.I., Sikes K.J., Hone J. *et al.*, "Temperature dependent transport in suspended graphene", *Phys. Rev. Lett.*, vol. 101, art. 096802, 2008.

[BON 10] Bonaccorso F., Sun Z., Hasan T. *et al.*, "Graphene photonics and optoelectronics", *Nat. Photonics*, vol. 4, pp. 611–622, 2010.

[BOU 12] Bourlinos A.B., Bakandritsos A., Kouloumpis A. *et al.*, "Gd(III)-doped carbon dots as a dual fluorescent-MRI probe", *J. Mater. Chem.*, vol. 22, pp. 23327–23330, 2012.

[BRI 99] Brissot C., Rosso M., Chazalviel J.-N. *et al.*, "Dendritic growth mechanisms in lithium polymer cells", *J. Power Sources*, vols 81–82, pp. 925–929, 1999.

[BRO 01] Bronikowski M.J., Willis P.A., Colbert D.T. *et al.*, "Gas-phase production of carbon single-walled nanotubes from carbon monoxide via the HiPco process: A parametric study", *J. Vac. Sci. Technol. A*, vol. 19, pp. 1800–1805, 2001.

[BRO 11] Brownson D.A.C., Kampouris D.K., Banks C.E., "An overview of graphene in energy production and storage applications", *J. Power Sources*, vol. 196, pp. 4873–4885, 2011.

[BRU 12] BRUCE P.G., FREUNBERGER S.A., HARDWICK L.J. *et al.*, "Li-O_2 and Li-S batteries with high energy storage", *Nat. Mater.*, vol. 11, pp. 19–29, 2012.

[BRY 18] BRYANS D., AMSTUTZ V., GIRAULT H.H. *et al.*, "Characterisation of a 200 kW/400 kWh vanadium redox flow battery", *Batteries*, vol. 4, no. 54, 2018.

[BUN 07] BUNDGAARD E., KREBS F.C., "Low band gap polymers for organic photovoltaics", *Sol. Energy. Sol. Cells*, vol. 91, pp. 954–985, 2007.

[BUS 17] BUSH K.A., PALMSTROM A.F., YU Z.J. *et al.*, "23.6%-efficient monolithic perovskite/silicon tandem solar cells with improved stability", *Nat. Energy*, vol. 2, no. 17009, 2017.

[BUT 08] BUTLER J.E., SUMANT A.V., "The CVD of nanodiamond materials", *Chem. Vap. Deposition*, vol. 14, pp. 145–160, 2008.

[CAI 19] CAI Y., HOU Y., LU Y. *et al.*, "Recent progress on catalysts for the positive electrode of aprotic lithium-oxygen batteries", *Inorganics*, vol.7, no. 69, 2019.

[CAL 16] CALLEJAS J.F., READ C.G., ROSKE C.W. *et al.*, "Synthesis, characterization, and properties of metal phosphide catalysts for the hydrogen-evolution reaction", *Chem. Mater.*, vol. 28, pp. 6017–6044, 2016.

[CAO 07] CAO L., WANG X., MEZIANI M.J. *et al.*, "Carbon dots for multiphoton bioimaging", *J. Am. Chem. Soc.*, vol. 129, pp. 11318–11319, 2007.

[CAO 15] CAO S., CHEN Y., WANG C.-J. *et al.*, "Spectacular photo-catalytic hydrogen evolution using metal-phosphide/CdS hybrid catalysts under sunlight irradiation", *Chem. Commun.*, vol. 51, pp. 8708–8711, 2015.

[CAO 17] CAO Y., SAYGILI Y., UMMADISINGU A. *et al.*, "11% efficiency solid-state dye-sensitized solar cells with copper (II/I) hole transport materials", *Nat. Commun.*, vol. 8, art. 15390, 2017.

[CAR 11] CARNE A., CARBONELL C., IMAZ I. *et al.*, "Nanoscale metal-organic frameworks", *Chem. Soc. Rev.*, vol. 40, pp. 291–305, 2011.

[CAR 13] CARMO M., FRITZ D.L., MERGEL J. *et al.*, "A comprehensive review on PEM water electrolysis", *Int. J. Hydrogen Energy*, vol. 38, pp. 4901–4934, 2013.

[CAR 18] CARELLA A., BORBONE F., CENTORE R., "Research progress on photo-sensitizers for DSSC", *Front. Chem.*, vol. 6, no. 481, 2018.

[CAS 17] CASTRO E., HERNANDEZ GARCIA A., ZAVALA G. *et al.*, "Fullerenes in biology and medicine", *J. Mater. Chem. B*, vol. 5, pp. 6523–6535, 2017.

[CHA 91] CHAI Y., GUO T., JIN C. *et al.*, "Fullerenes with metals inside", *J. Phys. Chem.*, vol. 95, pp. 7564–7568, 1991.

[CHA 08] CHANG Y.-R., LEE H.-Y., CHEN K. *et al.*, "Mass production and dynamic imaging of fluorescent nanodiamonds", *Nat. Nanotechnol.*, vol. 3, pp. 284–288, 2008.

[CHA 14] CHANG K., MEI Z., WANG T. *et al.*, "MoS_2/Graphene cocatalyst for efficient photocatalytic H_2 evolution under visible light irradiation", *ACS Nano*, vol. 8, pp. 7078–7087, 2014.

[CHA 17] CHAKRABORTY I., PRADEEP T., "Atomically precise clusters of noble metals: Emerging link between atoms and nanoparticles", *Chem. Rev.*, vol. 117, pp. 8208–8271, 2017.

[CHA 18] CHAN Z.M., KITCHAEV D.A., WEKER J.N. *et al.*, "Electrochemical trapping of metastable Mn^{3+} ions for activation of MnO_2 oxygen evolution catalysts", *Proc. Nat. Acad. Sci.*, vol. 115, pp. E5261–E5268, 2018.

[CHA 19] CHAND S., ELAHI S.M., PAL A. *et al.*, "Metal-organic frameworks and other crystalline materials for ultrahigh superprotonic conductivities of 10^{-2} Scm^{-1} or higher", *Chem. Eur. J.*, vol. 25, pp. 6259–6269, 2019.

[CHE 08] CHEN H., MÜLLER M.B., GILMORE K.J. *et al.*, "Mechanically strong, electrically conductive, and biocompatible graphene paper", *Adv. Mater.*, vol. 20, pp. 3557–3561, 2008.

[CHE 12] CHEN W.F., SASAKI K., MA C. *et al.*, "Hydrogen-evolution catalysts based on non-noble metal nickel-molybdenum nitride nanosheets", *Angew. Chem. Int. Ed.*, vol. 51, pp. 6131–6135, 2012.

[CHE 13a] CHEN J., XI J., WANG D. *et al.*, "Carrier mobility in graphyne should be even larger than that in graphene: A theoretical prediction", *J. Phys. Chem. Lett.*, vol. 4, pp. 1443–1448, 2013.

[CHE 13b] CHEN W.F., WANG C.-H., SASAKI K. *et al.*, "Highly active and durable nanostructured molybdenum carbide electrocatalysts for hydrogen production", *Energy Environ. Sci.*, vol. 6, pp. 943–951, 2013.

[CHE 14a] CHEN C., KANG Y., HUO Z. *et al.*, "Electro-catalytic surfaces highly crystalline multimetallic nanoframes with three-dimensional electrocatalytic surfaces", *Science*, vol. 343, pp. 1339–1343, 2014.

[CHE 14b] CHEN Y., SONG B., TANG X. *et al.*, "Ultrasmall Fe_3O_4 nano-particle/MoS_2 nanosheet composites with superior performances for Lithium ion batteries", *Small*, vol. 10, pp. 1536–1543, 2014.

[CHE 14c] CHEN Y., ZHANG J., "Chemical vapor deposition growth of single-walled carbon nanotubes with controlled structures for nanodevice applications", *Acc. Chem. Res.*, vol. 47, pp. 2273–2881, 2014.

[CHE 14d] CHEN Y., ZHANG Y., HU Y. *et al.*, "State of the art of single-walled carbon nanotube synthesis on surfaces", *Adv. Mater.*, vol. 26, pp. 5898–5922, 2014.

[CHE 14e] CHENG L., HUANG W., GONG Q. *et al.*, "Ultra-thin WS_2 nanoflakes as a high-performance electrocatalyst for the hydrogen evolution reaction", *Angew. Chem. Int. Ed.*, vol. 53, pp. 7860–7863, 2014.

[CHE 15a] CHEN L.-Y., WANG C.-W., YUAN Z. *et al.*, "Fluorescent gold nanoclusters: Recent advances in sensing and imaging", *Anal. Chem.*, vol. 87, pp. 216–229, 2015.

[CHE 15b] CHENG X., FABBRI E., NACHTEGAAL M. *et al.*, "Oxygen evolution reaction on La_{1-x} Sr_xCoO_3 perovskites: A combined experimental and theoretical study of their structural, electronic, and electrochemical properties", *Chem. Mater.*, vol. 27, pp. 7662–7672, 2015.

[CHE 16] CHENG H., YE M., ZHAO F. *et al.*, "A general and extremely simple remote approach toward graphene bulks with in situ multifunctionalization", *Adv. Mater.*, vol. 28, pp. 3305–3312, 2016.

[CHE 17] CHEN J., YAN L., "Recent advances in carbon nanotube-polymer composites", *Adv. in Mat.*, vol. 6, pp. 129–148, 2017.

[CHE 18a] CHEN C.-H., TIERCE N.T., LEUNG M.-K. *et al.*, "Efficient triplet–triplet annihilation upconversion in an electroluminescence device with a fluorescent sensitizer and a triplet-diffusion singlet-bloc-king layer", *Adv. Mater.*, vol. 30, art. 1804850, 2018.

[CHE 18b] CHEN L., LI Y., LI S.-P. *et al.*, "PEO/garnet composite electrolytes for solid-state lithium batteries: From 'ceramic-in-polymer' to 'polymer-in-ceramic'", *Nano Energy*, vol. 46, pp. 176–184, 2018.

[CHE 18c] CHEN Y., JI S., CHEN C. *et al.*, "Single-atom catalysts: Synthetic strategies and electrochemical applications", *Joule*, vol. 2, pp. 1242–1264, 2018.

[CHE 19] CHEN Y., WANG Y., LIN Y. *et al.*, "Nanomanufacturing of graphene nanosheets through nano-hole opening and closing", *Mat. Today*, vol. 24, pp. 27–32, 2019.

[CHH 01] CHHOWALLA M., TEO K.B.K., DUCATI C. *et al.*, "Growth process conditions of vertically aligned carbon nanotubes using plasma enhanced chemical vapor deposition", *J. Appl. Phys.*, vol. 90, pp. 5308–5317, 2001.

[CHH 13] CHHOWALLA M., SHIN H.S., EDA G. *et al.*, "The chemistry of two-dimensional layered transition metal dichalcogenide nanosheets", *Nat. Chem.*, vol. 5, pp. 263–275, 2013.

[CHO 10] CHOU H.-H., CHENG C.-H., "A highly efficient universal bipolar host for blue, green, and red phosphorescent OLEDs", *Adv. Mater.*, vol. 22, pp. 2468–2471, 2010.

[CHO 13] CHO J., SCHUBERT E.F., KIM J.K., "Efficiency droop in light-emitting diodes: Challenges and countermeasures", *Laser Photonics Rev.*, vol. 7, pp. 408–421, 2013.

[CHO 18a] CHOI M.K., YANG J., HYEON T. *et al.*, "Flexible quantum dot light-emitting diodes for next-generation displays", *npj Flexible Electron.*, vol. 2, no. 10, 2018.

[CHO 18b] CHOI M.K., YANG J., KIM D.C. *et al.*, "Extremely vivid, highly transparent, and ultrathin quantum dot light-emitting diodes", *Adv Mater.*, vol. 30, art. 1703279, 2018.

[CHU 12] CHUNG I., LEE B., HE J. *et al.*, "All-solid-state dye-sensitized solar cells with high efficiency", *Nature*, vol. 485, pp. 486–490, 2012.

[COL 06] COLEMAN J.N., KHAN U., BLAU W.J. *et al.*, "Small but strong: A review of the mechanical properties of carbon nanotube-polymer composites", *Carbon*, vol. 44, pp. 1624–1652, 2006.

[COL 19] COLLI A.N., GIRAULT H.H., BATTISTEL A., "Non-precious electrodes for practical alkaline water electrolysis", *Materials*, vol. 12, no. 1336, 2019.

[CÔT 07] CÔTÉ A.P., EL-KADERI H.M., FURUKAWA H. *et al.*, "Reticular synthesis of microporous and mesoporous 2D covalent organic frameworks", *J. Am. Chem. Soc.*, vol. 129, pp. 12914–12915, 2007.

[CRA 92] CRAFORD M.G., "LEDs challenge the incandescents", *IEEE Circuits Devices Mag.*, vol. 8, issue. 5, pp. 24–29, 1992.

[CUI 12] CUI Y., YUE Y., QIAN G. *et al.*, "Luminescent functional metal-organic frameworks", *Chem. Rev.*, vol. 112, pp. 1126–1162, 2012.

[DAI 14] DAI X., ZHANG Z., JIN Y. *et al.*, "Solution-processed, high-performance light-emitting diodes based on quantum dots", *Nature*, vol. 515, pp. 96–99, 2014.

[DAI 17] DAI X., DENG Y., PENG X. *et al.*, "Quantum-dot light-emitting diodes for large-area displays: Towards the dawn of commercialization", *Adv. Mater.*, vol. 29, art. 1607022, 2017.

[DAN 04a] DANIEL M.-C., ASTRUC D., "Gold nanoparticles: Assembly, supramolecular chemistry, quantum-size-related properties, and applications toward biology, catalysis, and nanotechnology", *Chem. Rev.*, vol. 104, pp. 293–346, 2004.

[DAN 04b] DANILENKO V.V., "On the history of the discovery of nanodiamond synthesis", *Phys. Solid State*, vol. 46, pp. 595–599, 2004.

[DAN 05] DANILENKO V.V., "Nanocarbon phase diagram and conditions for detonation nanodiamond formation", *NATO Adv. Res. Workshop*, pp. 181–198, 2005.

[DAS 13] DAS T.K., PRUSTY S., "Graphene-based polymer composites and their applications", *Polymer-Plastics Technol. Eng.*, vol. 52, pp. 319–331, 2013.

[DEG 17] DEGRAFF J., LIANG R., LE M.Q. *et al.*, "Printable low-cost and flexible carbon nanotube buckypaper motion sensors", *Mat. Design*, vol. 133, pp. 47–53, 2017.

[DEK 18] DEKEL D.R., "Review of cell performance in anion exchange membrane fuel cells", *J. Power Sources*, vol. 375, pp. 158–169, 2018.

[DEM 13] DEMCHENKO P., DEKALIUK M.O., "Novel fluorescent carbonic nanomaterials for sensing and imaging", *Methods and Applications in Fluorescence*, vol. 1, 2013.

[DEN 15a] DENG B., HSU P.-C., CHEN G. *et al.*, "Roll-to-roll encapsulation of metal nanowires between graphene and plastic substrate for high-performance flexible transparent electrodes", *Nano Lett.*, vol. 15, pp. 4206–4213, 2015.

[DEN 15b] DENG J., LI H., XIAO J. *et al.*, "Triggering the electrocatalytic hydrogen evolution activity of the inert two-dimensional MoS_2 surface via single-atom metal doping", *Energy Environ. Sci.*, vol. 8, pp.1594–1601, 2015.

[DEV 13] DE VOLDER M.F.L., TAWFICK S.H., BAUGHMAN R.H. *et al.*, "Carbon nanotubes present and future commercial applications", *Science*, vol. 239, pp. 535–539, 2013.

[DIA 15] DIAC A., DEMETER D., ALLAIN M. *et al.*, "Simple and versatile molecular donors for organic photovoltaics prepared by metal-free synthesis", *Chem. Eur. J.*, vol. 21, pp. 1598–1608, 2015.

[DIK 07] DIKIN D.A., STANKOVICH S., ZIMNEY E.J. *et al.*, "Preparation and characterization of graphene oxide paper", *Nature*, vol. 448, pp. 457–460, 2007.

[DOE 16] DEPARTMENT of ENERGY (U.S.), Fuel cells multi-year research development and demonstration plan, Fuel Cell Technologies Office, 2016.

[DOE 19] DEPARTMENT OF ENERGY (U.S.), Energy savings forecast of solid-state lighting in general illumination applications, Office of Energy Efficiency and Renewable Energy, 2019.

[DOL 02] DOLGAEV S.I., SIMAKIN A.V., VORONOV V.V., SHAFEEV G.A., BOZON-VERDURAZ F., "Nanoparticles produced by laser ablation of solids in liquid environment", *Appl. Surf. Sci.*, no. 186, pp. 546–551, 2002.

[DON 12] DONG Z., JIANG C., CHENG H. *et al.*, "Facile fabrication of light, flexible and multifunctional graphene fibers", *Adv. Mater.*, vol. 24, pp. 1856–1861, 2012.

[DON 16] DONG L., XU C., LI Y. *et al.*, "Flexible electrodes and supercapacitors for wearable energy storage: A review by category", *J. Mater. Chem. A*, vol. 4, pp. 4659–4685, 2016.

[DOU 17] DOUGLAS A., PINTA C.L., "Review – Electrochemical growth of carbon nanotubes and graphene from ambient carbon dioxide: Synergy with conventional gas-phase growth mechanisms", *ECS J. Solid State Sci. Technol.*, vol. 6, pp. M3084–M3089, 2017.

[DOU 18] DOUGLAS A., CARTER R., LI M. *et al.*, "Toward small-diameter carbon nanotubes synthesized from captured carbon dioxide: Critical role of catalyst coarsening", *ACS Appl. Mater. Interfaces*, vol. 10, pp. 19010–19018, 2018.

[DU 14] DU F.-H., LI B., FU W. *et al.*, "Surface binding of polypyrrole on porous silicon hollow nanospheres for li-ion battery anodes with high structure stability", *Adv. Mater.*, vol. 26, pp. 6145–6150, 2014.

[DUA 15] DUAN J., CHEN S., JARONIEC M. *et al.*, "Porous C_3N_4 nanolayers @N-Graphene films as catalyst electrodes for highly efficient hydrogen evolution", *ACS Nano*, vol. 9, pp. 931–940, 2015.

[DUG 97] DUGAN L.L., TURETSKY D.M., DU C. *et al.*, "Carboxyfullerenes as neuroprotective agents", *Proc. Natl. Acad. Sci. USA*, vol. 94, pp. 9434–9439, 1997.

[DUN 16] DUNLEAVY R., LU L., KIELY C.J. *et al.*, "Single-enzyme biomineralization of cadmium sulfide nanocrystals with controlled optical properties", *Proc. Natl. Acad. Sci. USA.*, vol. 113, pp. 5275–5280, 2016.

[DÜR 04] DÜRKOP T., GETTY S.A., COBAS E. *et al.*, "Extraordinary mobility in semiconducting carbon nanotubes", *Nano Lett.*, vol. 4, pp. 35–39, 2004.

[EFR 16] EFROS A.L., NESBITT D.J., "Origin and control of blinking in quantum dots", *Nat. Nanotechnol.*, vol. 11, pp. 661–671, 2016.

[EIG 13] EIGLER S., ENZELBERGER-HEIM M., GRIMM S. *et al.*, "Wet chemical synthesis of graphene", *Adv. Mater.*, vol. 25, pp. 3583–3587, 2013.

[EMT 09] EMTSEV K.V., BOSTWICK A., HORN K. *et al.*, "Towards wafer-size graphene layers by atmospheric pressure graphitization of silicon carbide", *Nat. Mater.*, vol. 8, pp. 403–407, 2009.

[EUS 06] EUSTIS S., EL-SAYED M.A., "Why gold nanoparticles are more precious than pretty gold: Noble metal surface plasmon resonance and its enhancement of the radiative and nonradiative properties of nanocrystals of different shapes", *Chem. Soc. Rev.*, vol. 35, pp. 209–217, 2006.

[FAB 11] FABREGAT-SANTIAGO F., GARCIA-BELMONTE G., MORA-SERO I. *et al.*, "Characterization of nanostructured hybrid and organic solar cells by impedance spectroscopy", *Phys. Chem. Chem. Phys.*, vol. 13, pp. 9083–9118, 2011.

[FAL 14] FALCARO P., RICCO R., DOHERTY C.M. *et al.*, "MOF positioning technology and device fabrication", *Chem. Soc. Rev.*, vol. 43, pp. 5513–5560, 2014.

[FAN 13] FANG H., TOSUN M., SEOL G. *et al.*, "Degenerate n-doping of few-layer transition metal dichalcogenides by potassium", *Nano Lett.*, vol. 13, pp. 1991–1995, 2013.

[FAN 18a] FAN X., CHEN L., JI X. *et al.*, "Highly fluorinated interphases enable high-voltage Li-metal batteries", *Chem*, vol. 4, pp. 174–185, 2018.

[FAN 18b] FANG Y., XIAO L., CHEN Z. *et al.*, "Recent advances in sodium-ion battery materials", *Electrochem. Energy Rev.*, vol. 1, pp. 294–323, 2018.

[FAR 12] FARHA O.K., ERYAZICI I., JEONG N.C. *et al.*, "Metal-organic framework materials with ultrahigh surface areas: Is the sky the limit?", *J. Am. Chem. Soc.*, vol. 134, pp. 15016–15021, 2012.

[FEI 15] FEI H., DONG J., ARELLANO-JIMENEZ M.J. *et al.*, "Atomic cobalt on nitrogen-doped graphene for hydrogen generation", *Nature Commun.*, vol. 6, no. 8668, 2015.

[FEI 18] FEI H., DONG J., FENG Y. *et al.*, "General synthesis and definitive structural identification of MN4C4 single-atom catalysts with tunable electrocatalytic activities", *Nat. Catal.*, vol. 1, pp. 63–72, 2018.

[FEL 15] FELGENHAUER M., HAMACHER T., "State-of-the-art of commercial electrolyzers and on-site hydrogen generation for logistic vehicles in South Carolina", *Int. J. Hydrogen Energy*, vol. 40, pp. 2084–2090, 2015.

[FER 06] FERRARI A.C., MEYER J.C., SCARDACI V. *et al.*, "Raman spectrum of graphene and graphene layers", *Phys. Rev. Lett.*, vol. 97, 2006.

[FÉR 08a] FÉREY G., "Hybrid porous solids: Past, present, future", *Chem. Soc. Rev.*, vol. 37, pp. 191–214, 2008.

[FER 08b] FERNANDEZ J.A., MORISHITA T., TOYODA M. *et al.*, "Performance of mesoporous carbons derived from poly(vinyl alcohol) in electrochemical capacitors", *J. Power Sources*, vol. 157, pp. 675–679, 2008.

[FER 15] FERNANDO K.A.S., SAHU S., LIU Y. *et al.*, "Carbon quantum dots and applications in photocatalytic energy conversion", *ACS Appl. Mater. Interfaces*, vol. 7, pp. 8363–8376, 2015.

[FOT 17] FOTOUHI A., AUGER D.J., O'NEILL L. *et al.*, "Lithium-sulfur battery technology readiness and applications. A review", *Energies*, vol. 10, no. 1937, 2017.

[FRA 08] FRANTSUZOV P., KUNO M., JANKO B. *et al.*, "Universal emission intermittency in quantum dots, nanorods, and nanowires", *Nature Physics*, vol. 4, pp. 519–522, 2008.

[FRA 20] FRAUNHOFER ISE, Photovoltaics report, available at: https://www.ise.fraunhofer.de/content/dam/ise/de/documents/publications/studies/Photovoltaics-Report.pdf, Freiburg, 2020.

[FUJ 72] FUJISHIMA A., HONDA K., "Electrochemical photolysis of water at a semiconductor electrode", *Nature*, vol. 238, pp. 37–38, 1972.

[FUR 13] FURUKAWA H., CORDOVA K.E., O'KEEFFE M. *et al.*, "The chemistry and applications of metal-organic frameworks", *Science*, vol. 341, art. 1230444, 2013.

[GAL 16] GALL D., "Electron mean free path in elemental metals", *J. Appl. Physics*, vol. 119, 2016.

[GAO 19] GAO X., LIU H., WANG D. *et al.*, "Graphdiyne: Synthesis, properties, and applications", *Chem. Soc. Rev.*, vol. 48, pp. 908–936, 2019.

[GAR 15a] GARCIA-SEGURA S., DOS SANTOS E.V., MARTINEZ-HUITLE C.A., "Role of sp^3/sp^2 ratio on the electrocatalytic properties of boron-doped diamond electrodes: A mini review", *Electrochem. Commun.*, vol. 59, pp. 52–55, 2015.

[GAR 15b] GARCIA-SEGURA S., KELLER J., BRILLAS E. *et al.*, "Removal of organic contaminants from secondary effluent by anodic oxidation with a boron-doped diamond anode as tertiary treatment", *J. Hazard. Mat.*, vol. 283, pp. 551–557, 2015.

[GAV 15] GAVGANI J.N., DEHSARI H.S., HASANI A. *et al.*, "A room temperature volatile organic compound sensor with enhanced performance, fast response and recovery based on N-doped graphene quantum dots and poly(3,4-ethylenedioxythiophene) poly(styrenesulfonate) nanocomposite", *RSC Adv.*, vol. 5, pp. 57559–57567, 2015.

[GE 19] GE C., CHEN J., TANG S. *et al.*, "Review of the electronic, optical, and magnetic properties of graphdiyne: From theories to experiments", *ACS Appl. Mater. Interfaces*, vol. 11, pp. 2707–2716, 2019.

[GEI 07] GEIM A.K., NOVOSELOV K.S., "The rise of graphene", *Nat. Mat.*, vol. 6, pp. 183–191, 2007.

[GEO 15] GEORGAKILAS V., PERMAN J.A., TUCEK J. *et al.*, "Broad family of carbon nanoallotropes: Classification, chemistry, and applications of fullerenes, carbon dots, nanotubes, graphene, nanodiamonds, and combined superstructures", *Chem. Rev.*, vol. 115, pp. 4744–4822, 2015.

[GEW 18] GEWIRTH A.A., VARNELL J.A., DIASCRO A.M., "Non precious metal catalysts for oxygen reduction in heterogeneous aqueous systems", *Chem. Rev.*, vol. 118, pp. 2313–2339, 2018.

[GHA 05] GHARBI N., PRESSAC M., HADCHOUEL M. *et al.*, "[60]Fullerene is a powerful antioxidant in vivo with no acute or subacute toxicity", *Nano Lett.*, vol. 5, pp. 2578–2585, 2005.

[GHO 10] GHOSH S., BACHILO S.M., WEISMAN R.B., "Advanced sorting of single-walled carbon nanotubes by nonlinear density-gradient ultracentrifugation", *Nat. Nanotechnol.*, vol. 5, pp. 443–450, 2010.

[GIL 16] GILROY K.D., RUDITSKIY A., PENG H.-C. *et al.*, "Bimetallic nanocrystals: Syntheses, properties, and applications", *Chem. Rev.*, vol. 116, pp. 10414–10472, 2016.

[GOL 14] GOLUBKOV A.W., FUCHS D., WAGNER J. *et al.*, "Thermal-runaway experiments on consumer Li-ion batteries with metal-oxide and olivin-type cathodes", *RSC Adv.*, vol. 4, pp. 3633–3642, 2014.

[GOO 13] GOODENOUGH J.B., PARK K.-S., "The Li-Ion rechargeable battery: A perspective", *J. Am. Chem. Soc.*, vol. 135, 1167–1176, 2013.

[GOO 17] GOODARZI S., DA ROS T., CONDE J. *et al.*, "Fullerene: Biomedical engineers get to revisit an old friend", *Mater. Today*, vol. 20, pp. 460–480, 2017.

[GOR 11] GORE J.P., SANE A., "Flame synthesis of carbon nanotubes, carbon tubes – Synthesis, characterization, applications", in YELLAMPALLI S. (ed.), *Intech. Open*, available at: http://www.intechopen.com/books/carbon-nanotubes-synthesis-characterization-applications/flame-synthesis-of-carbon-nanotubes, 2011.

[GRA 07] GRAF D., MOLITOR F., ENSSLIN K. *et al.*, "Spatially resolved Raman spectroscopy of single- and few-layer graphene", *Nano Lett.*, vol. 7, pp. 238–242, 2007.

[GRÄ 09] GRÄTZEL M., "Recent advances in sensitized mesoscopic solar cells", *Acc. Chem. Res.*, vol. 42, pp. 1788–1798, 2009.

[GRA 17a] GRANCINI G., ROLDÀN-CARMONA C., ZIMMERMANN I. *et al.*, "One-year stable perovskite solar cells by 2D/3D interface engineering", *Nat. Commun.*, vol. 8, art. 15684, 2017.

[GRÄ 17b] GRÄTZEL M., "The rise of highly efficient and stable perovskite solar cells", *Acc. Chem. Res.*, vol. 50, pp. 487–491, 2017.

[GRE 88] GREINER N.R., PHILLIPS D.S., JOHNSON J.D. *et al.*, "Diamonds in detonation soot", *Nature*, vol. 333, pp. 440–441, 1988.

[GRE 11] GREEN A.A., HERSAM M.C., "Nearly single-chirality single-walled carbon nanotubes produced via orthogonal iterative density gradient ultracentrifugation", *Adv. Mater.*, vol. 23, pp. 2185–2190, 2011.

[GRE 17] GREY C.P., TARASCON J.M., "Sustainability and *in situ* monitoring in battery development", *Nat. Mater.*, vol. 16, pp. 45–56, 2017.

[GRE 20] GREEN M.A., DUNLOP E.D., HOHL-EBINGER J. *et al.*, "Solar cell efficiency tables (Version 55)", *Prog. Photovolt. Res. Appl.*, vol. 28, pp. 3–15, 2020.

[GRI 16] GRIMAUD A., HONG W.T., SHAO-HORN Y. *et al.*, "Anionic redox processes for electrochemical devices", *Nat. Mater.*, vol. 15, pp. 121–126, 2016.

[GRI 17] GRIMAUD A., DIAZ-MORALES O., HAN B. *et al.*, "Activating lattice oxygen redox reactions in metal oxides to catalyse oxygen evolution", *Nat. Chem.*, vol. 9, pp. 457–465, 2017.

[GRO 07] GROBERT N., "Carbon nanotubes becoming clean", *Mater. Today*, vol. 10, pp. 28–35, 2007.

[GU 16] GU J., ZHANG X., PANG A. *et al.*, "Facile synthesis and photoluminescence characteristics of blue-emitting nitrogen-doped graphene quantum dots", *Nanotechnology*, vol. 27, art. 165704, 2016.

[GUN 17] GUNEY M.S., TEPE Y., "Classification and assessment of energy storage systems", *Renewable Sustainable Energy Rev.*, vol. 75, pp. 1187–1197, 2017.

[GUO 10] GUO W., ZHANG M., BANERJEE A. *et al.*, "Catalyst-free InGaN/GaN nanowire light emitting diodes grown on (001) silicon by molecular beam epitaxy", *Nano Lett.*, vol. 10, pp. 3355–3359, 2010.

[GUO 14] GUO X., FACCHETTI A., MARKS T.J., "Imide- and amide-functionalized polymer semiconductors", *Chem. Rev.*, vol. 114, pp. 8943–9021, 2014.

[GUO 17a] GUO X., ZHENG S., ZHANG G. *et al.*, "Nanostructured graphene-based materials for flexible energy storage", *Energy Storage Mater.*, vol. 9, pp. 150–169, 2017.

[GUO 17b] GUO Y., LI H., ZHAI T., "Reviving lithium-metal anodes for next-generation high-energy batteries", *Adv. Mater.*, vol. 29, art. 1700007, 2017.

[HAG 10] HAGFELDT A., BOSCHLOO G., SUN L.C. *et al.*, "Dye-sensitized solar cells", *Chem. Rev.*, vol. 110, pp. 6595–6663, 2010.

[HAL 95] HALLS J.J.M., WALSH C.A., GREENHAM N.C. *et al.*, "Efficient photodiodes from interpenetrating polymer networks", *Nature*, vol. 376, pp. 498–500, 1995.

[HAL 08] HALEY M.M., "Synthesis and properties of annulenic subunits of graphyne and graphdiyne nanoarchitectures", *Pure Appl. Chem.*, vol. 80, pp. 519–532, 2008.

[HAL 18a] HALAWA M.I., LAI J., XU G., "Gold nanoclusters: Synthetic strategies and recent advances in fluorescent sensing", *Mater. Today Nano*, vol. 3, pp. 9–27, 2018.

[HAL 18b] HALAWA M.I., WUA F., FEREJA T.H. *et al.*, "One-pot green synthesis of supramolecular β-cyclodextrin functionalized gold nanoclusters and their application for highly selective and sensitive fluorescent detection of dopamine", *Sensors and Actuators B*, vol. 254, pp. 1017–1024, 2018.

[HAN 18] HAN X., TONG X., LIU X. *et al.*, "Hydrogen evolution reaction on hybrid catalysts of vertical MoS_2 nanosheets and hydrogenated Graphene", *ACS Catal.*, vol. 8, pp. 1828–1836, 2018.

[HAN 20] HAN L., GALÁN-MASCARÓS J.R., "The positive effect of iron doping in the electrocatalytic activity of cobalt hexacyanoferrate", *Catalysts*, vol. 10, no. 130, 2020.

[HAR 12] HARDIN B.E., SNAITH H.J., MCGEHEE M.D., "The renaissance of dye-sensitized solar cells", *Nature Photonics*, vol. 6, pp. 162–169, 2012.

[HAR 21] HARIYANI S., BRGOCH J., "Advancing human-centric LED lighting using $Na_2MgPO_4F:Eu^{2+}$", *ACS Appl. Mater. Interfaces*, vol.13, pp. 16669–16676, 2021.

[HAT 04] HATA K., FUTABA D.N., MIZUNO K. *et al.*, "Water-assisted highly efficient synthesis of impurity-free single walled carbon nanotubes", *Science*, vol. 306, pp. 1362–1364, 2004.

[HE 17] HE R., HUA J., ZHANG A. *et al.*, "Molybdenum disulfide-black phosphorus hybrid nanosheets as a superior catalyst for electrochemical hydrogen evolution", *Nano Lett.*, vol. 17, pp. 4311–4316, 2017.

[HE 18] HE R., XU D., CHENG B. *et al.*, "Review on nanoscale Bi-based photocatalysts", *Nanoscale Horiz.*, vol. 3, pp. 464–504, 2018.

[HE 19] HE Y., LINA H., GUOB Z. *et al.*, "Recent developments and advances in boron-doped diamond electrodes for electrochemical oxidation of organic pollutants", *Sep. Purif. Technol.*, vol. 212, pp. 802–821, 2019.

[HEA 85] HEATH J.R., O'BRIEN S.C., ZHANG Q. *et al.*, "Small lanthanum complexes of spheroidal carbon shells", *J. Am. Chem. Soc.*, vol. 107, pp. 7779–7780, 1985.

[HEA 08] HEAVEN M.W., DASS A., WHITE P.S. *et al.*, "Crystal structure of the gold nanoparticle $[N(C_8H_{17})_4][Au_{25}(SCH_2CH_2Ph)_{18}]$", *J. Am. Chem. Soc.*, vol. 130, pp. 3754–3755, 2008.

[HEE 14] HEEGER A.J., "25th anniversary article: Bulk heterojunction solar cells: Understanding the mechanism of operation", *Adv. Mater.*, vol. 26, pp. 10–28, 2014.

[HER 08] HERSAM M.C., "Progress towards monodisperse single-walled carbon nanotubes", *Nature Nanotechn.*, vol. 3, pp. 387–394, 2008.

[HER 13] HERRAIZ-CARDONA I., GONZÁLEZ-BUCH C., VALERO-VIDAL C. *et al.*, "Co-modification of Ni-based type Raney electrodeposits for hydrogen evolution reaction in alkaline media", *J. Power Sources*, vol. 240, pp. 698–704, 2013.

[HIG 16] HIGGINS T.M., PARK S.-H., KING P.J. *et al.*, "A commercial conducting polymer as both binder and conductive additive for silicon nanoparticle-based lithium-ion battery negative electrodes", *ACS Nano*, vol. 10, pp. 3702–3713, 2016.

[HIG 18] HIGAKI T., ZHOU M., LAMBRIGHT K.J. *et al.*, "Sharp transition from nonmetallic Au_{246} to metallic Au_{279} with nascent surface plasmon resonance", *J. Am. Chem. Soc.*, vol. 140, pp. 5691–5695, 2018.

[HIG 19] HIGAKI T., LI Y., ZHAO S. *et al.*, "Atomically tailored gold nanoclusters for catalytic application", *Angew. Chem. Int. Ed.*, vol. 58, pp. 8291–8302, 2019.

[HIN 15] HINE T., "Transition metal chalcogenides: Ultrathin inorganic materials with tunable electronic properties", *Acc. Chem. Res.*, vol. 48, pp. 65–72, 2015.

[HOF 05] HOFMANN S., CSANYI G., FERRARI A.C. *et al.*, "Surface diffusion: The low activation energy path for nanotube growth", *Phys. Rev. Lett.*, vol. 95, art. 036101, 2005.

[HOF 12] HOFMANN S., Exciton dynamics in white organic light-emitting diodes comprising triplet harvesting, Dissertation, Dresden University of Technology, Dresden, 2012.

[HON 12] HONG G., ROBINSON J.T., ZHANG Y. *et al.*, "In vivo fluorescence imaging with Ag_2S quantum dots in the second near-infrared region", *Angew. Chem.*, vol. 124, pp. 9956–9959, 2012.

[HOR 10] HORCAJADA P., CHALATI T., SERRE C. *et al.*, "Porous metal-organic-framework nanoscale carriers as a potential platform for drug delivery and imaging", *Nat. Mater.*, vol. 9, pp. 172–178, 2010.

[HSI 16] HSIAO W.W.-W., HUI Y.Y., TSAI P.-C. *et al.*, "Fluorescent nanodiamond: A versatile tool for long-term cell tracking, super-resolution imaging, and nanoscale temperature sensing", *Acc. Chem. Res.*, vol. 49, pp. 400–407, 2016.

[HU 09] HU S.-L., NIU K.-Y., SUN J. *et al.*, "One-step synthesis of fluorescent carbon nanoparticles by laser irradiation", *J. Mater. Chem.*, vol. 19, pp. 484–488, 2009.

[HU 14a] HU K., KULKARNI D.D., CHOI I. *et al.*, "Graphene-polymer nano-composites for structural and functional applications", *Prog. Polym. Sci.*, vol. 39, pp. 1934–1972, 2014.

[HU 14b] HU Z., DEIBERT B.J., LI J., "Luminescent metal-organic frameworks for chemical sensing and explosive detection", *Chem. Soc. Rev.*, vol. 43, pp. 5815–5840, 2014.

[HU 17] HU C., DAI L., "Multifunctional carbon-based metal-free electrocatalysts for simultaneous oxygen reduction, oxygen evolution, and hydrogen evolution", *Adv. Mater.*, art. 1604942, 2017.

[HU 19] HU C., LI M., QIU J. *et al.*, "Design and fabrication of carbon dots for energy conversion and storage", *Chem. Soc. Rev.*, vol. 48, pp. 2315–2337, 2019.

[HUA 03] HUANG S., CAI X., LIU J., "Growth of millimeter-long and horizontally aligned single-walled carbon nanotubes on flat substrates", *J. Am. Chem. Soc.*, vol. 125, pp. 5636–5637, 2003.

[HUA 09] HUANG X., NERETINA S., EL-SAYED M.A., "Gold nanorods: From synthesis and properties to biological and biomedical applications", *Adv. Mater.*, vol. 21, pp. 4880–4910, 2009.

[HUA 13a] HUANG X., ZENG Z., ZHANG H., "Metal dichalcogenide nanosheets: Preparation, properties and applications", *Chem. Soc. Rev.*, vol. 42, pp. 1934–1946, 2013.

[HUA 13b] HUANG X., ZENG Z., BAO S. *et al.*, "Solution-phase epitaxial growth of noble metal nanostructures on dispersible single-layer molybdenum disulfide nanosheets", *Nat. Commun.*, vol. 4, art. 1444, 2013.

[HUA 15] HUANG X., ZHAO Z., CAO L. *et al.*, "High-performance transition metal-doped Pt_3Ni octahedra for oxygen reduction reaction", *Science*, vol. 348, pp. 1230–1234, 2015.

[HUA 17] HUANG S., CHEONG L.-Z., WANG D. *et al.*, "Nanostructured phosphorus doped silicon/graphite composite as anode for high-performance lithium-ion batteries", *ACS Appl. Mater. Interfaces*, vol. 9, pp. 23672–23678, 2017.

[HUF 19] HUFNAGEL A.G., HÄRINGER S., BEETZ M. *et al.*, "Carbon-templated conductive oxide supports for oxygen evolution catalysis", *Nanoscale*, vol. 11, pp. 14285–14293, 2019.

[HUM 58] HUMMERS Jr., W.S. OFFEMAN R.E., "Preparation of graphite oxide", *J. Am. Chem. Soc.*, vol. 80, pp. 1339–1339, 1958.

[HUO 19] HUO J., WANG H., LI S. *et al.*, "Design and development of highly efficient light-emitting layers in OLEDs with dimesitylboranes: An updated review", *Chem. Rec.*, vol. 19, pp. 1–15, 2019.

[HUY 14] HUYNH M., BEDIAKO D.K., NOCERA D.G., "A functionally stable manganese oxide oxygen evolution catalyst in acid", *J. Am. Chem. Soc.*, vol. 136, pp. 6002–6010, 2014.

[HWA 17] HWANG J.-Y., MYUNG S.-T., SUN Y.-K., "Sodium-ion batteries: Present and future", *Chem. Soc. Rev.*, vol. 46, pp. 3529–3614, 2017.

[IEA 11] INTERNATIONAL ENERGY AGENCY, "Technology roadmap: Electric and plug-in hybrid electric vehicles", *OECDiLibrary – IEA Technology Roadmaps*, available at: http://www.oecd-ilibrary.org/energy/technology-roadmap-electric-and-plug-in-hybrid-electric-vehicles_9789264088177-en, 2011.

[IEA 17] INTERNATIONAL ENERGY AGENCY, Key world energy statistics – Data and statistics, available at: http://www.iea.org/statistics/, 2017.

[INA 19] INAGAKI M., TSUMURA T., KINUMOTO T. *et al.*, "Graphitic carbon nitrides (g-C_3N_4) with comparative discussion to carbon materials", *Carbon*, vol. 141, pp. 580–607, 2019.

[INS 19] INSTITUT FRANÇAIS DU PÉTROLE ET DES ÉNERGIES NOUVELLES, "La synthèse – table ronde", *Captage-stockage-valorisation du CO_2, un levier pour décarboner l'industrie*, Paris, France, available at: https://www.ifpenergiesnouvelles.fr/article/captage-stockage-valorisation-du-co2-levier-decarboner-lindustrie-synthese, November 15, 2019.

[INT 18] INTERNATIONAL TECHNOLOGY ROADMAP PHOTOVOLTAIC, Results 2017 including maturity report 2018, 9th edition, September 2018.

[IRI 15] IRISARRI E., PONROUCH A., PALACIN M.R., "Review-hard carbon negative electrode materials for sodium-ion batteries", *J. Electrochem. Soc.*, vol. 162, pp. A2476–A2482, 2015.

[ISH 17] ISHIDA Y., CORPUZ R.D., YONEZAWA T., "Matrix sputtering method: A novel physical approach for photoluminescent noble metal nanoclusters", *Acc. Chem. Res.*, vol. 50, pp. 2986–2995, 2017.

[ISL 14] ISLAM M.S., FISHER C.A.J., "Lithium and sodium battery cathode materials: Computational insights into voltage, diffusion and nanostructural properties", *Chem. Soc. Rev.*, vol. 43, pp. 185–204, 2014.

[ISL 17] ISLAMOGLU T., GOSWAMI S., LI Z. *et al.*, "Postsynthetic tuning of metal-organic frameworks for targeted applications", *Acc. Chem. Res.*, vol. 50, pp. 805–813, 2017.

[IWA 11] IWASE A., NG Y.H., ISHIGURO Y. *et al.*, "Reduced graphene oxide as a solid-state electron mediator in z-scheme photocatalytic water splitting under visible light", *J. Am. Chem. Soc.*, vol. 133, pp. 11054–11057, 2011.

[JAD 07] JADZINSKY P.D., CALERO G., ACKERSON C.J. *et al.*, "Structure of a thiol monolayer-protected gold nanoparticle at 1.1: A resolution", *Science*, vol. 318, pp. 430–433, 2007.

[JÄG 14] JÄGER K., ISABELLA O., SMETS A.H.M. *et al.*, Solar energy, fundamentals, technology, and systems, massive open online course on solar energy, Delft University of Technology, Delft, 2014.

[JAN 16] JANAS D., KOZIOL K.K., "Carbon nanotube fibers and films: Synthesis, applications and perspectives of the direct spinning method", *Nanoscale*, vol. 8, pp. 19475–19490, 2016.

[JAR 07] JARAMILLO T.F., JORGENSEN K.P., BONDE J. *et al.*, "Identification of active edge sites for electrochemical H_2 evolution from MoS_2 nanocatalysts", *Science*, vol. 317, pp. 100–102, 2007.

[JEA 15] JEAN J., BROWN P.R., JAFFE R.L. *et al.*, "Pathways for solar photovoltaics", *Energy Environ. Sci.*, vol. 8, pp. 1200–1219, 2015.

[JEN 19] JENA A.K., KULKARNI A., MIYASAKA T., "Halide perovskite photovoltaics: Background, status, and future prospects", *Chem. Rev.*, vol. 119, pp. 3036–3103, 2019.

[JI 09] JI X., LEE K.T., NAZAR L.F., "A highly ordered nanostructured carbon-sulphur cathode for lithium-sulphur batteries", *Nat. Mater.*, vol. 8, pp. 500–506, 2009.

[JIA 09] JIANG D., WHETTEN R.L., LUO W. *et al.*, "The smallest thiolated gold superatom complexes", *J. Phys. Chem. C*, vol. 113, pp. 17291–17295, 2009.

[JIA 18a] JIAN J., JIANG G., VAN DE KROL R. *et al.*, "Recent advances in rational engineering of multinary semiconductors for photoelectrochemical hydrogen generation", *Nano Energy*, vol. 51, pp. 457–480, 2018.

[JIA 18b] JIANG K., WANG H., "Electrocatalysis over graphene-defect-coordinated transition-metal single-atom catalysts", *Chem.*, vol. 4, pp. 191–198, 2018.

[JIN 07] JIN Z., CHU H., WANG J. *et al.*, "Ultralow feeding gas flow guiding growth of large-scale horizontally aligned single-walled carbon nanotube arrays", *Nano Lett.*, vol. 7, pp. 2073–2079, 2007.

[JIN 10] JIN R., "Quantum sized, thiolate-protected gold nanoclusters", *Nanoscale*, vol. 2, pp. 343–362, 2010.

[JIN 15] JIN R., "Atomically precise metal nanoclusters: Stable sizes and optical properties", *Nanoscale*, vol. 7, pp. 1549–1565, 2015.

[JIN 16] JIN R., ZENG C., ZHOU M. *et al.*, "Atomically precise colloidal metal nanoclusters and nanoparticles: Fundamentals and opportunities", *Chem. Rev.*, vol. 116, pp. 10346–10413, 2016.

[JOL 14] JOLY D., PELLEJÀ L., NARBEY S. *et al.*, "A robust organic dye for dye sensitized solar cells based on iodine/iodide electrolytes combining high efficiency and outstanding stability", *Sci. Rep.*, vol. 4, no. 4033, 2014.

[JUD 18] JUDEZ X., ZHANG H., LI C. *et al.*, "Review – Solid electrolytes for safe and high energy density lithium-sulfur batteries: Promises and challenges", *Journal of the Electrochemical Society*, vol. 165, pp. A6008–A6016, 2018.

[JUN 15] JUNG H.S., PARK N.-G., "Perovskite solar cells: From materials to devices", *Small*, vol. 11, pp. 10–25, 2015.

[KAL 18] KALANOOR B.S., SEO H., KALANUR S.S., "Recent developments in photo-electrochemical water-splitting using $WO_3/BiVO_4$ heterojunction photoanode: A review", *Mater. Sci. Energy Techn.*, vol. 1, pp. 49–62, 2018.

[KAM 11] KAMAYA N., HOMMA K., YAMAKAWA Y. *et al.*, "A lithium superionic conductor", *Nat. Mater.*, vol. 10, pp. 682–686, 2011.

[KAN 01] KANNO R., MURAYAMA M., "Lithium ionic conductor thio-LISICON: The Li_2S GeS_2 P_2S_5 System", *J. Electrochem. Soc.*, vol. 148, pp. A742–A746, 2001.

[KAN 10] KANDIAH M., NILSEN M.H., USSEGLIO S. *et al.*, "Synthesis and stability of tagged UiO-66 Zr-MOFs", *Chem. Mater.*, vol. 22, pp. 6632–6640, 2010.

[KAN 14] KANG S., ARAYA-SECCHI R., WANG D. *et al.*, "Dual inhibitory pathways of metallofullerenol Gd@C82(OH)22 on matrix metal-loproteinase-2: Molecular insight into drug-like nanomedicine", *Sci. Rep.*, vol. 4, vol. 4775, 2014.

[KAN 17] KANG J., KIM J., LEE S. *et al.*, "Breathable carbon-free electrode: Black TiO_2 with hierarchically ordered porous structure for stable Li–O_2 battery", *Adv. Energy Mat.*, vol. 1700814, 2017.

[KAN 18] KANG X., CHONG H., ZHU M., "$Au_{25}(SR)_{18}$: The captain of the great nanocluster ship", *Nanoscale*, vol. 10, pp. 10758–10834, 2018.

[KAN 19a] KANG J., WEI Z., LI J., "Graphyne and its family: Recent theoretical advances", *ACS Appl. Mater. Interfaces*, vol. 11, pp. 2692–2706, 2019.

[KAN 19b] KANTAREDDY S.N.R., MATHEWS I., SUN S. *et al.*, "Perovskite PV-powered RFID: Enabling low-cost self-powered IoT sensors", *IEEE Sensors Journal*, vol. 99, 2019.

[KAS 07] KASAVAJJULA U., WANG C., APPLEBY A.J., "Nano- and bulk-silicon-based insertion anodes for lithium-ion secondary cells", *J. Power Sources*, vol. 163, pp. 1003–1039, 2007.

[KAT 99] KATAURA H., KUMAZAWA Y., MANIWA Y. *et al.*, "Optical properties of single-wall carbon nanotubes", *Synth. Met.*, vol. 103, pp. 2555–2558, 1999.

[KAU 18] KAUR M., KAUR M., SHARMA V.K., "Nitrogen-doped graphene and graphene quantum dots: A review on synthesis and applications in energy, sensors and environment", *Adv. in Colloids and Interface Sci.*, vol. 259, pp. 44–64, 2018.

[KE 16] KE Q., WANG J., "Graphene-based materials for supercapacitor electrodes – A review", *J. Materiomics*, vol. 2, pp. 37–54, 2016.

[KEL 15] KELLER J., BRILLAS E., RADJENOVIC J., "Removal of organic contaminants from secondary effluent by anodic oxidation with a boron-doped diamond anode as tertiary treatment", *J. Hazard. Mater.*, vol. 283, pp. 551–557, 2015.

[KHA 17] KHANDELWAL P., PODDAR P., "Fluorescent metal quantum clusters: An updated overview of the synthesis, properties, and biological applications", *J. Mater. Chem. B*, vol. 5, pp. 9055–9084, 2017.

[KID 94] KIDO J., HONGAWA K., OKUYAMA K. *et al.*, "White light-emitting organic electroluminescent devices using the poly(Nvinylcarbazole) emitter layer doped with three fluorescent dyes", *Applied Physics Letters*, vol. 64, pp. 815–817, 1994.

[KIM 09] KIM K.S., ZHAO Y., JANG H. *et al.*, "Large-scale pattern growth of graphene films for stretchable transparent electrodes", *Nature*, vol. 457, pp. 706–710, 2009.

[KIM 11] KIM J.E., HAN T.H., LEE S.H. *et al.*, "Graphene oxide liquid crystals", *Angew. Chem. Int. Ed.*, vol. 50, pp. 3043–3047, 2011.

[KIM 12] KIM J., COTE L.J., HUANG J., "Two-dimensional soft material: New faces of graphene oxide", *Acc. Chem. Res.*, vol. 45, pp. 1356–1364, 2012.

[KIM 15] KIM B.K., SY S., YU A. *et al.*, "Electrochemical supercapacitors for energy storage and conversion", *Handbook of Clean Energy Systems*, John Wiley & Sons, Hoboken, 2015.

[KIM 16] KIM J.-H., YANG H., "High-efficiency Cu–In–S quantum-dot-light-emitting device exceeding 7%", *Chem. Mater.*, vol. 28, pp. 6329–6335, 2016.

[KIM 17] KIM T.-H., LEE C.-S., KIM S. *et al.*, "Fully stretchable optoelectronic sensors based on colloidal quantum dots for sensing photople-thysmographic signals", *ACS Nano*, vol. 11, pp. 5992–6003, 2017.

[KIM 18] KIM S., JOARDER B., HURD J.A. *et al.*, "Achieving superprotonic conduction in metal-organic frameworks through iterative design advances", *J. Am. Chem. Soc.*, vol. 140, pp. 1077–1082, 2018.

[KOB 13] KOBAYASHI T., BANDO M., KIMURA N. *et al.*, "Production of a 100-m-long high-quality graphene transparent conductive film by roll-to-roll chemical vapor deposition and transfer process", *Appl. Phys. Lett.*, vol. 102, art. 023112, 2013.

[KOJ 09] KOJIMA A., TESHIMA K., SHIRAI Y. *et al.*, "Organometal halide perovskites as visible-light sensitizers for photovoltaic cells", *J. Am. Chem. Soc.*, vol. 131, pp. 6050–6051, 2009.

[KOM 10] KOMATSU N., WANG F., "A comprehensive review on separation methods and techniques for single-walled carbon nanotubes", *Materials*, vol. 3, pp. 3818–3844, 2010.

[KON 09] KONDAKOV D.Y., PAWLIK T.D., HATWAR T.K. *et al.*, "Triplet annihilation exceeding spin statistical limit in highly efficient fluorescent organic light-emitting diodes", *J. Appl. Physics*, vol. 106, art. 124510, 2009.

[KOR 90] KORTAN R., HULL R., OPILA R. *et al.*, "Nucleation and growth of CdSe on ZnS quantum crystallite seeds, and vice versa, in inverse micelle media", *J. Am. Chem. Soc.*, vol. 112, pp. 1327–1332, 1990.

[KOR 16] KORTSHAGEN U.R., SANKARAN R.M., PEREIRA R.N. *et al.*, "Non thermal plasma synthesis of nanocrystals: Fundamental principles, materials, and applications", *Chem. Rev.*, vol. 116, pp. 11061–11127, 2016.

[KOU 15] KOU L., CHEN C., SMITH S.C., "Phosphorene: Fabrication, properties, and applications", *J. Phys. Chem. Lett.*, vol. 6, pp. 2794–2805, 2015.

[KOU 20] KOUR R., ARYA S., VERMA S. *et al.*, "Review – Recent advances and challenges in indium gallium nitride ($In_xGa_{1-x}N$) materials for solid state lighting", *ECS J. Solid State Sci. Technol.*, vol. 9, art. 015011, 2020.

[KOV 11] KOVALENKO I., ZDYRKO B., MAGASINSKI A. *et al.*, "A major constituent of brown algae for use in high-capacity Li-ion batteries", *Sci.*, vol. 334, pp. 75–79, 2011.

[KOV 15] KOVALENKO M.V., MANNA L., CABOT A. *et al.*, "Prospects of nanoscience with nanocrystals", *ACS Nano*, vol. 9, pp. 1012–1057, 2015.

[KOZ 17] KOZIOL K.K., JANAS D., BROWN E. *et al.*, "Thermal properties of continuously spun carbon nanotube fibres", *Physica E: Low-dimensional Systems and Nanostructures*, vol. 88, pp. 104–108, 2017.

[KRI 20] KRISHNAN S., FAIRLIE M., ANDRES P. *et al.*, "Power to gas (H_2): Alkaline electrolysis", in JUNGINGER M., LOUWEN A. (eds), *Technological Learning in the Transition to a Low-Carbon Energy System*, Academic Press, Cambridge, 2020.

[KRO 85] KROTO H.W., HEATH J.R., O'BRIEN S.C. *et al.*, "C_{60}: Buckminsterfullerene", *Nature*, vol. 318, pp. 162–163, 1985.

[KUD 99] KUDO A., OMORI K., KATO H., "A novel aqueous process for preparation of crystal form-controlled and highly crystalline $BiVO_4$ powder from layered vanadates at room temperature and its photocatalytic and photophysical properties", *J. Am. Chem. Soc.*, vol. 121, pp. 11459–11467, 1999.

[KUH 08] KUHN P., ANTONIETTI M., THOMAS A. "Porous, covalent triazine-based frameworks prepared by ionothermal synthesis", *Angew. Chem. Int. Ed.*, vol. 47, pp. 3450–3453, 2008.

[KUL 18] KULKARNI A., SIAHROSTAMI S., PATEL A. *et al.*, "Understanding catalytic activity trends in the oxygen reduction reaction", *Chem. Rev.*, vol. 118, pp. 2302–2312, 2018.

[KUM 06] KUMAR S., GANDHI K.S., KUMAR R., "Modeling of formation of gold nanoparticles by citrate method", *Industrial & Engineering Chemistry Research*, vol. 46, pp. 3128–3136, 2006.

[KUS 19] KUSIAK-NEJMAN E., MORAWSKI A.W., "TiO_2/graphene-based nanocomposites for water treatment: A brief overview of charge carrier transfer, antimicrobial and photocatalytic performance", *Appl. Catalysis B: Environm.*, vol. 253, pp. 179–186, 2019.

[LAG 12] LAGUNA-BERCERO M.A., "Recent advances in high temperature electrolysis using solid oxide fuel cells: A review", *J. Power Sources*, vol. 203, pp. 4–16, 2012.

[LAL 13] LALWANI G., SITHARAMAN B., "Multifunctional fullerene and metallo fullerene-based nanobiomaterials", *Nano Life*, vol. 3, art. 1342003, 2013.

[LAN 09] LAN A., LI K., WU H. *et al.*, "A luminescent microporous metal-organic framework for the fast and reversible detection of high explosives", *Angew. Chem. Int. Ed.*, vol. 48, pp. 2334–2338, 2009.

[LAR 08] LARGEOT C., PORTET C., CHMIOLA J. *et al.*, "Relation between the ion size and pore size for an electric double-layer capacitor", *J. Am. Chem. Soc.*, vol. 130, pp. 2730–2731, 2008.

[LAT 17] LATZEL M., BÜTTNER P., SARAU G. *et al.*, "Significant performance enhancement of InGaN/GaN nanorod LEDs with multi-layer graphene transparent electrodes by alumina surface passivation", *Nanotechn.*, vol. 28, art. 055201, 2017.

[LAU 07] LAUS S., SITHARAMAN B., TOTH E. *et al.*, "Understanding paramagnetic relaxation phenomena for water-soluble gadofullerenes", *J. Phys. Chem. C*, vol. 111, pp. 5633–5639, 2007.

[LEE 12] LEE Y., SUNTIVICH J., MAY K.J. *et al.*, "Synthesis and activities of rutile IrO_2 and RuO_2 nanoparticles for oxygen evolution in acid and alkaline solutions", *J. Phys. Chem. Lett.*, vol. 3, pp. 399–404, 2012.

[LEY 14] LEY M.B., JEPSEN L.H., LEE Y.-S. *et al.*, "Complex hydrides for hydrogen storage – New perspectives", *Mater. Today*, vol. 17, pp. 122–128, 2014.

[LI 99] LI H., EDDAOUDI M., O'KEEFFE M. *et al.*, "Design and synthesis of an exceptionally stable and highly porous metal-organic framework", *Nature*, vol. 402, pp. 276–279, 1999.

[LI 09a] LI J.-R., KUPPLER R.J., ZHOU H.-C., "Selective gas adsorption and separation in metal-organic frameworks", *Chem. Soc. Rev.*, vol. 38, pp. 1477–1504, 2009.

[LI 09b] LI X., CAI W., AN J. *et al.*, "Large-area synthesis of high-quality and uniform graphene films on copper foils", *Science*, vol. 324, pp. 1312–1314, 2009.

[LI 10a] LI G., LI Y., LIU H. *et al.*, "Architecture of graphdiyne nano-scale films", *Chem. Commun.*, vol. 46, pp. 3256–3258, 2010.

[LI 10b] LI H., HE X., KANG Z. *et al.*, "Water-soluble fluorescent carbon quantum dots and photocatalyst design", *Angew. Chem. Int. Ed.*, vol. 49, pp. 4430–4434, 2010.

[LI 14] LI H., WU J., YIN Z. *et al.*, "Preparation and applications of mechanically exfoliated single-layer and multilayer MoS_2 and WSe_2 nanosheets", *Acc. Chem. Res.*, vol. 47, pp. 1067–1075, 2014.

[LI 15] LI Z., LIU Z., SUN H. *et al.*, "Superstructured assembly of nanocarbons: Fullerenes, nanotubes, and graphene", *Chem. Rev.*, vol. 115, pp. 7046–7117, 2015.

[LI 16a] LI M., ZHAO Z., CHENG T. *et al.*, "Ultrafine jagged platinum nanowires enable ultrahigh mass activity for the oxygen reduction reaction", *Science*, vol. 354, pp. 1414–1419, 2016.

[LI 16b] LI N.-W., YIN Y.-X., YANG C.-P. *et al.*, "An artificial solid electrolyte interphase layer for stable lithium metal anodes", *Adv. Mater.*, vol. 28, pp. 1853–1858, 2016.

[LI 16c] LI R., LIU Y., LI Z. *et al.*, "Bottom-up fabrication of single-layered nitrogen-doped graphene quantum dots through intermolecular carbonization arrayed in a 2d plane", *Chem. Eur. J.*, vol. 22, pp. 272–278, 2016.

[LI 16d] LI X., YU J., WAGEH S. *et al.*, "Graphene in photo-catalysis: A review", *Small*, vol. 12, pp. 6640–6696, 2016.

[LI 17a] LI B., LIU J., "Progress and directions in low-cost redox-flow batteries for large-scale energy storage", *Natl. Sci. Rev.*, vol. 4, pp. 91–105, 2017.

[LI 17b] LI M., "World energy 2017–2050: Annual report", *Seeking Alpha*, available at: https://seekingalpha.com/article/4083393-world-energy-2017minus-2050-annual-report, 2017.

[LI 18a] LI H., WANG K., SUN Y. *et al.*, "Recent advances in gas storage and separation using metal-organic frameworks", *Mater. Today*, vol. 21, pp. 108–121, 2018.

[LI 18b] LI N.-W., SHI Y., YIN Y.-X. *et al.*, "A flexible solid electrolyte interphase layer for long-life lithium metal anodes", *Angew. Chem. Int. Ed.*, vol. 57, pp. 1505–1509, 2018.

[LI 18c] LI P., ZHAO G., ZHENG X. *et al.*, "Recent progress on silicon-based anode materials for practical lithium-ion battery applications", *Energy Storage Mater.*, vol. 15, pp. 422–446, 2018.

[LI 18d] LI Z., KLEIN T.R., KIM D.H. *et al.*, "Scaling up perovskite photovoltaics: Progress, challenges, and outlook of a transformational technology", *Nat. Rev. Mater.*, vol. 3, art. 18017, 2018.

[LI 19a] LI A., OOKA H., BONNET N. *et al.*, "Stable potential windows for long-term electrocatalysis by manganese oxides under acidic conditions", *Angew. Chem. Int. Ed.*, vol. 58, pp. 5054–5058, 2019.

[LI 19b] LI F., LI J., ZHU F. *et al.*, "Atomically intimate contact between solid electrolytes and electrodes for Li batteries", *Matter*, vol. 1, pp. 1001–1016, 2019.

[LI 19c] LI Y., LI Q., WANG H. *et al.*, "Recent progresses in oxygen reduction reaction electrocatalysts for electrochemical energy applications", *Electrochem. Energy Rev.*, vol. 2, pp. 518–538, 2019.

[LIA 14] LIAO L., WANG S., XIAO J. *et al.*, "A nanoporous molybdenum carbide nanowire as an electrocatalyst for hydrogen evolution reaction", *Energy Environ. Sci.*, vol. 7, pp. 387–392, 2014.

[LIJ 91] LIJIMA S., "Helical microtubules of graphitic carbon", *Nature*, vol. 354, pp. 56–58, 1991.

[LIM 03] LIM S., CIUPARU D., PAK C. *et al.*, "Synthesis and characterization of highly ordered Co-MCM-41 for production of aligned single walled carbon nanotubes (SWNT)", *J. Phys. Chem. B*, vol. 107, pp. 11048–11056, 2003.

[LIM 15] LIM S.Y., SHEN W., GAO Z., "Carbon quantum dots and their applications", *Chem. Soc. Rev.*, vol. 44, pp. 362–381, 2015.

[LIN 02] LINK S., BEEBY A., FITZGERALD S. *et al.*, "Visible to infrared luminescence from a 28-atom gold cluster", *J. Phys. Chem. B*, vol. 106, pp. 3410–3415, 2002.

[LIN 17a] LIN B.-Y., EASLEY C.J., CHEN C.-H. *et al.*, "Exciplex-sensitized triplet-triplet annihilation in heterojunction organic thin-film", *ACS Appl. Mater. Interfaces*, vol. 9, pp. 10963–10970, 2017.

[LIN 17b] LIN D., LIU Y., CUI Y., "Reviving the lithium metal anode for high-energy batteries", *Nat. Nanotechnol.*, vol. 12, pp. 194–206, 2017.

[LIU 10a] LIU J., HERSAM M.C., "Recent developments in carbon nanotube sorting and selective growth", *MRS Bull.*, vol. 35, pp. 315–321, 2010.

[LIU 10b] LIU X., ZHANG W., UCHIDA S. *et al.*, "An efficient organic-dye-sensitized solar cell with in situ polymerized poly(3,4-ethy-lene dioxythiophene) as a hole-transporting material", *Adv. Mater.*, vol. 22, pp. E150–E155, 2010.

[LIU 12] LIU N., WU H., MCDOWELL M.T., YAO Y., WANG C., CUI Y., "A yolk-shell design for stabilized and scalable li-ion battery alloy anodes", *Nano Lett.*, no. 12, pp. 3315–3321, 2012.

[LIU 16a] LIU C., NEALE Z.G., CAO G., "Understanding electrochemical potentials of cathode materials in rechargeable batteries", *Mater. Today*, vol. 19, pp. 109–123, 2016.

[LIU 16b] LIU Y., XU Z., ZHAN J. *et al.*, "Superb electrically conductive graphene fibers via doping strategy", *Adv. Mater.*, vol. 28, pp. 7941–7947, 2016.

[LIU 17a] LIU W., LOH K.P., "Two-dimensional conjugated polymers based on c-c coupling", *Acc. Chem. Res.*, vol. 50, pp. 522–526, 2017.

[LIU 17b] LIU Y., LIN D., YUEN P.Y. *et al.*, "An artificial solid electrolyte interphase with high Li-ion conductivity, mechanical strength, and flexibility for stable lithium metal anodes", *Adv. Mater.*, vol. 1605531, 2017.

[LIU 17c] LIU Y., LIANG H., XU Z. *et al.*, "Superconducting continuous graphene fibers via calcium intercalation", *ACS Nano*, vol. 11, pp. 4301–4306, 2017.

[LIU 18] LIU Q., LIU X., ZHENG L. *et al.*, "The solid-phase synthesis of an Fe-N-C electrocatalyst for high-power proton-exchange membrane fuel cells", *Angew. Chem. Int. Ed.*, vol. 57, pp. 1204–1208, 2018.

[LIU 19] LIU J., BAO Z., CUI Y. *et al.*, "Pathways for practical high-energy long-cycling lithium metal batteries", *Nat. Energy*, vol. 4, pp. 180–186, 2019.

[LU 14] LU Z., ZHU W., YU X. *et al.*, "Ultrahigh hydrogen evolution performance of underwater 'superaerophobic' MoS_2 nanostructured electrodes", *Adv. Mater.*, vol. 26, pp. 2683–2687, 2014.

[LU 20] LU H., LIU Y., AHLAWAT P. *et al.*, "Vapor-assisted deposition of highly efficient, stable black-phase $FAPbI_3$ perovskite solar cells", *Science*, vol. 370, pp. 74–81, 2020.

[LUO 01] LUO J., XIE Z., LAM J.W.Y. *et al.*, "Aggregation-induced emission of 1-methyl-1,2,3,4,5-pentaphenylsilole", *Chem. Commun.*, pp. 1740–1741, 2001.

[LUO 12] LUO Z., YUAN X., YU Y. *et al.*, "From aggregation-induced emission of Au(I)-thiolate complexes to ultrabright Au(0)@ Au(I)-thiolate core-shell nanoclusters", *J. Am. Chem. Soc.*, vol. 134, pp. 16662–16670, 2012.

[LUO 14] LUO Z., CASTLEMAN A.W., "Special and general superatoms", *Acc. Chem. Res.*, vol. 47, pp. 2931–2940, 2014.

[LUO 15] LUO X, WANG J., DOONER M. *et al.*, "Overview of current development in electrical energy storage technologies and the application potential in power system operation", *Appl. Energy*, vol. 137, pp. 511–536, 2015.

[LUO 16] LUO Z., QI G., CHEN K. *et al.*, "Microwave-assisted preparation of white fluorescent graphene quantum dots as a novel phosphor for enhanced white-light-emitting diodes", *Adv. Funct. Mat.*, vol. 26, pp. 2739–2744, 2016.

[LÜS 13] LÜSSEM B., RIEDE M., LEO K., "Doping of organic semiconductors", *Phys. Status Solidi*, vol. A 210, pp. 9–43, 2013.

[LV 15] LV R., ROBINSON J.A., SCHAAK R.E. *et al.*, "Transition metal dichalcogenides and beyond: Synthesis, properties, and applications of single- and few-layer nanosheets", *Acc. Chem. Res.*, vol. 48, pp. 56–64, 2015.

[LV 16] LV L., ZHANG P., CHENG H. *et al.*, "Solution-processed ultraelastic and strong air-bubbled graphene foams", *Small*, vol. 12, pp. 3229–3234, 2016.

[MA 05] MA W., YANG C., GONG X. *et al.*, "Thermally stable, efficient polymer solar cells with nanoscale control of the interpenetrating network morphology", *Adv. Funct. Mater.*, vol. 15, pp. 1617–1622, 2005.

[MA 19] MA R., LIN G., ZHOU Y. *et al.*, "A review of oxygen reduction mechanisms for metal-free carbon-based electrocatalysts", *npj Comput. Mater.*, vol. 78, 2019.

[MAE 10] MAEDA K., DOMEN K., "Photocatalytic water splitting: Recent progress and future challenges", *J. Phys. Chem. Lett.*, vol. 1, pp. 2655–2661, 2010.

[MAG 19] MAGGIO G., NICITA A., SQUADRITO G., "How the hydrogen production from RES could change energy and fuel markets: A review of recent literature", *Int. J. of Hydrogen Energy*, vol. 44, pp. 11371–11384, 2019.

[MAH 08] MAHLER B., SPINICELLI P., BUIL S. *et al.*, "Towards non-blinking colloidal quantum dots", *Nat. Mater.*, vol. 7, pp. 661–664, 2008.

[MAH 17] MAHMOOD J., LI F., JUNG S.M. *et al.*, "An efficient and pH-universal ruthenium-based catalyst for the hydrogen evolution reaction", *Nat. Nanotechnol.*, vol. 12, pp. 441–447, 2017.

[MAH 18] MAHMOOD N., YAO Y., ZHANG J.-W. *et al.*, "Electrocatalysts for hydrogen evolution in alkaline electrolytes: Mechanisms, challenges, and prospective solutions", *Adv. Sci.*, vol. 5, art. 1700464, 2018.

[MAL 09] MALARD L.M., PIMENTA M.A., DRESSELHAUS G. *et al.*, "Raman spectroscopy in graphene", *Phys. Rep.*, vol. 473, pp. 51–87, 2009.

[MAL 17] MALUCELLI G., "Graphene-based polymer nanocomposites: Recent advances and still open challenges", *Curr. Graphene Sci.*, vol. 1, pp. 16–25, 2017.

[MAL 18] MALATHI A., MADHAVAN J., ASHOKKUMAR M. *et al.*, "A review on $BiVO_4$ photocatalyst: Activity enhancement methods for solar photo-catalytic applications", *Applied Catalysis A, General*, vol. 555, pp. 47–74, 2018.

[MAR 18] MARBELLA L.E., EVANS M.L., GROH M.F. *et al.*, "Sodiation and desodiation via helical phosphorus intermediates in high-capacity anodes for sodium-ion batteries", *J. Am. Chem. Soc.*, vol. 140, pp. 7994–8004, 2018.

[MAT 16] MATTEOCCI F., CINÀ L., LAMANNA E. *et al.*, "Encapsulation for long-term stability enhancement of perovskite solar cells", *Nano Energy*, vol. 30, pp. 162–172, 2016.

[MAU 05] MAULTZSCH J., TELG H., REICH S. *et al.*, "Radial breathing mode of single-walled carbon nanotubes: Optical transition energies and chiral-index assignment", *Phys. Rev. B*, vol. 72, art. 205438, 2005.

[MAY 18] MAY G.J., DAVIDSON A., MONAHOV B., "Lead batteries for utility energy storage: A review", *J. of Energy Storage*, vol. 15, pp. 145–157, 2018.

[MCC 13] MCCRORY C.C.L., JUNG S., PETERS J.C. *et al.*, "Benchmarking heterogeneous electrocatalysts for the oxygen evolution reaction", *J. Am. Chem. Soc.*, vol. 135, pp. 16977–16987, 2013.

[MCC 15] MCCRORY C.C.L, JUNG S., FERRER I.M. *et al.*, "Benchmarking hydrogen evolving reaction and oxygen evolving reaction electrocatalysts for solar water splitting devices", *J. Am. Chem. Soc.*, vol. 137, pp. 4347–4357, 2015.

[MCD 12] MCDONALD T.M., LEE W.R., MASON J.A. *et al.*, "Capture of carbon dioxide from air and flue gas in the alkylamine-appended metal-organic framework mmen-Mg2(dobpdc)", *J. Am. Chem. Soc.*, vol. 134, pp. 7056–7065, 2012.

[MCD 15] MCDONALD T.M., MASON J.A., Kong X. *et al.*, "Cooperative insertion of CO_2 in diamine appended metal-organic frameworks", *Nature*, vol. 519, pp. 303–308, 2015.

[MED 19] MEDVEDEV D., "Trends in research and development of protonic ceramic electrolysis cells", *Int. J. Hydrogen Energy*, vol. 44, pp. 26711–26740, 2019.

[MEN 18] MENG L., ZHANG Y., WAN X. *et al.*, "Organic and solution-processed tandem solar cells with 17.3% efficiency", *Science*, vol. 361, pp. 1094–1098, 2018.

[MER 11] MERLE G., WESSLING M., NIJMEIJER K., "Anion exchange membranes for alkaline fuel cells: A review", *J. Membr. Sci.*, vol. 377, pp. 1–35, 2011.

[MIZ 80] MIZUSHIMA K., JONES P.C., WISEMAN P.J. *et al.*, "Li_xCoO_2 ($0<x<1$): A new cathode material for batteries of high energy density", *Mater. Res. Bull.*, vol. 15, pp. 783–789, 1980.

[MIZ 09] MIZUNO K., ISHII J., KISHIDA H. *et al.*, "A black body absorber from vertically aligned single-walled carbon nanotubes", *Proc. Natl. Acad. Sci. USA*, vol. 106, pp. 6044–6047, 2009.

[MOC 12] MOCHALIN V.N., SHENDEROVA O., HO D. *et al.*, "The properties and applications of nanodiamonds", *Nat. Nanotechn.*, vol. 7, pp. 11–23, 2012.

[MON 13] MONDLOCH J.E., BURY W., FAIREN-JIMENEZ D. *et al.*, "Vapor-phase metalation by atomic layer deposition in a metal-organic framework", *J. Am. Chem. Soc.*, vol. 135, pp. 10294–10297, 2013.

[MON 16] MONTOYA J.H., SEITZ L.C., CHAKTHRANONT P. *et al.*, "Materials for solar fuels and chemicals", *Nat. Mater.*, vol. 16, pp. 70–81, 2016.

[MON 17] MONDSCHEIN J.S., CALLEJAS J.F., READ C.G. *et al.*, "Crystalline cobalt oxide films for sustained electrocatalytic oxygen evolution under strongly acidic conditions", *Chem. Mater.*, vol. 29, pp. 950–957, 2017.

[MOR 11] MOROZAN A., JOUSSELME B., PALACIN S., "Low-platinum and platinum-free catalysts for the oxygen reduction reaction at fuel cell cathodes", *Energy Environ. Sci.*, vol. 4, pp. 1238–1254, 2011.

[MOR 16] MORALES-GUIO C.G., LIARDET L., HU X., "Oxidatively electrodeposited thin-film transition metal (oxy)hydroxides as oxygen evolution catalysts", *J. Am. Chem. Soc.*, vol. 138, pp. 8946–8957, 2016.

[MOS 15] MOSELEY P.T., RAND D.A.J., PETERS K., "Enhancing the performance of lead acid batteries with carbon. In pursuit of an understanding", *J. Power Sources*, vol. 295, pp. 268–274, 2015.

[MUK 19] MUKHOPADHYAY S., DEBGUPTA J., SINGH C. *et al.*, "Designing UiO-66-based superprotonic conductor with the highest metal-organic framework-based proton conductivity", *ACS Appl. Mater. Interfaces*, vol. 11, pp. 13423–13432, 2019.

[MUR 93] MURRAY C.B., NORRIS D.J., BAWENDI M.G., "Synthesis and characterization of nearly monodisperse CdE (E = S, Se, Te) semiconductor nanocrystallites", *J. Am. Chem. Soc.*, vol. 115, pp. 8706–8715, 1993.

[MUR 13] MURAWSKI C., LEO K., GATHER M.C., "Efficiency roll-off in organic light-emitting diodes", *Adv. Mater.*, vol. 25, pp. 6801–6827, 2013.

[NAG 11] NAGUIB M., KURTOGLU M., PRESSER V. *et al.*, "Two-dimensional nanocrystals produced by exfoliation of Ti_3AlC_2", *Adv. Mater.*, vol. 23, pp. 4248–4253, 2011.

[NAK 93] NAKAMURA S., SENOH M., MUKAI T., "p-GaN/n-InGaN/n-GaN double-heterostructure blue-light-emitting diodes", *Jpn. J. Appl. Phys.*, vol. 32, pp. L8–L11, 1993.

[NAK 95] NAKAMURA S., SENOH M., IWASA N. *et al.*, "High-brightness InGaN blue, green and yellow light-emitting diodes with quantum well structures", *Jpn. J. Appl. Phys.*, vol. 34, pp. L797–L799, 1995.

[NAK 14] NAKAMURA S., Background story of the invention of efficient blue InGaN light emitting diodes, Nobel lecture, December 2014.

[NAT 15] NATARAJAN G., MATHEW A., NEGISHI Y. *et al.*, "A unified framework for understanding the structure and modifications of atomically precise monolayer protected gold clusters", *J. Phys. Chem. C*, vol. 119, pp. 27768–27785, 2015.

[NAZ 01] NAZEERUDDIN M.K., PÉCHY P., RENOUARD T. *et al.*, "Engineering of efficient panchromatic sensitizers for nanocrystalline TiO_2-based solar cells", *J. Am. Chem. Soc.*, vol. 123, pp. 1613–1624, 2001.

[NEG 05] NEGISHI Y., NOBUSADA K., TSUKUDA T., "Glutathione-protected gold clusters revisited: Bridging the gap between gold(i)-thiolate complexes and thiolate-protected gold nanocrystals", *J. Am. Chem. Soc.*, vol. 127, pp. 5261–5270, 2005.

[NIE 18] NIE K., HONG Y., QIU J. *et al.*, "Interfaces between cathode and electrolyte in solid state lithium batteries: Challenges and perspectives", *Front. Chem.*, vol. 6, no. 616, 2018.

[NIR 96] NIRMAL M., DABBOUSI B., BAWENDI M. *et al.*, "Fluorescence intermittency in single cadmium selenide nanocrystals", *Nature*, vol. 383, pp. 802–804, 1996.

[NIT 15] NITTA N., WU F., LEE J.T. *et al.*, "Li-ion battery materials: Present and future", *Mater. Today*, vol. 18, pp. 252–264, 2015.

[NOH 16] NOH T.H., JUNG O.-S., "Recent advances in various metal-organic channels for photochemistry beyond confined spaces", *Acc. Chem. Res.*, vol. 49, pp. 1835–1843, 2016.

[NOV 04] NOVOSELOV K.S., GEIM A.K., MOROZOV S.V. *et al.*, "Electric field effect in atomically thin carbon films", *Science*, vol. 306, pp. 666–669, 2004.

[NOV 05a] NOVOSELOV K.S., JIANG D., SCHEDIN F. *et al.*, "Two-dimensional atomic crystals", *Proc. Nat. Acad. Sci. USA*, vol. 10, pp. 10451–10453, 2005.

[NOV 05b] NOVOSELOV K.S., GEIM A.K., MOROZOV S.V. *et al.*, "Two-dimensional gas of massless Dirac fermions in graphene", *Nature*, vol. 438, pp. 197–200, 2005.

[ODO 98] ODOM T.W., HUANG J.-L., KIM P. *et al.*, "Atomic structure and electronic properties of single-walled carbon nanotubes", *Nature*, vol. 391, pp. 62–64, 1998.

[OHN 15] OHNSORG M.L., BEAUDOIN C.K., ANDERSON M.E., "Fundamentals of MOF thin film growth via liquid-phase epitaxy: Investigating the initiation of deposition and the influence of temperature", *Langmuir*, vol. 31, pp. 6114–6121, 2015.

[OKU 07] OKUBO M., HOSONO E., KIM J. *et al.*, "Nanosize effect on high-rate li-ion intercalation in $LiCoO_2$ electrode", *J. Am. Chem. Soc.*, vol. 129, pp. 7444–7452, 2007.

[OMA 11] OMACHI H., SEGAWA Y., ITAMI K., "Synthesis and racemization process of chiral carbon nanorings: A step toward the chemical synthesis of chiral carbon nanotubes", *Org. Lett.*, vol. 13, pp. 2480–2483, 2011.

[OMA 13] OMACHI H., NAKAYAMA T., TAKAHASHI E. *et al.*, "Initiation of carbon nanotube growth by well-defined carbon nanorings", *Nat. Chem.*, vol. 5, pp. 572–576, 2013.

[OMA 18] OMASTA T.J., WANG L., PENG X. *et al.*, "Importance of balancing membrane and electrode water in anion exchange membrane fuel cells", *J. Power Sources*, vol. 375, pp. 205–213, 2018.

[ORE 91] O'REGAN B., GRÄTZEL M., "A low-cost, high-efficiency solar cell based on dye-sensitized colloidal TiO_2 films", *Nature*, vol. 353, pp. 737–740, 1991.

[OWE 17] OWEN J., BRUS L., "Chemical synthesis and luminescence applications of colloidal semiconductor quantum dots", *J. Am. Chem. Soc.*, vol. 139, pp. 10939–10943, 2017.

[PAG 15] PAGE A.J., DING F., IRLE S. *et al.*, "Insights into carbon nano-tube and graphene formation mechanisms from molecular simulations: A review", *Rep. Prog. Phys.*, vol. 78, art. 03650, 2015.

[PAN 05] PANIZZA M., CERISOLA G., "Application of diamond electrodes to electrochemical processes", *Electrochimica Acta*, vol. 51, pp. 191–199, 2005.

[PAN 18] PAN C., BAO L., YU X. *et al.*, "Facile access to Y_2C_{2n} (2n = 92–130) and crystallographic characterization of Y_2C_2@C1(1660) -C108: A giant nanocapsule with a linear carbide cluster", *ACS Nano*, vol. 12, pp. 2065–2069, 2018.

[PAR 09] PARK S.H., ROY A., BEAUPRE S. *et al.*, "Bulk heterojunction solar cells with internal quantum efficiency approaching 100%", *Nature Photonics*, vol. 3, pp. 297–303, 2009.

[PAR 15] PARK J., ZHANG L., CHOI S.-I. *et al.*, "Atomic layer-by-layer deposition of platinum on palladium octahedra for enhanced catalysts toward the oxygen reduction reaction", *ACS Nano*, vol. 9, pp. 2635–2647, 2015.

[PAR 18] PARK J.-B., LEE S.H., JUNG H.-G. *et al.*, "Redox mediators for Li-O_2 batteries: Status and perspectives", *Adv. Mater.*, vol. 30, art. 1704162, 2018.

[PAT 14] PATON K.R., VARRLA E., BACKES C. *et al.*, "Scalable production of large quantities of defect-free few-layer graphene by shear exfoliation in liquids", *Nat. Mater.*, vol. 13, pp. 624–630, 2014.

[PEA 18] PEARSE A., SCHMITT T., SAHADEO E. *et al.*, "Three-dimensional solid-state lithium-ion batteries fabricated by conformal vapor-phase chemistry", *ACS Nano*, vol. 12, pp. 4286–4294, 2018.

[PEC 10] PECH D., BRUNET M., DUROU H. *et al.*, "Ultrahigh-power micrometre-sized supercapacitors based on onion-like carbon", *Nat. Nanotechn.*, vol. 5, pp. 651–654, 2010.

[PEI 19] PEI Y., WANG P., MA Z. *et al.*, "Growth-rule-guided structural exploration of thiolate-protected gold nanoclusters", *Acc. Chem. Res.*, vol. 52, pp. 23–33, 2019.

[PEN 12] PENG Z., FREUNBERGER S.A., CHEN Y. *et al.*, "A reversible and higher-rate Li-O_2 battery", *Science*, vol. 337, pp. 563–566, 2012.

[PEU 03] PEUMANS P., UCHIDA S., FORREST S.R., "Efficient bulk heterojunction photovoltaic cells using small molecular-weight organic thin films", *Nature*, vol. 425, pp. 158–162, 2003.

[PHA 20] PHAM-TRUONG T.-N., WANG Q., GHILANE J. *et al.*, "Recent advances in the development of organic and organometallic redox shuttles for lithium-ion redox flow batteries", *ChemSusChem*, vol. 13, pp. 2142–2159, 2020.

[PIH 15] PIHOSH Y., TURKEVYCH I., MAWATARI K. *et al.*, "Photocatalytic generation of hydrogen by core-shell $WO_3/BiVO_4$ nanorods with ultimate water splitting efficiency", *Scientific Reports*, vol. 5, art. 11141, 2015.

[PIL 03] PILENI M.P., "The role of soft colloidal templates in controlling the size and shape of inorganic nanocrystals", *Nat. Mater.*, vol. 2, pp. 145–149, 2003.

[PIL 17] PILLOT C., "The rechargeable battery market and main trends 2016–2025", *34th Annual International Battery Seminar and Exhibit*, Fort Lauderdale, USA, March 20–23, 2017.

[POL 12] POLMAN A., ATWATER H.A. "Photonic design principles for ultra-high efficiency photovoltaics", *Nat. Mater.*, vol. 11, pp. 174–177, 2012.

[PON 08] PONOMARENKO L.A., SCHEDIN F., KATSNELSON M.I. *et al.*, "Chaotic Dirac billiard in graphene quantum dots", *Science*, vol. 320, pp. 356–358, 2008.

[PON 12] PONROUCH A., MARCHANTE E., COURTY M. *et al.*, "In search of an optimized electrolyte for Na-ion batteries", *Energy Environ. Sci.*, vol. 5, pp. 8572–8583, 2012.

[POP 06] POP E., MANN D., WANG Q. *et al.*, "Thermal conductance of an individual single-wall carbon nanotube above room temperature", *Nano Lett.*, vol. 6, pp. 96–100, 2006.

[POP 13] POPCZUN E.J., MCKONE J.R., READ C.G. *et al.*, "Nanostructured nickel phosphide as an electrocatalyst for the hydrogen evolution reaction", *J. Am. Chem. Soc.*, vol. 135, pp. 9267–9270, 2013.

[POR 18] PORTEHAULT D., DELACROIX S., GOUGET G. *et al.*, "Beyond the compositional threshold of nanoparticle-based materials", *Acc. Chem. Res.*, vol. 51, pp. 930–939, 2018.

[POT 11] POTTS J.R., DREYER D.R., BIELAWSKI C.W. *et al.*, "Graphene-based polymer nanocomposites", *Polymer*, vol. 52, pp. 5–25, 2011.

[PRO 15] PROTESESCU L., YAKUNIN S., BODNARCHUK M.I. *et al.*, "Nanocrystals of Cesium Lead Halide Perovskites ($CsPbX_3$, X = Cl, Br, and I): Novel optoelectronic materials showing bright emission with wide color gamut", *Nano Lett.*, vol. 15, pp. 3692–3696, 2015.

[PU 18] PU Y., CAI F., WANG D. *et al.*, "Colloidal synthesis of semiconductor quantum dots toward large-scale production: A review", *Ind. Eng. Chem. Res.*, vol. 57, pp. 1790–1802, 2018.

[PUM 11] PUMERA M., "Graphene in biosensing", *Mater. Today*, vol. 14, pp. 308–315, 2011.

[QIA 12a] QIAN H., ZHU M., WU Z. *et al.*, "Quantum sized gold nanoclusters with atomic precision", *Acc. Chem. Res.*, vol. 45, pp. 1470–1479, 2012.

[QIA 12b] QIAN X.M., NING Z.Y., LI Y.L. *et al.*, "Construction of Graphdiyne nanowires with high-conductivity and mobility", *Dalton T*, vol. 41, pp. 730–733, 2012.

[QIA 18a] QIAN D., ZHENG Z., YAO H. *et al.*, "Design rules for minimizing voltage losses in high efficiency organic solar cells", *Nat. Mater.*, vol. 17, pp. 703–709, 2018.

[QIA 18b] QIAO Y., HE Y., WU S. *et al.*, "MOF-based separator in an Li-O_2 battery: An effective strategy to restrain the shuttling of dual redox mediators", *ACS Energy Lett.*, vol. 3, pp. 463–468, 2018.

[QIU 15] QIU H.J., ITO Y., CONG W. *et al.*, "Nanoporous graphene with single atom nickel dopants: An efficient and stable catalyst for electrochemical hydrogen production", *Angew. Chem. Int. Ed.*, vol. 54, pp. 14031–14035, 2015.

[QIU 17] QIU N., ZHANG H., WAN X. *et al.*, "A new nonfullerene electron acceptor with a ladder type backbone for high-performance organic solar cells", *Adv. Mat.*, vol. 29, art. 1604964, 2017.

[QU 14] QU B., MA C., JI G. *et al.*, "Layered SnS_2-reduced graphene oxide composite – A high-capacity, high-rate, and long-cycle life sodium-ion battery anode material", *Adv. Mater.*, vol. 26, pp. 3854–3859, 2014.

[QU 15] QU D., ZHENG M., LI J. *et al.*, "Tailoring color emissions from N-doped graphene quantum dots for bioimaging applications", *Light: Science & Applications*, vol. 4, art. e364, 2015.

[RAD 11] RADISAVLJEVIC B., RADENOVIC A., BRIVIO J. *et al.*, "Single-layer MoS2 transistors", *Nat. Nanotechn.*, vol. 6, pp. 147–150, 2011.

[RAM 16] RAMANUJAM J., VERMA A., GONZÁLEZ-DÍAZ B. *et al.*, "Inorganic photovoltaics – Planar and nanostructured devices", *Progr. Mater. Sci.*, vol. 82, pp. 294–404, 2016.

[REN 15a] REN C., "Polarisation fields in III-nitrides: Effects and control", *Materials Science and Technology*, vol. 32, 2015.

[REN 15b] REN Y.Y., CHEN K., CHEN R.J. *et al.*, "Oxide electrolytes for lithium batteries", *J. Am. Ceram. Soc.*, vol. 98, pp. 3603–3623, 2015.

[ROH 86] ROHRER H., BINNIG G., "Scanning tunneling microscopy. From birth to adolescence", *Nobel Lecture*, 1986.

[RUD 94] RUDGE A., DAVEY J., RAISTRICK I. *et al.*, "Conducting polymers as active materials in electrochemical capacitors", *J. Power Sources*, vol. 47, pp. 89–107, 1994.

[RUS 86] RUSKA E., "The development of the electron microscope and of early electron microscopy", *Nobel Lecture*, 1986.

[RUS 14] RUSSIER-ANTOINE I., BERTORELLE F., VOJKOVIC M. *et al.*, "Non-linear optical properties of gold quantum clusters. The smaller the better", *Nanoscale*, vol. 6, pp. 13572–13578, 2014.

[RYU 04] RYU J.H., KIM J.W., SUNG Y.-E. *et al.*, "Failure modes of silicon powder negative electrode in lithium secondary batteries", *Electrochemical and Solid-State Letters*, vol. 7, pp. A306–A309, 2004.

[SAB 15] SABIHUDDIN S., KIPRAKIS A.E., MUELLER M., "A numerical and graphical review of energy storage technologies", *Energies*, vol. 8, pp. 172–216, 2015.

[SAD 09] SADAKIYO M., YAMADA T., KITAGAWA H., "Rational designs for highly proton-conductive metal-organic frameworks", *J. Am. Chem. Soc.*, vol. 131, pp. 9906–9907, 2009.

[SAK 13] SAKAUSHI K., HOSONO E., NICKER G. *et al.*, "Aromatic porous-honeycomb electrodes for a sodium-organic energy storage device", *Nat. Commun.*, vol. 4, pp. 1485–1492, 2013.

[SAL 09] SALOT R., MARTIN S., OUKASSI S. *et al.*, "Microbattery technology overview and associated multilayer encapsulation process", *Appl. Surface Sci.*, vol. 256, pp. S54–S57, 2009.

[SAL 16] SALIBA M., MATSUI T., SEO J.-Y. *et al.*, "Cesium-containing triple cation perovskite solar cells: Improved stability, reproducibility and high efficiency", *Energy Environ. Sci.*, vol. 9, pp. 1989–1997, 2016.

[SAN 20] SANCHEZ S., JERONIMO-RENDON O., SALIBA M. *et al.*, "Highly efficient and rapid manufactured perovskite solar cells via Flash InfraRed Annealing", *Mater. Today*, vol. 35, pp. 9–15, 2020.

[SAR 92] SARICIFTCI N.S., SMILOWITZ L., HEEGER A.J. *et al.*, "Photoinduced electron transfer from a conducting polymer to Buckminsterfullerene", *Science*, vol. 258, pp. 1474–1476, 1992.

[SAR 13] SARAVANAN K., MASON C.W., RUDOLA A. *et al.*, "The first report on excellent cycling stability and superior rate capability of Na_3V_2 $(PO_4)_3$ for sodium ion batteries", *Adv. En. Mat.*, vol. 3, pp. 444–450, 2013.

[SAT 11] SATHIYA M., PRAKASH A.S., RAMESHA K. *et al.*, "V_2O_5-anchored carbon nanotubes for enhanced electrochemical energy storage", *J. Am. Chem. Soc.*, vol. 133, pp. 16291–16299, 2011.

[SAU 04] SAU T.K., MURPHY C.J., "Seeded high yield synthesis of short Au nanorods in aqueous solution", *Langmuir*, vol. 20, pp. 6414–6420, 2004.

[SCH 06] SCHUBERT F.E., *Light Emitting Diodes*, 2nd edition, Cambridge University Press, Cambridge, 2006.

[SCH 09] SCHWARTZ G., REINEKE S., ROSENOW T.C. *et al.*, "Triplet harvesting in hybrid white organic light-emitting diodes", *Adv. Funct. Mater.*, vol. 19, pp. 1319–1333, 2009.

[SCH 10] SCHWIERZ F., "Graphene transistors", *Nat. Nanotechnol.*, vol. 5, pp. 487–496, 2010.

[SCH 11] SCHNORR J.M., SWAGER T.M., "Emerging applications of carbon nanotubes", *Chem. Mater.*, vol. 23, pp. 646–657, 2011.

[SCH 14] SCHIRHAGL R., CHANG K., LORETZ M. *et al.*, "Nitrogen-vacancy centers in Diamond: Nanoscale sensors for physics and biology", *Annu. Rev. Phys. Chem.*, vol. 65, pp. 83–105, 2014.

[SCH 16] SCHALENBACH M., TJARKS G., CARMO M. *et al.*, "Acidic or alkaline? Towards a new perspective on the efficiency of water electrolysis", *J. Electrochem. Soc.*, vol. 163, pp. F3197–F3208, 2016.

[SEH 13] SEH W.Z., LI W., CHA J.J. *et al.*, "Sulphur-TiO2 yolk-shell nanoarchitecture with internal void space for long-cycle lithium-sulphur batteries", *Nat. Commun.*, vol. 4, no. 1331, 2013.

[SEI 14] SEINO Y., OTA T., TAKADA K. *et al.*, "A sulphide lithium super ion conductor is superior to liquid ion conductors for use in rechargeable batteries", *Energy Environ. Sci.*, vol. 7, pp. 627–631, 2014.

[SEO 08] SEO D., YOO C., CHUNG I.S. *et al.*, "Shape adjustment between multiply twinned and single-crystalline polyhedral gold nano-crystals: Decahedra, icosahedra, and truncated tetrahedral", *J. Phys. Chem. C*, vol. 112, pp. 2469–2475, 2008.

[SEO 09] SEO D., PARK J.H., JUNG J. *et al.*, "One-dimensional gold nanostructures through directed anisotropic overgrowth from gold decahedrons", *J. Phys. Chem. C*, vol. 113, pp. 3449–3454, 2009.

[SER 04] SERRE C., MILLANGE F., THOUVENOT C. *et al.*, "Synthesis, characterisation and luminescent properties of a new three-dimensional lanthanide trimesate: $M((C_6H_3)$–$(CO_2)_3)$ (M ~ Y, Ln) or MIL-78", *J. Mat. Chem.*, vol. 14, pp. 1540–1543, 2004.

[SHA 15] SHAO Y., EL-KADY M.F., WANG L.J. *et al.*, "Graphene-based materials for flexible supercapacitors", *Chem. Soc. Rev.*, vol. 44, pp. 3639–3665, 2015.

[SHA 16] SHAO Y., EL-KADY M.F., LIN C.-W. *et al.*, "3D freeze-casting of cellular graphene films for ultrahigh-power-density supercapacitors", *Adv. Mater.*, vol. 28, pp. 6719–6726, 2016.

[SHE 09] SHEKHAH O., WANG H., PARADINAS M. *et al.*, "Controlling interpenetration in metal-organic frameworks by liquid-phase epitaxy", *Nat. Mat.*, vol. 8, pp. 481–484, 2009.

[SHE 13a] SHEN L., ZHANG L., CHEN M. *et al.*, "The production of pH-sensitive photoluminescent carbon nanoparticles by the carbonization of polyethylenimine and their use for bioimaging", *Carbon*, vol. 55, pp. 343–349, 2013.

[SHE 13b] SHENG W., MYINT M., CHEN J.G. *et al.*, "Correlating the hydrogen evolution reaction activity in alkaline electrolytes with the hydrogen binding energy on monometallic surfaces", *Energy Environ. Sci.*, vol. 6, pp. 1509–1512, 2013.

[SHE 19] SHEN H., THOMAS T., RASAKI S.A. *et al.*, "Oxygen reduction reactions of Fe-N-C catalysts: Current status and the way forward", *Electrochem. Energy Rev.*, vol. 2, pp. 252–276, 2019.

[SHI 15] SHIN S., JIN Z., KWON D.H. *et al.*, "High turnover frequency of hydrogen evolution reaction on amorphous MoS2 thin film directly grown by atomic layer deposition", *Langmuir*, vol. 31, pp. 1196–1202, 2015.

[SIM 08] SIMON P., GOGOTSI Y., "Materials for electrochemical capacitors", *Nat. Mater.*, vol. 7, pp. 845–854, 2008.

[SIM 14] SIMON P., GOGOTSI Y., DUNN B., "Where do batteries end and supercapacitors begin?", *Sci. Magazine*, vol. 343, pp. 1210–1211, 2014.

[SIN 03] SINHA S., ROTHE C., GUNTNER R. *et al.*, "Electrophosphorescence and delayed electroluminescence from pristine poly fluorene thin-film devices at low temperature", *Phys. Rev. Lett.*, vol. 90, art. 127402, 2003.

[SIN 05] SINHA S., BARJAMI S., IANNACCHIONE G. *et al.*, "Off-axis thermal properties of carbon nanotube films", *J. Nanoparticle Res.*, vol. 7, pp. 651–657, 2005.

[SNA 13] SNAITH H.J., "Perovskites: The emergence of a new era for low-cost, high-efficiency solar cells", *J. Phys. Chem. Lett.*, vol. 4, pp. 3623–3630, 2013.

[SON 14] SONG F., HU X., "Ultrathin cobalt-manganese layered double hydroxide is an efficient oxygen evolution catalyst", *J. Am. Chem. Soc.*, vol. 136, pp. 16481–16484, 2014.

[SON 20] SONG J., WEI C., HUANG Z.-F. *et al.*, "A review on fundamentals for designing oxygen evolution electrocatalysts", *Chem. Soc. Rev.*, vol. 49, pp. 2196–2214, 2020.

[SPI 10] SPITALSKY Z., TASIS D., PAPAGELIS K. *et al.*, "Carbon nanotube-polymer composites: Chemistry, processing, mechanical and electrical properties", *Prog. Polymer Sci.*, vol. 35, pp. 357–401, 2010.

[STA 06] STANKOVICH S., DIKIN D.A., DOMMETT G.H.B. *et al.*, "Graphene-based composite materials", *Nature*, vol. 442, pp. 282–286, 2006.

[STE 99] STEVENSON S., RICE G., GLASS T. *et al.*, "Small-bandgap endohedral metallofullerenes in high yield and purity", *Nature*, vol. 401, pp. 55–57, 1999.

[STR 12] STRONG V., DUBIN S., EL-KADY M.F. *et al.*, "Patterning and electronic tuning of laser scribed graphene for flexible all-carbon devices", *ACS Nano*, vol. 6, pp. 1395–1403, 2012.

[SU 14] SU Y., KRAVETS V.G., WONG S.L. *et al.*, "Impermeable barrier films and protective coatings based on reduced graphene oxide", *Nat. Commun.*, vol. 5, no. 4843, 2014.

[SUB 11] SUBBARAMAN R., TRIPKOVIC D., STRMCNIK D. *et al.*, "Enhancing hydrogen evolution activity in water splitting by tailoring Li+-Ni $(OH)_2$-Pt interfaces", *Science*, vol. 334, pp. 1256–1260, 2011.

[SUE 17] SUEN N.-T., HUNG S.F., QUAN Q. *et al.*, "Electrocatalysis for the oxygen evolution reaction: Recent development and future perspectives", *Chem. Soc. Rev.*, vol. 46, pp. 337–365, 2017.

[SUM 12] SUMIDA K., ROGOW D.L., MASON J.A. *et al.*, "Carbon dioxide capture in metal-organic frameworks", *Chem. Rev.*, vol. 112, pp. 724–781, 2012.

[SUN 06] SUN Y.-P., ZHOU B., LIN Y. *et al.*, "Quantum-sized carbon dots for bright and colorful photoluminescence", *J. Am. Chem. Soc.*, vol. 128, pp. 7756–7757, 2006.

[SUN 10] SUN C., HUI R., ROLLER J., "Cathode materials for solid oxide fuel cells: A review", *J. Solid State Electrochem.*, vol. 14, pp. 1125–1144, 2010.

[SUN 13] SUN X., SUN H., LI H. *et al.*, "Developing polymer composite materials: Carbon nanotubes or graphene?", *Adv. Mater.*, vol. 25, pp. 5153–5176, 2013.

[SUR 17] SURE R., HANSEN A., SCHWERDTFEGER P. *et al.*, "Comprehensive theoretical study of all 1812 C_{60} isomers", *Phys. Chem. Chem. Phys.*, vol. 19, pp. 14296–14305, 2017.

[TAB 06] TABERNA P.L., MITRA S., POIZOT P. *et al.*, "High rate capabilities Fe_3O_4-based Cu nano-architectured electrodes for lithium-ion battery applications", *Nat. Mater.*, vol. 5, pp. 567–573, 2006.

[TAN 86] TANG C.W., "Two-layer organic photovoltaic cell", *Appl. Phys. Lett.*, vol. 48, pp. 183–185, 1986.

[TAN 87] TANG C.W., VANSLYKE S.A., "Organic electroluminescent diodes", *Appl. Phys. Lett.*, vol. 51, pp. 913–915, 1987.

[TAN 15] TAN C., ZHANG H., "Two-dimensional transition metal dichalcogenide nanosheet-based composites", *Chem. Soc. Rev.*, vol. 44, pp. 2713–2731, 2015.

[TAO 17] TAO L., LIN C.-Y., DOU S. *et al.*, "Creating coordinatively unsaturated metal sites in metal-organic frameworks as efficient electrocatalysts for the oxygen evolution reaction: Insights into the active centers", *Nano Energy*, vol. 41, pp. 417–425, 2017.

[TAR 18] TARASCON J.M., "The Li-ion battery: 25 years of exciting and enriching experiences", *Electrochem. Soc. Interface*, pp. 79–83, 2018.

[TAY 93] TAYLOR R., WALTON D.R.M., "The chemistry of fullerenes", *Nature*, vol. 363, pp. 685–693, 1993.

[TES 11] TESSONNIER J.P., SU D.S., "Recent progress on the growth mechanism of carbon nanotubes: A review", *ChemSusChem*, vol. 4, pp. 824–847, 2011.

[TIA 18] TIAN P., TANG L., TENG K.S. *et al.*, "Graphene quantum dots from chemistry to applications", *Mater. Today Chem.*, vol. 10, pp. 221–258, 2018.

[TIK 16] TIKEKAR M.D., CHOUDHURY S., TU Z. *et al.*, "Design principles for electrolytes and interfaces for stable lithium-metal batteries", *Nat. Energy*, vol. 1, no. 16114, 2016.

[TON 19] TONG J., SONG Z., KIM D.H. *et al.*, "Carrier lifetimes of >1 ms in Sn-Pb perovskites enable efficient all-perovskite tandem solar cells", *Science*, vol. 364, pp. 475–479, 2019.

[TOS 14] TOSUN M., CHUANG S., FANG H. *et al.*, "High-gain inverters based on WSe_2 complementary field-effect transistors", *ACS Nano*, vol. 8, pp. 4948–4953, 2014.

[TSA 14] TSAO J.Y., CRAWFORD M.H., COLTRIN M.E. *et al.*, "Toward smart and ultra-efficient solid-state lighting", *Adv. Optical Mater.*, vol. 2, pp. 809–836, 2014.

[TUN 11] TUNG V.C., KIM J., COTE L.J. *et al.*, "Sticky interconnect for solution-processed tandem solar cells", *J. Am. Chem. Soc.*, vol. 133, pp. 9262–9265, 2011.

[TUR 51] TURKEVICH J., STEVENSON P.C., HILLIER J., "A study of the nucleation and growth processes in the synthesis of colloidal gold", *Discuss. Faraday Soc.*, vol. 11, pp. 55–75, 1951.

[UME 19] UMESHBABU E., ZHENG B., YANG Y., "Recent progress in all-solid-state lithium-sulfur batteries using high Li-ion conductive solid electrolytes", *Electrochem. Energy Rev.*, vol. 2, pp. 199–230, 2019.

[UNI 18] UNITED NATIONS, Energy statistics pocket book, Statistic Division, available at: https://unstats.un.org/unsd/energy, 2018.

[UOY 12] UOYAMA H., GOUSHI K., SHIZU K. *et al.*, "Highly efficient organic light-emitting diodes from delayed fluorescence", *Nature*, vol. 492, pp. 234–240, 2012.

[VAL 17] VALAPPIL M.O., PILLAI V.K., ALWARAPPAN S., "Spotlighting graphene quantum dots and beyond: Synthesis, properties and sensing applications", *Appl. Mater. Today*, vol. 9, pp. 350–371, 2017.

[VAN 17] VAN AKEN K.L., MALESKI K., MATHIS T.S. *et al.*, "Processing of onion-like carbon for electrochemical capacitors", *ECS J. Solid State Sci. Technol.*, vol. 6, pp. M3103–M3108, 2017.

[VLA 13] VLASSIOUK I., FULVIO P., MEYER H. *et al.*, "Large scale atmospheric pressure chemical vapor deposition of graphene", *Carbon*, vol. 54, pp. 58–67, 2013.

[VRU 12] VRUBEL H., HU X.L., "Molybdenum boride and carbide catalyze hydrogen evolution in both acidic and basic solutions", *Angew. Chem. Int. Ed.*, vol. 51, pp. 12703–12706, 2012.

[WAD 14] WADA T., ICHITSUBO T., YUBUTA K. *et al.*, "Bulk-nanoporous-silicon negative electrode with extremely high cyclability for lithium-ion batteries prepared using a top-down process", *Nano Lett.*, vol. 14, pp. 4505–4510, 2014.

[WAD 19] WADSWORTH A., MOSER M., MARKS A. *et al.*, "Critical review of the molecular design progress in non-fullerene electron acceptors towards commercially viable organic solar cells", *Chem. Soc. Rev.*, vol. 48, pp. 1596–1625, 2019.

[WAL 11] WALL M., The Raman spectroscopy of graphene and the determination of layer thickness, Note 52252, Madison, 2011.

[WAL 17] WALKOWIAK-KULIKOWSKA J., WOLSKA J., KORONIAK H., "Polymers application in proton exchange membranes for fuel cells (PEMFCs)", *Physics Sci. Rev.*, vol. 2, art. 20170018 2017.

[WAN 10] WANG X., CAO L., YANG S.T. *et al.*, "Bandgap-like strong fluorescence in functionalized carbon nanoparticles", *Angew. Chem., Int. Ed.*, vol. 49, pp. 5310–5314, 2010.

[WAN 11] WANG T., JEPPSON K., YE L. *et al.*, "Carbon-nanotube through-silicon via interconnects for three-dimensional integration", *Small*, vol. 7, pp. 2313–2317, 2011.

[WAN 12a] WANG C., WANG Y., XU L. *et al.*, "Facile aqueous-phase synthesis of biocompatible and fluorescent Ag_2S nanoclusters for bioimaging: Tunable photoluminescence from red to near infrared", *Small*, vol. 8, pp. 3137–3142, 2012.

[WAN 12b] WANG Q.H., KALANTAR-ZADEH K., KIS A. *et al.*, "Electronics and optoelectronics of two-dimensional transition metal dichalcogenides", *Nat. Nanotechnol.*, vol. 7, pp. 699–712, 2012.

[WAN 14a] WANG S., MENG X., DAS A. *et al.*, "A 200-fold quantum yield boost in the photoluminescence of silver-doped Ag_xAu_{25-x} nanoclusters: The 13th silver atom matters", *Angew. Chem. Int. Ed.*, vol. 53, pp. 2376–2380, 2014.

[WAN 14b] WANG Y., HU A., "Carbon quantum dots: Synthesis, properties and applications", *J. Mater. Chem. C*, vol. 2, pp. 6921–6939, 2014.

[WAN 15a] WANG H., YUAN Y., WEI L. *et al.*, "Catalysts for chirality selective synthesis of single-walled carbon nanotubes", *Carbon*, vol. 81, pp. 1–19, 2015.

[WAN 15b] WANG W., FAVORS Z., IONESCU R. *et al.*, "Monodisperse porous silicon spheres as anode materials for lithium ion batteries", *Sci. Rep.*, vol. 5, no. 8781, 2015.

[WAN 15c] WANG X., SUN L., HU X. *et al.*, "Ni-Si nanosheet network as high performance anode for Li ion batteries", *J. Power Sources*, vol. 280, pp. 393–396, 2015.

[WAN 16a] WANG D., WRIGHT M., ELUMALAI N.K. *et al.*, "Stability of perovskite solar cells", *Sol. Energy Mater. Sol. Cells*, vol. 147, pp. 255–275, 2016.

[WAN 16b] WANG K., LIU C., MENG T. *et al.*, "Inverted organic photovoltaic cells", *Chem. Soc. Rev.*, vol. 45, pp. 2937–2975, 2016.

[WAN 17] WANG N., HE J., TU Z. *et al.*, "Synthesis of chlorine-substituted graphdiyne and applications for lithium-ion storage", *Angew. Chem. Int. Ed.*, vol. 56, pp. 10740–10745, 2017.

[WAN 18] WANG H., YANG L., SUI X. *et al.*, "Selective synthesis of single walled carbon nanotubes on metal (iron, nickel or cobalt) sulfate-based catalysts", *Carbon*, vol. 129, pp. 128–136, 2018.

[WAN 19a] WANG D., MU X., HE P. *et al.*, "Materials for advanced Li-O_2 batteries: Explorations, challenges and prospects", *Mater. Today*, vol. 26, pp. 87–99, 2019.

[WAN 19b] WANG L., CHEN M., KUNGAS R. *et al.*, "Power-to-fuels via solid-oxide electrolyzer: Operating window and technoeconomics", *Renewable Sustainable Energy Rev.*, vol. 110, pp. 174–187, 2019.

[WAN 19c] WANG X., LI Z., QU Y. *et al.*, "Review of metal catalysts for oxygen reduction reaction: From nanoscale engineering to atomic design", *Chem*, vol. 5, pp. 1486–1511, 2019.

[WEI 01] WEI B.Q., VAJTAI R., AJAYAN P.M., "Reliability and current carrying capacity of carbon nanotubes", *Appl. Phys. Lett.*, vol. 79, pp. 1172–1174, 2001.

[WEI 15] WEISBUCH C., PICCARDO M., MARTINELLI L. *et al.*, "The efficiency challenge of nitride light-emitting diodes for lighting", *Phys. Status Solidi A*, vol. 212, pp. 899–913, 2015.

[WEI 18] WEISBUCH C., "Historical perspective on the physics of artificial lighting", *C. R. Physique*, vol. 19, pp. 89–112, 2018.

[WEI 20] WEISBUCH C., "Review – On the search for efficient solid state light emitters: Past, present, future", *ECS J. Solid State Sci. Technol.*, vol. 9, art. 016022, 2020.

[WEN 10] WEN Q., QIAN W., NIE J. *et al.*, "100 mm long, semiconducting triple-walled carbon nanotubes", *Adv. Mater.*, vol. 22, pp. 1867–1871, 2010.

[WHE 96] WHETTEN R.L., KHOURY J.T., ALVAREZ M.M. *et al.*, "Nanocrystal gold molecules", *Adv. Mater.*, vol. 8, pp. 428–433, 1996.

[WHE 19] WHETTEN R.L., WEISSKER H.-C., PELAYO J.J. *et al.*, "Chiral-icosahedral (i) symmetry in ubiquitous metallic cluster compounds (145A,60X): Structure and bonding principles", *Acc. Chem. Res.*, vol. 52, pp. 34–43, 2019.

[WHI 76] WHITTINGHAM M.S., "Electrical energy storage and intercalation chemistry", *Science*, vol. 192, pp. 1126–1127, 1976.

[WIL 15] WILD M., O'NEILL L., ZHANG T. *et al.*, "Lithium sulfur batteries, a mechanistic review", *Energy Environ. Sci.*, vol. 8, pp. 3477–3494, 2015.

[WU 10] WU Z., JIN R., "On the ligand's role in the fluorescence of gold nanoclusters", *Nano Lett.*, vol. 10, pp. 2568–2573, 2010.

[WU 12a] WU H., CHAN G., CHOI J.W. *et al.*, "Stable cycling of double-walled silicon nanotube battery anodes through solid-electrolyte interphase control", *Nature Nanotechnol.*, vol. 7, pp. 310–315, 2012.

[WU 12b] WU Z.S., YANG S., SUN Y. *et al.*, "3D nitrogen-doped graphene aerogel-supported Fe_3O_4 nanoparticles as efficient electrocatalysts for the oxygen reduction reaction", *J. Am. Chem. Soc.*, vol. 134, pp. 9082–9085, 2012.

[WU 13] WU J., YANG H., "Platinum-based oxygen reduction electrocatalysts", *Acc. Chem. Research*, vol. 46, pp. 1848–1857, 2013.

[WU 18] WU W., PU J., WANG J. *et al.*, "Biomimetic bipolar microcapsules derived from *Staphylococcus aureus* for enhanced properties of lithium-sulfur battery cathodes", *Adv. Energy Mater.*, vol. 8, art. 1702373, 2018.

[XIA 12] XIANG Q., YU J., JARONIEC M., "Synergetic effect of MoS_2 and graphene as cocatalysts for enhanced photocatalytic H_2 production activity of TiO_2 nanoparticles", *J. Am. Chem. Soc.*, vol. 134, pp. 6575–6578, 2012.

[XIA 17a] XIA Y., GILROY K.D., PENG H.-C. *et al.*, "Seed-mediated growth of colloidal metal nanocrystals", *Angew. Chem. Int. Ed.*, vol. 56, pp. 60–95, 2017.

[XIA 17b] XIA Y., YANG X., "Toward cost-effective and sustainable use of precious metals in heterogeneous catalysts", *Acc. Chem. Res.*, vol. 50, pp. 450–454, 2017.

[XIA 18a] XIA S., WANG Y., LIU Y. *et al.*, "Ultrathin MoS_2 nanosheets tightly anchoring onto nitrogen-doped graphene for enhanced lithium storage properties", *Chem. Eng. J.*, vol. 332, pp. 431–439, 2018.

[XIA 18b] XIAO B., SOTO F.A., GU M. *et al.*, "Lithium-pretreated hard carbon as high-performance sodium-ion battery anodes", *Adv. Energy Mater.*, vol. 8, art. 1801441, 2018.

[XIA 19a] XIA S., WU X., ZHANG Z. *et al.*, "Practical challenges and future perspectives of all-solid-state lithium-metal batteries", *Chem*, vol. 5, pp. 1–33, 2019.

[XIA 19b] XIAO B., ROJO T., LI X., "Hard carbon as sodium-ion battery anodes: Progress and challenges", *ChemSusChem*, vol. 12, pp. 133–144, 2019.

[XIE 92] XIE Q., PÉREZ-CORDERO E., ECHEGOYEN L., "Electrochemical detection of $C_{60}{}^{6-}$ and $C_{70}{}^{6-}$: Enhanced stability of fullerides in solution", *J. Am. Chem. Soc.*, vol. 114, pp. 3978–3980, 1992.

[XIE 14] XIE S., CHOI S.-I., LU N. *et al.*, "Atomic layer-by-layer deposition of Pt on Pd nanocubes for catalysts with enhanced activity and durability toward oxygen reduction", *Nano Lett.*, vol. 14, pp. 3570–3576, 2014.

[XIN 17] XIN S., YOU Y., WANG S. *et al.*, "Solid-state lithium metal batteries promoted by nanotechnology: Progress and prospects", *ACS Energy Lett.*, vol. 2, pp. 1385–1394, 2017.

[XIN 18] XIN H., LI W., "A review on high throughput roll-to-roll manufacturing chemical vapor deposition graphene", *Appl. Phys. Rev.*, vol. 5, art. 031105, 2018.

[XIO 19] XIONG P., MA R., WANG G. *et al.*, "Progress and perspective on two-dimensional unilamellar metal oxide nanosheets and tailored nanostructures from them for electrochemical energy storage", *Energy Storage Mater.*, vol. 19, pp. 281–298, 2019.

[XU 04] XU X.Y., RAY R., GU Y.L. *et al.*, "Electrophoretic analysis and purification of fluorescent single-walled carbon nanotube fragments", *J. Am. Chem. Soc.*, vol. 126, pp. 12736–12737, 2004.

[XU 11] XU Z., GAO C., "Aqueous liquid crystals of graphene oxide", *ACS Nano*, vol. 5, pp. 2908–2915, 2011.

[XU 16] XU Z., LIU Y., ZHAO X. *et al.*, "Ultrastiff and strong graphene fibers via full-scale synergetic defect engineering", *Adv. Mater.*, vol. 28, pp. 6449–6456, 2016.

[XU 18] XU Y., ZHOU M., LEI Y., "Organic materials for rechargeable sodium-ion batteries", *Materials Today*, vol. 21, pp. 60–78, 2018.

[XU 19a] XU C., ANUSUYADEVI P.R., AYMONIER C. *et al.*, "Nano-structured materials for photocatalysis", *Chem. Soc. Rev.*, vol. 48, pp. 3868–3902, 2019.

[XU 19b] XU H., CHIEN P.-H., SHI J. *et al.*, "High-performance all-solid-state batteries enabled by salt bonding to perovskite in poly(ethylene oxide)", *Proc. Natl. Acad. Sci.*, vol. 116, pp. 18815–18821, 2019.

[XUE 18] XUE Y., HUANG B., YI Y. *et al.*, "Anchoring zero valence single atoms of nickel and iron on graphdiyne for hydrogen evolution", *Nat. Commun.*, vol. 9, no. 1460, 2018.

[YAB 14] YABUUCHI N., KUBOTA K., DAHBI M. *et al.*, "Research development on sodium-ion batteries", *Chem. Rev.*, vol. 114, pp. 11636–11682, 2014.

[YAG 03] YAGHI O.M., O'KEEFFE M., OCKWIG N.W. *et al.*, "Reticular synthesis and the design of new materials", *Nature*, vol. 423, pp. 705–714, 2003.

[YAM 13] YAMADA T., ISHIHARA M., HASEGAWA M., "Large area coating of graphene at low temperature using a roll-to-roll microwave plasma chemical vapor deposition", *Thin Solid Films*, vol. 532, pp. 89–93, 2013.

[YAN 05] YANG Y.A., WU H., WILLIAMS K.R. *et al.*, "Synthesis of CdSe and CdTe nanocrystals without precursor injection", *Angew. Chem. Int. Ed.*, vol. 44, pp. 6712–6715, 2005.

[YAN 13a] YANG H.-Y., ZHAO Y.-W., ZHANG Z.-Y. *et al.*, "One-pot synthesis of water-dispersible Ag_2S quantum dots with bright fluorescent emission in the second near-infrared window", *Nanotechnology*, vol. 24, art. 055706, 2013.

[YAN 13b] YANG X.-F., WANG A., QIAO B. *et al.*, "Single-atom catalysts: A new frontier in heterogeneous catalysis", *Acc. Chem. Res.*, vol. 46, pp. 1740–1748, 2013.

[YAN 13c] YANG Z., LIU M., ZHANG C. *et al.*, "Carbon nanotubes bridged with graphene nanoribbons and their use in high-efficiency dye-sensitized solar cells", *Angew. Chem. Int. Ed.*, vol. 52, pp. 3996–3999, 2013.

[YAN 15] YANG F., WANG X., ZHANG D. *et al.*, "Growing zigzag (16,0) carbon nanotubes with structure-defined catalysts", *J. Am. Chem. Soc.*, vol. 137, pp. 8688–8691, 2015.

[YAN 16] YANG J., CHEN F., LI C. *et al.*, "A free-standing sulfur-doped microporous carbon interlayer derived from luffa sponge for high performance lithium-sulfur batteries", *J. Mater. Chem. A*, vol. 4, pp. 14324–14333, 2016.

[YAN 17] YANG S., WEI T., JIN F., "When metal clusters meet carbon cages: Endohedral clusterfullerenes", *Chem. Soc. Rev.*, vol. 46, pp. 5005–5058, 2017.

[YAN 18] YANG L., CHENG D., XU H. *et al.*, "Unveiling the high-activity origin of single-atom iron catalysts for oxygen reduction reaction", *Proc. Nat. Acad. Sci. USA*, vol. 115, pp. 6626–6631, 2018.

[YAO 14] YAO J., YANG M., DUAN Y., "Chemistry, biology, and medicine of fluorescent nanomaterials and related systems: New insights into biosensing, bioimaging, genomics, diagnostics, and therapy", *Chem. Rev.*, vol. 114, pp. 6130–6178, 2014.

[YAO 19] YAO H., CUI Y., QIAN D. *et al.*, "14.7% efficiency organic photovoltaic cells enabled by active materials with a large electrostatic potential difference", *J. Am. Chem. Soc.*, vol. 141, pp. 7743–7750, 2019.

[YAR 19] YARTS V.A., LOTOTSKYY M.V., AKIBA E. *et al.*, "Magnesium based materials for hydrogen based energy storage: Past, present and future", *Int. J. Hydrogen Energy*, vol. 44, pp. 7809–7859, 2019

[YAZ 83] YAZAMI R., TOUZAIN P., "A reversible graphite-lithium negative electrode for electrochemical generators", *J. Power Sources*, vol. 9, pp. 365–371, 1983.

[YE 15a] YE M., WEN X., WANG M. *et al.*, "Recent advances in dye-sensitized solar cells: From photoanodes, sensitizers and electrolytes to counter electrodes", *Mater. Today*, vol. 18, pp. 155–162, 2015.

[YE 15b] YE S., ZHOU T., CHENG K. *et al.*, "Carboxylic acid fullerene (C_{60}) derivatives attenuated neuroinflammatory responses by modulating mitochondrial dynamics", *Nanoscale Res. Lett.*, vol. 10, no. 246, 2015.

[YEL 11] YELLA A., LEE H.-W., TSAO H.N. *et al.*, "Porphyrin-sensitized solar cells with cobalt (II/III)–based redox electrolyte exceed 12 percent efficiency", *Science*, vol. 334, pp. 629–635, 2011.

[YEL 13] YELLA A., HUMPHRY-BAKER R., CURCHOD B.F.E. *et al.*, "Molecular engineering of a fluorene donor for dye-sensitized solar cells", *Chem. Mater.*, vol. 25, pp. 2733–2739, 2013.

[YIN 20] YIN H., YAN C., HU H. *et al.*, "Recent progress of all-polymer solar cells: From chemical structure and device physics to photovoltaic performance", *Materials Science & Engineering*, vol. 140, art. 100542, 2020.

[YOS 87] YOSHINO A., SANECHIKA K., NAKAJIMA T., United States Patent, no. 4668595, May 26, 1987.

[YU 95] YU G., HEEGER A.J., "Charge separation and photovoltaic conversion in polymer composites with internal donor/acceptor heterojunctions", *J. Appl. Phys.*, vol. 78, pp. 4510–4515, 1995.

[YU 14] YU J., ZHENG Y., HUANG J., "Towards high performance organic photovoltaic cells: A review of recent development in organic photovoltaics", *Polymers*, vol. 6, pp. 2473–2509, 2014.

[YU 17] YU X., MANTHIRAM A., "Electrode-electrolyte interfaces in lithium-sulfur batteries with liquid or inorganic solid electrolytes", *Acc. Chem. Res.*, vol. 50, pp. 2653–2660, 2017.

[YU 19] YU H., XUE Y., HUANG B. *et al.*, "Ultrathin nanosheet of graphdiyne-supported palladium atom catalyst for efficient hydrogen production", *iScience*, vol. 11, pp. 31–41, 2019.

[YUA 19] YUAN H., LIU T., LIU Y. *et al.*, "A review of biomass materials for advanced lithium-sulfur batteries", *Chem. Sci.*, vol. 10, pp. 7484–7495, 2019.

[YUN 18] YUN S., QIN Y., UHL A.R. *et al.*, "New-generation integrated devices based on dye-sensitized and perovskite solar cells", *Energy Environ. Sci.*, vol. 11, pp. 476–526, 2018.

[ZEI 16] ZEIGER M., JÄCKEL N., MOCHALIN V.N. *et al.*, "Review: Carbon onions for electrochemical energy storage", *J. Mater. Chem. A*, vol. 4, pp. 3172–3196, 2016.

[ZEN 14] ZENG M., TAN L., WANG J. *et al.*, "Liquid metal: An innovative solution to uniform graphene films", *Chem. Mater.*, vol. 26, pp. 3637–3643, 2014.

[ZEN 15] ZENG M., LI Y., "Recent advances in heterogeneous electrocatalysts for the hydrogen evolution reaction", *J. Mater. Chem. A*, vol. 3, pp. 14942–14962, 2015.

[ZEN 16] ZENG M., TAN L., WANG L. *et al.*, "Isotropic growth of graphene toward smoothing stitching", *ACS Nano*, vol. 10, pp. 7189–7196, 2016.

[ZEN 18] ZENG M., FU L., "Controllable fabrication of graphene and related two-dimensional materials on liquid metals via chemical vapor deposition", *Acc. Chem. Res.*, vol. 51, pp. 2839–2847, 2018.

[ZHA 08] ZHANG H., CAO G., WANG Z. *et al.*, "Growth of manganese oxide nanoflowers on vertically aligned carbon nanotube arrays for high-rate electrochemical capacitive energy storage", *Nano Lett.*, vol. 8, pp. 2664–2668, 2008.

[ZHA 11] ZHANG R., WEN Q., QIAN W. *et al.*, "Superstrong ultralong carbon nanotubes for mechanical energy storage", *Adv. Mater.*, vol. 23, pp. 3387–3391, 2011.

[ZHA 12] ZHAI X., ZHANG P., LIU C. *et al.*, "Highly luminescent carbon nanodots by microwave-assisted pyrolysis", *Chem. Commun.*, vol. 48, pp. 7955–7957, 2012.

[ZHA 13] ZHANG Y., ZHANG L., ZHOU C., "Review of chemical vapor deposition of graphene and related applications", *Acc. Chem. Res.*, vol. 46, pp. 2329–2339, 2013.

[ZHA 14] ZHANG L., WANG E., "Metal nanoclusters: New fluorescent probes for sensors and bioimaging", *Nano Today*, vol. 9, pp. 132–157, 2014.

[ZHA 15] ZHANG Q., JIE J., DIAO S. *et al.*, "Solution-processed graphene quantum dot deep-UV photodetectors", *ACS Nano*, vol. 9, pp. 1561–1570, 2015.

[ZHA 16a] ZHANG S., CAI Y., HE H. *et al.*, "Heteroatom doped graphdiyne as efficient metal free electrocatalyst for oxygen reduction reaction in alkaline medium", *J. Mater. Chem. A*, vol. 4, pp. 4738–4744, 2016.

[ZHA 16b] ZHAO S., WANG Y., DONG J. *et al.*, "Ultrathin metal-organic framework nanosheets for electrocatalytic oxygen evolution", *Nat. Energy*, vol. 1, no. 16184, 2016.

[ZHA 16c] ZHAO Y., ZHU H., ZHU Q. *et al.*, "Three-in-one: Sensing, self-assembly and cascade catalysis of cyclodextrin modified gold nanoparticles", *J. Am. Chem. Soc.*, vol. 138, pp. 16645–16654, 2016.

[ZHA 17a] ZHANG K., FENG Y., WANG F. *et al.*, "Two dimensional hexagonal boron nitride (2D-hBN): Synthesis, properties and applications", *J. Mater. Chem. C*, vol. 5, pp. 11992–12022, 2017.

[ZHA 17b] ZHANG L., HAN L., LIU H. *et al.*, "Potential-cycling synthesis of single platinum atoms for efficient hydrogen evolution in neutral media", *Angew. Chem. Int. Ed.*, vol. 56, pp. 13694–13698, 2017.

[ZHA 17c] ZHANG R., ZHANG Y., WEI F., "Controlled synthesis of ultralong carbon nanotubes with perfect structures and extraordinary properties", *Acc. Chem. Res.*, vol. 50, pp. 179–189, 2017.

[ZHA 17d] ZHANG X., LIU T., ZHANG S. *et al.*, "Synergistic coupling between $Li_{6.75}La_3Zr_{1.75}Ta_{0.25}O_{12}$ and poly(vinylidene fluoride) induces high ionic conductivity, mechanical strength, and thermal stability of solid composite electrolytes", *J. Am. Chem. Soc.*, vol. 139, pp. 13779–13785, 2017.

[ZHA 17e] ZHANG Z., ZHAO Y., CHEN S. *et al.*, "An advanced construction strategy of all-solid state lithium batteries with excellent interfacial compatibility and ultralong cycle life", *J. Mater. Chem. A*, vol. 5, pp. 16984–16993, 2017.

[ZHA 17f] ZHAO W., LI S., YAO H. *et al.*, "Molecular optimization enables over 13% efficiency in organic solar cells", *J. Am. Chem. Soc.*, vol. 139, pp. 7148–7151, 2017.

[ZHA 18a] ZHANG L., JIA Y., GAO G. *et al.*, "Graphene defects trap atomic Ni species for hydrogen and oxygen evolution reactions", *Chem*, vol. 4, pp. 1–13, 2018.

[ZHA 18b] ZHANG P., ZHAO Y., ZHANG X., "Functional and stability orientation synthesis of materials and structures in aprotic Li-O_2 batteries", *Chem. Soc. Rev.*, vol. 47, pp. 2921–3004, 2018.

[ZHA 18c] ZHANG Y., LIU T., ZHANG Q. *et al.*, "High-performance all-solid-state lithium-sulfur batteries with sulfur/carbon nano-hybrids in a composite cathode", *J. Mater. Chem. A*, vol. 6, pp. 23345–23356, 2018.

[ZHA 18d] ZHANG Y., MURTAZA I., MENG H., "Development of fullerenes and their derivatives as semiconductors in field-effect transistors: Exploring the molecular design", *J. Mater. Chem. C*, vol. 6, pp. 3514–3537, 2018.

[ZHA 18e] ZHANG Z., SHAO Y., LOTSCH B. *et al.*, "New horizons for inorganic solid state ion conductors", *Energy Environ. Sci.*, vol. 11, pp. 1945–1976, 2018.

[ZHA 19a] ZHANG C., PARK S.-H., SERAL-ASCASO A. *et al.*, "High capacity silicon anodes enabled by MXene viscous aqueous ink", *Nat. Commun.*, vol. 10, no. 849, 2019.

[ZHA 19b] ZHAO W., YI J., HE P. *et al.*, "Solid-state electrolytes for lithium-ion batteries: Fundamentals, challenges and perspectives", *Electrochem. Energy Rev.*, vol. 2, pp. 574–605, available at: https://doi.org/10.1007/s41918-019-00048-02019, 2019.

[ZHA 21] ZHANG J., LAI Y.Y., LI P. *et al.*, "Boosting rate and cycling performance of K-doped $Na_3V_2(PO_4)_2F_3$ cathode for high-energy-density sodium-ion batteries", *Green Energy & Environ.*, February 1, 2021.

[ZHE 04] ZHENG J., ZHANG C., DICKSON R.M., "Highly fluorescent, water-soluble, size-tunable gold quantum dots", *Phys. Rev. Lett.*, vol. 93, art. 077402, 2004.

[ZHE 12] ZHENG J., ZHOU C., YU M. *et al.*, "Different sized luminescent gold nanoparticles", *Nanoscale*, vol. 4, pp. 4073–4083, 2012.

[ZHE 14a] ZHENG M., LIU S., LI J. *et al.*, "Integrating oxaliplatin with highly luminescent carbon dots: An unprecedented theranostic agent for personalized medicine", *Adv. Mater.*, vol. 26, pp. 3554–3560, 2014.

[ZHE 14b] ZHENG Y., ZHONG X., LI Z. *et al.*, "Successive, seed-mediated growth for the synthesis of single-crystal gold nanospheres with uniform diameters controlled in the range of 5 – 150 nm", *Part. Part. Syst. Char.*, vol. 31, pp. 266–273, 2014.

[ZHE 16] ZHENG Y., JIAO Y., ZHU Y. *et al.*, "High electrocatalytic hydrogen evolution activity of an anomalous ruthenium catalyst", *J. Am. Chem. Soc.*, vol. 138, pp. 16174–16181, 2016.

[ZHE 17] ZHENG Y., LAI L., LIU W. *et al.*, "Recent advances in bio-medical applications of fluorescent gold nanoclusters", *Adv. Colloid Interface Sci.*, vol. 242, pp. 1–16, 2017.

[ZHE 18] ZHENG F., KOTOBUKI M., SONG S. *et al.*, "Review on solid electrolytes for all-solid-state lithium-ion batteries", *J. Power Sources*, vol. 389, pp.198–213, 2018.

[ZHO 14] ZHOU H., CHEN Q., LI G. *et al.*, "Interface engineering of highly efficient perovskite solar cells", *Science*, vol. 345, pp. 542–546, 2014.

[ZHO 18] ZHOU K.-G., VASU K.S., CHERIAN C.T. *et al.*, "Electrically controlled water permeation through graphene oxide membranes", *Nature*, vol. 559, pp. 236–240, 2018.

[ZHU 08] ZHU M., LANNI E., GARG N. *et al.*, "Kinetically controlled, high-yield synthesis of Au_{25} clusters", *J. Am. Chem. Soc.*, vol. 130, pp. 1138–1139, 2008.

[ZHU 13] ZHU S.J., MENG Q.N., WANG L. *et al.*, "Highly photoluminescent carbon dots for multicolor patterning, sensors, and bioimaging", *Angew. Chem. Int. Ed.*, vol. 52, pp. 3953–3957, 2013.

[ZHU 14a] ZHU C., MU X., VAN NAKEN P.A. *et al.*, "Single-layered ultrasmall nanoplates of MoS_2 embedded in carbon nanofibers with excellent electro-chemical performance for lithium and sodium storage", *Angew. Chem. Int. Ed.*, vol. 53, pp. 2152–2156, 2014.

[ZHU 14b] ZHU S., WANG L., LI B. *et al.*, "Investigation of photoluminescence mechanism of graphene quantum dots and evaluation of their assembly into polymer dots", *Carbon*, vol. 77, pp. 462–472, 2014.

[ZON 11] ZONG J., ZHU Y., YANG X. *et al.*, "Synthesis of photoluminescent carbogenic dots using mesoporous silica spheres as nanoreactors", *Chem. Commun.*, vol. 47, pp. 764–766, 2011.

[ZUO 17a] ZUO W., TANG L., XIANG J. *et al.*, "Functionalization of graphene quantum dots by fluorine: Preparation, properties, application, and their mechanisms", *Appl. Phys. Lett.*, vol. 110, art. 221901, 2017.

[ZUO 17b] ZUO X., ZHU J., MÜLLER-BUSCHBAUM P. *et al.*, "Silicon based lithium-ion battery anodes: A chronicle perspective review", *Nano Energy*, vol. 31, pp. 113–143, 2017.

[ZOU 20] ZOU S.-J., SHEN Y., XIE F.-M. *et al.*, "Recent advances in organic light-emitting diodes: Toward smart lighting and displays", *Mater. Chem. Front.*, issue 3, 2020.

Index

A, B

adsorbate evolution mechanism (AEM), 280
atomic layer deposition (ALD), 224, 293
band
  conduction, 197
  valence, 197
batteries
  Li/Air, 171
  Li/S, 165
  Li-Co (LCO), 142
  Li-ion (LIB), 76
  rechargeable, 137, 139
  sodium ion (SIB), 175
blocking layer
  electron (EBL), 234
  hole (HBL), 234
boron-doped diamonds (BDD), 15
bulk heterojunction (BHJ), 204

C

carbon
  nanotube (CNT), 21, 170, 187, 190, 258
  quantum dots (CQD), 16
carboxyfullerenes, 7
cell(s)
  dye sensitized solar (DSSC), 212
  perovskite solar (PSC), 218
  photovoltaic
    perovskite, 196
    silicon, 200
  tandem, 210
chemical vapor deposition (CVD), 11, 22, 41
CIGS (copper-indium-gallium-selenium) semi-conductors, 196
core-shell, 84, 99, 152, 167
composite solid electrolytes (CSE), 160

D, E

density gradient ultracentrifugation (DGU), 29
double heterojunction (DH), 226
electroluminescence, 226
electron
  acceptor, 203
  donor, 203
endocluster fullerenes (ECF), 8
endometallofullerenes (EMF), 8
excitons, 203, 204

F

fluorescence, 235
properties of carbon quantum dots (CQD), 18
thermally assisted delayed (TADF), 242
fluorescent nanodiamonds (FND), 13
fullerenes, 4, 207

G

graphdiyne (GDY), 72, 270
graphene (GR), 37, 187, 190
fibers (GF), 57
oxide (GO), 16, 47, 52
paper, 54
polymer, 60
quantum dots (GQD)
photoluminescence, 66
reduced graphene oxide (r-GO), 48, 187

H

h-graphene (holey graphene), 55
hexagonal boron nitride (hBN), 104
hole transport layer (HTL), 216
horizontally aligned carbon nanotube (HACNT), 32
hydrogen
evolution reaction (HER), 255
acidic, 257
alkaline, 274
oxidation reaction (HOR), 256

I, L

Internet of Things (IoT), 211
lattice oxygen-mediated mechanism (LOM), 281
Li-manganese oxide (LMO), 143
light emitting diodes (LED), 98, 225
QDots (QLED), 226
white (WLED), 228
lithium ion battery, 139
luminescence, 123
properties of Au NCs, 91

M

metal
nanoclusters (MNC), 87
organic framework (MOF), 113, 284
microbatteries (MB), 181
MoS_2, 103, 104, 108, 130, 259, 267
multi-wall carbon nanotube (MWCNT), 24, 169, 190
MXenes, 157

N

nanocluster
gold (Au NC), 87
silver (Ag NC), 87
nanodiamonds (ND), 11
nanoparticles
gold (Au NP), 81
graphene (NP-Gs), 60
graphene oxide (NP-GO), 60
nitrogen-doped GQDs (N-GQDs), 64
non-fullerene acceptors (NFA), 207

O

onion-like fullerenes (OLF), 10
organic
photovoltaic cells (OPV), 203
semiconductor materials, 196
sodium ion batteries (OSIB), 178
organic light-emitting diodes (OLED), 64, 225, 233
fluorescence
delayed, 239
thermally assisted delayed fluorescence (TADF), 243

triplet-triplet annihilation (TTA), 240
phosphorescent, 237
oxygen
evolution reaction (OER), 255, 279
acidic, 288
alkaline, 279
reduction reaction (ORR), 74, 256

P

phosphorescence, 235
photovoltaic, 6, 45, 98, 136, 196, 296
proton exchange membrane/ polymer electrolyte membrane (PEM), 253
pseudocapacitors, 189

Q

quantum dots (QDots)
graphene, 63
light-emitting diodes (QLEDs), 244
semiconductor, 97, 215, 244
quantum well (QW), 229, 230

R, S

redox mediators (RM), 172
roll-to-roll process (R2R), 212
ruthenium composite (Ru@C_2N), 276
solid electrolyte interphase (SEI), 147, 150
silicon
anode, 151
monocrystalline, 200
p-n junction, 196
polycrystalline, 200
single atom catalyst (SAC), 266
single wall carbon nanotube (SWCNT), 21, 24
solid state
electrolyte, 159
lighting (SSL), 225
solidified graphene foam (SGF), 50
supercapacitors, 21, 68, 137, 184
superprotonic conductors, 127

T, U

transition metal dichalcogenides (TMD), 103
triplet-triplet annihilation (TTA), 240
turnover frequency (TOF), 263, 276
UV photodetector, 71

V, W, Y

vertically aligned carbon nanotube (VACNT), 31
volatile organic compounds (VOC), 69
water splitting, 136, 252
white organic light-emitting diodes (WOLED), 233, 236, 237
WS_2, 103, 130, 264
WSe_2, 103
yolk-shell nanoparticles, 167

Other titles from

iSTE

in

Energy

2020

BOISGIBAULT Louis, AL KABBANI Fahad
Energy Transition in Metropolises, Rural Areas and Deserts

SOUALHI Abdenour, RAZIK Hubert
Electrical Systems 1: From Diagnosis to Prognosis
Electrical Systems 2: From Diagnosis to Prognosis

2019

BENALLOU Abdelhanine
Energy Transfers by Convection
(Energy Engineering Set – Volume 3)
Energy Transfers by Radiation
(Energy Engineering Set – Volume 4)
Mass Transfers and Physical Data Estimation
(Energy Engineering Set – Volume 5)

LACHAL Bernard
Energy Transition

ROBYNS Benoît, DAVIGNY Arnaud, BARRY Hervé, KAZMIERCZAK Sabine, SAUDEMONT Christophe, ABBES Dhaker, FRANÇOIS Bruno
Electrical Energy Storage for Buildings in Smart Grids

2018

BENALLOU Abdelhanine
Energy and Mass Transfers: Balance Sheet Approach and Basic Concepts (Energy Engineering Set – Volume 1)
Energy Transfers by Conduction (Energy Engineering Set – Volume 2)

JEMEÏ Samir
Hybridization, Diagnostic and Prognostic of Proton Exchange Membrane Fuel Cells: Durability and Reliability

RUFFINE Livio, BROSETA Daniel, DESMEDT Arnaud
Gas Hydrates 2: Geoscience Issues and Potential Industrial Applications

VIALLET Virginie, FLEUTOT Benoit
Inorganic Massive Batteries
(Energy Storage – Batteries, Supercapacitors Set – Volume 4)

2017

BROSETA Daniel, RUFFINE Livio, DESMEDT Arnaud
Gas Hydrates 1: Fundamentals, Characterization and Modeling

LA SCALA Massimo
From Smart Grids to Smart Cities: New Challenges in Optimizing Energy Grids

MOLINA Géraldine, MUSY Marjorie, LEFRANC Margot
Building Professionals Facing the Energy Efficiency Challenge

SIMON Patrice, BROUSSE Thierry, FAVIER Frédéric
Supercapacitors Based on Carbon or Pseudocapacitive Materials
(Energy Storage – Batteries, Supercapacitors Set – Volume 3)

2016

ALLARD Bruno
Power Systems-on-Chip: Practical Aspects of Design

ANDRE Michel, SAMARAS Zissis
Energy and Environment

DUFOUR Anthony
Thermochemical Conversion of Biomass for the Production of Energy and Chemicals

2015

CROGUENNEC Laurence, MONCONDUIT Laure, DEDRYVÈRE Rémi
Electrodes for Li-ion Batteries
(Energy Storage – Batteries, Supercapacitors Set – Volume 2)

LEPRINCE-WANG Yamin
Piezoelectric ZnO Nanostructure for Energy Harvesting
(Nanotechnologies for Energy Recovery Set – Volume 1)

ROBYNS Benoît, FRANÇOIS Bruno, DELILLE Gauthier, SAUDEMONT Christophe
Energy Storage in Electric Power Grids

ROSSI Carole
Al-based Energetic Nanomaterials
(Nanotechnologies for Energy Recovery Set – Volume 2)

TARASCON Jean-Marie, SIMON Patrice
Electrochemical Energy Storage
(Energy Storage – Batteries, Supercapacitors Set – Volume 1)

2013

LALOUI Lyesse, DI DONNA Alice
Energy Geostructures: Innovation in Underground Engineering

2012

BECKERS Benoit
Solar Energy at Urban Scale

ROBYNS Benoît, DAVIGNY Arnaud, FRANÇOIS Bruno, HENNETON Antoine, SPROOTEN Jonathan
Electricity Production from Renewable Energies

2011

GAO Fei, BLUNIER Benjamin, MIRAOUI Abdellatif
Proton Exchange Membrane Fuel Cell Modeling

MULTON Bernard
Marine Renewable Energy Handbook

2010

BRUNET Yves
Energy Storage

2009

SABONNADIÈRE Jean-Claude
Low Emission Power Generation Technologies and Energy Management

SABONNADIÈRE Jean-Claude
Renewable Energy Technologies

Printed and bound by CPI Group (UK) Ltd, Croydon, CR0 4YY

15/09/2026

14962352-0004